Distributed Generation:
The Power Paradigm
for the New Millennium

T0174720

CRC Press
Taylor & Francis Group
6000 Broken Sound Parkway NW, Suite 300
Boca Raton, FL 33487-2742

First issued in paperback 2019

© 2001 by Taylor & Francis Group, LLC
CRC Press is an imprint of Taylor & Francis Group, an Informa business

No claim to original U.S. Government works

ISBN-13: 978-0-8493-0074-5 (hbk)
ISBN-13: 978-0-367-39719-7 (pbk)

Library of Congress Cataloging-in-Publication Data

Distributed generation: the power paradigm for the new millennium / edited by
Anne-Marie Borbely and Jan F. Krieder.
    p. cm.
    Includes bibliographical references and index.
    ISBN 0-8493-0074-6
      1. Power resources. 2. Energy development. 3. Environmental protection. 4.
Distributed generation of electric power. I. Borbely, Anne-Marie. II. Kreider, Jan F.,
1942-

    TJ163.2 .D56 2001
    333.79—dc21   2001025250

Visit the CRC Press Web site at www.crcpress.com

Library of Congress Card Number 2001025250

**Visit the Taylor & Francis Web site at
http://www.taylorandfrancis.com**

**and the CRC Press Web site at
http://www.crcpress.com**

Preface

Behind the public eye a quiet revolution is taking place, one that will permanently alter our relationship with energy — the building block of our industrial, digital society. Most people today have heard about deregulation of the electric utility industry. A smaller but significant portion of people have joined the stock-buying frenzy surrounding fuel cell developers and other darlings of the energy IPO world. But there's much more at stake here, and this book is a first step in understanding the myriad issues facing both homeowners and businesses.

Distributed generation is not a new concept. Originally, all energy was produced and consumed at or near the process that required it. A fireplace, wood stove, and candle are all forms of "distributed" — small scale, demand-sited — energy. So is a pocket watch, alarm clock, or car battery. The key to today's energy revolution, however, involves turning the resource clock backwards (from large power plants hundreds or thousands of miles away to a "heat engine" in the building) by riding the rapidly accelerating technology wave forward.

For that reason, this book describes not only the technologies being developed today — fuel cells, microturbines, Stirling engines, photovoltaics — but also the communications and control systems that will populate the new energy landscape. The new energy world has as many questions attendant upon its birth as answers. What regulatory issues are at stake? What are the financial and economic implications? How do the installation and operation affect the building owner? What fuels can be used, and what does this all mean for the existing electric distribution system? All these questions are addressed here as well. And, finally, the profound resource and air-quality implications of combined heat and power, an old idea also given new life by a suite of technical advances, are discussed.

The editors invited the developers to represent their respective technologies, with regulatory personnel, research scientists, economists, and financial advisors all providing their unique perspectives. The final product is intended to give the engineer or energy business developer a broad understanding of the distributed generation world as it is emerging today.

The editors wish to thank several individuals, in addition to the expert chapter authors, whose contributions were critical to this book. Christian Yoder contributed to the financial sections, sharing his experience with electricity and gas trading. Craig Moorhead shared his wisdom on the future of natural gas markets. Lois Arena and Peter Curtiss undertook critical readings of the final manuscript. William Reinert first stimulated the interest of both editors in the new energy paradigm for the 21st century.

Editors

Anne-Marie Borbely is a technology policy and planning manager with the Pacific Northwest National Laboratory in Richland, Washington. Ms. Borbely works with both private industry and the public sector to develop the distributed generation industry in the United States. Clients include manufacturers and electric or gas utilities, who contract for market analysis, distribution channel development, sales/customer support, and systems integration. Her 10 years of publishing and editorial experience include over 60 popular and industry articles published in English, Hungarian, and Romanian, as well as editorship of nine popular and technical texts. Ms. Borbely frequently lectures on distributed generation at industry conferences and popular forums.

Jan F. Kreider, Ph.D., P.E. is professor of engineering and founding director of the University of Colorado's (CU) Joint Center for Energy Management. He is cofounder of the Building Systems Program at CU and has written seven textbooks on alternative energy, two books on building systems and other energy related topics, and more than 175 technical papers. He is president of Kreider and Associates, LLC, an energy consulting company. For 10 years he was a technical editor of the *ASME Transactions Journal of Solar Energy Engineering*.

During the past decade, Dr. Kreider has directed more than $5,000,000 in energy related research and development. His work on energy systems for buildings, building performance monitoring, building diagnostics, and renewable energy research is known all over the world. Among his major accomplishments are the first applications of neural networks to building control, energy management, and systems identification, as well as the development of applied artificial intelligence approaches for building design and operation. He has assisted governments and universities worldwide in establishing renewable energy and energy efficiency programs since the 1970s. In 1980 Dr. Kreider and his colleagues connected the first wind-powered distributed generation system to the grid of the Public Service Company of Colorado. He has designed numerous systems since then. He is a Fellow of the American Society of Mechanical Engineers, a registered professional engineer, and a member of several honorary and professional societies. Dr. Kreider received ASHRAE's E. K. Campbell Award of Merit, CU's Outstanding Researcher Award (College of Engineering, Boulder), and the Distinguished Engineering Alumnus Award, CU's highest honor.

Dr. Kreider earned his B.S. degree (magna cum laude) from Case Institute of Technology, and his M.S. and Ph.D. degrees in engineering from the University of Colorado. He was employed by General Motors for several years, where he was involved in the design and testing of automotive heating and air conditioning systems.

Contributors

Anne-Marie Borbely	Pacific Northwest National Laboratory, Richland, Washington
Richard Brent	Solar Turbines Incorporated, Washington, D.C.
Jacob Brouwer	University of California, National Fuel Cell Research Center, Irvine, California
Sunil Cherian	Sixth Dimension, Inc., Fort Collins, Colorado
Peter S. Curtiss	Kreider and Associates, LLC, Boulder, Colorado
Jeffrey Dagle	Pacific Northwest National Laboratory, Richland, Washington
Paul Dailey	Stirling Technology, Inc., Kennewick, Washington
James Daley	ASCO Power Technologies, Florham Park, New Jersey
Peter Fusaro	Global-Change Associates, Inc., New York, NY
Michael Godec	ICF Consulting Group, Inc., Fairfax, Virginia
Yogi Goswami	University of Florida, Department of Mechanical Engineering, Gainesville, Florida
Bruce Hedman	Onsite Energy Corporation, Washington, D.C.
Tina Kaarsberg	U.S. Department of Energy, Washington, D.C.
Jan F. Kreider	Kreider and Associates, LLC, Boulder, Colorado
Ken Nichols	Barber-Nichols, Inc., Arvada, Colorado
Ari Rabl	Ecole des Mines, Paris, France
Colin Rodgers	San Diego, California

Branch Russell	Syntroleum Corporation, Tulsa, Oklahoma
Lawrence Schienbein	Pacific Northwest National Laboratory, Richland, Washington
David Shearer	AVES, Sausalito, California
Don Stevens	Battelle Northwest Laboratory, Richland, Washington
Dan Thoren	Barber-Nichols, Inc., Arvada, Colorado
James Watts	Ingersoll Rand Energy Systems, Portsmouth, New Hampshire
Herb Whitall	EGSA Headquarters, Boca Raton, Florida
Morey Wolfson	National Renewable Energy Laboratory, Golden, Colorado
Eric Wong	Caterpillar, Inc., Sacramento, California

Contents

This book is dedicated

to Billy, who had the vision.

1

Distributed Generation: An Introduction

Anne-Marie Borbely and Jan F. Kreider

CONTENTS

This chapter is designed to provide a brief overview of the converging trends in the energy industry today — regulatory, economic, and technical — that make distributed generation (DG) such a compelling business case. Due to the pervasive, entrenched position of electric utilities in the United States (and other developed countries), however, this is not always an open playing field. The forces that created the contemporary U.S. electricity system continue to exert a profound influence on our perceptions of what is possible as both energy providers and consumers.

The term distributed generation is defined in this book as power generation technologies below 10 MW electrical output that can be sited at or near the load they serve. For this reason, not all small-scale technologies are included here. Hydro- and wind-powered generators are too fuel-dependent (i.e., their location is dictated by the availability of moving water or wind) to be considered truly load-sited or distributed generation.

1.1 Electricity Production:
Distributed — Centralized — Distributed Again

Distributed generation is not a new concept. Indeed, until electricity was introduced as a commercial alternative for the energy historically provided by steam, hydraulics, direct heating and cooling, and light, all energy was produced near the device or service requiring that energy.

From its inception as an industry, electricity has competed against gas for customers. Indeed, electric arc lighting, a mid-19th century stand-alone system located on the customer's premises, attempted to replace the less expensive but volatile gas lamps supplied with "town gas," a mixture of hydrogen and carbon monoxide. Gas production and delivery was the first centralized element in the modern energy industry, produced initially on the customer's premises and later in large gasifiers. By the 1870s, town gas was piped throughout virtually every major city in the U.S. and Europe. An early power plant is shown in Figure 1.1.

The economies of scale that made widespread municipal lighting possible, however, did nothing for the drawbacks of the product. The light was poor, tremendous waste heat made rooms smoky and hot, and the noxious elements of town gas mentioned earlier left room for a cleaner, cooler alternative in the marketplace, i.e., electricity.

Thomas Edison created the first electric utility system, mimicking the gas lighting industry but supplying energy through virtual "mains" to light filaments instead of via gas burners. The same reduction in capital cost per unit of power generated applied to electricity as it did to gas, and the inexorable trend toward centralized power generation, distribution, and system management began.

Initially, electric utilities were established in open territories without service, granting them de facto monopolies. The systems were isolated, without connection to other utilities. By the end of the 1920s, however, utility grids adjoined one another, and interconnection brought obvious benefits (e.g., sharing peak load coverage, and backup power). Through the 1920s, financing such private investor-backed ventures was relatively easy, until the Wall Street crash in 1929. The *Public Utility Holding Companies Act of 1934* recognized the public goods element of electric, gas, water, and telephone companies, and outlined restrictions to and regulatory oversight of corporations that provided such services. The "Golden Age of Regulation" (described in Chapter 9) was ushered in and is only now undergoing substantial change for the first time in over 60 years.

Technological advances were not confined, however, to large-scale operations. Fuel cells were first developed for space flight, and aeroderivative gas turbines that powered jet aircraft found a market in stationary power. The pursuit of soft path, environmentally sustainable economies produced solar engineering and photovoltaic systems. Parallel advances in communications

FIGURE 1.1
Early power generation system.

and microprocessing — the digital age — created the monitoring and dispatching architectures requisite for this new generation of energy providers. Key to it all, the economies of mass production replacing those of scale may mean that centralized power generation and distribution systems are about to give way to a new energy landscape.

1.1.1   Regulatory Restructuring

The massive shift in the U.S. regulatory system for electric (and natural gas) utilities began with the *Energy Policy Act* (EPAct) of 1992, which required interstate transmission line owners to allow all electric generators access to their lines. In effect, transmission lines became common carriers. In its Order 888 in 1996, the Federal Energy Regulatory Commission (FERC) implemented EPAct with respect to electric transmission lines, ordering all transmission line owners to post open-access tariffs. Every qualified generator of electricity — utility-affiliated or independent — was given access to transmission lines to transport electric output. The objective of these initiatives at the federal level was to create competitive wholesale electric markets.

This initial deregulation of the wholesale market, though it spawned hundreds of new power marketers, did not affect the individual end user.

Another movement for deregulation of the retail market — consisting of the distribution system, the contracted energy provider for each individual energy customer, billing, metering, and energy efficiency services — turned the U.S. electric utility industry on its end.

Regional differences in electric rates (as much as a $0.11-0.13/kWh range) drive the restructuring agenda among policymakers, but technological advances make it possible. Dramatic improvements in the efficiency of gas turbine power plants have reduced the cost of producing electricity as well as the size of the plants needed to obtain these cost reductions. Scale economies are no longer a justification for monopoly power production. States with high prices relative to other states are more inclined to restructure with the expectation of improving efficiency, lowering costs, and then lowering prices. Lower energy prices attract investment and industry. Conversely, high energy prices can drive an industry to relocate.

Table 1.1 shows the status of state-level restructuring as of the year 2000, dividing states whose legislatures have enacted restructuring legislation and those whose regulatory commissions have issued sweeping restructuring orders. Fifteen states have enacted legislation restructuring their electric industries. Three state regulatory commissions have issued orders requiring restructuring of their electric industries. Most remaining states are investigating restructuring. Because of the dynamic state of restructuring legislation in the U.S., changes in the contents of Table 1.1 will occur often into the first decade of the 21st century.

Thus, after five years of intense debate, 30 of the 48 states under the FERC's jurisdiction still have not adopted legislation or commission orders to restructure their electric industries. In anticipation of a "patchwork" industry in which generally higher-priced states restructure while lower-priced ones

TABLE 1.1

Status of State Electric Restructuring (2000)

State Legislation		State Regulatory Order	
State	**Date Enacted**	**State**	**Date Ordered**
Arizona	May 1998	Michigan	Jun. 1997
California	Sept. 1996	New York	May 1996
Connecticut	Apr. 1998	Vermont	Jan. 1997
Illinois	Dec. 1997		
Maine	May 1997		
Maryland	Apr. 1999		
Massachusetts	Nov. 1997		
Montana	May 1997		
Nevada	Jul. 1997		
New Hampshire	May 1996		
New Jersey	Feb. 1999		
Oklahoma	May 1997		
Pennsylvania	Nov. 1996		
Rhode Island	Aug. 1996		
Texas	Jun. 1999		

delay, several federal bills mandating national restructuring have been introduced in the U.S. Congress since 1996. The *Comprehensive Electricity Competition Act*, proposed by the Clinton Administration in June 1998 and amended in April 1999, required retail competition — i.e., direct access to sources of electric supply — for all customers by January 1, 2003. However, states would be allowed to opt out of provisions of the legislation if it was felt that consumers would be harmed (Hill, 1995).

The outcome of this legislative/regulatory morass on retail restructuring is difficult to predict. However, there are three certain and undeniable trends in the electric industry: (1) de-integration, (2) convergence, and (3) globalization.

1.1.2   De-Integration of Vertical Stages

The electric industry is becoming increasingly vertically de-integrated by state legislation, commission orders, and state policymaking. Utilities are more often required to sell their generating assets. Given these developments at the state level and the FERC's requirement of open access to transmission lines, electric markets and the electric utility of the future will look dramatically different than they do today. Electric markets will become more like gas markets, with a futures market, a spot market, and a variety of financial arrangements (described in Chapter 10). Many electric utilities will become national and international energy companies, diversifying into a broad arena of products and services.

These new electric utilities will be selling the energy form preferred by their customers. Electricity's share of total U.S. energy use is now approaching 40%, nearly double its share in the late 1950s. That percentage is expected to increase markedly in the future because of the information revolution's reliance on electricity and the introduction of new electricity-using technologies in other sectors. The latter include developments in space conditioning, industrial processes, and transportation.

1.1.3   Convergence of Utility Companies

Electric distribution utilities will face intense competition from marketers in purchasing reliable supplies of electricity and getting and maintaining customers. Technological progress in the development of the distributed generation technologies will provide incentives for local-distribution electric firms to merge with gas distribution firms. Besides the obvious advantage of economies in joint operation such as those realized from joint metering, a merged electric–gas distribution utility can use the gas infrastructure as a storage system for its electric operations. Although the merged utility will be indifferent as to how energy reaches the customer, electricity is the energy of choice for customers and, as discussed above, is likely to remain so as new electrotechnologies enter the consumer market.

Widespread use of advanced communications technologies such as fiber optics (which many electric utilities currently use in their transmission systems) is an incentive for the convergence of more than just electric and gas utilities. A drawback of using fiber-optic cable for retail communications by electric distribution companies is that only a small portion — roughly 2% — of the bandwidth will be used for transmitting information. This provides an additional incentive for mergers of existing utility companies, however. There are incentives for the convergence of telephone, internet, and cable utilities with the energy industries jointly using fiber-optic cable. Because of their use of meters, water utilities may even be a major part of this trend toward convergence.

1.1.4 Globalization

EPAct broke down the barriers to globalizing the electric industry. U.S. electric utilities that were formerly restricted to single service territory — or holding companies in multiple service territories in contiguous states — can now purchase utility assets anywhere in the world. By the same token, foreign interests can purchase U.S. electric utilities.

The globalization of the electric industry in the aftermath of EPAct's enactment is impressive. Many U.S. electric utilities have purchased foreign utility assets and many have also been sold to foreign interests. Recently, there were two significant foreign purchases of U.S. electric assets. National Grid in the United Kingdom, the world's largest privately owned transmission company, purchased the New England Electric system for $3.2 billion. Scottish Power paid $12.8 billion for Pacific Energy, creating one of the ten largest utilities in the world.

The final structure of electric markets will depend on state legislation or state regulatory commission rulings. The FERC has jurisdiction over interstate wholesale electric markets; states have jurisdiction over retail sales. State policymakers have two options:

- Allow the industry's structure to evolve, ensuring open access to transmission lines for any potential entrant in the generating stage of the industry and letting competitive forces emerge over time. Modest policy changes such as substituting performance-based for cost-of-service ratemaking could also be considered.
- Restructure the industry more radically, (1) facilitating competition in the generating stage of the industry by creating a spot market and independent system operator, and (2) allowing retail customers direct access to electric suppliers of their choice.

The paragraphs below address physical elements of the U.S. electric utility industry relevant to the rise of distributed generation today. Patterson (1999) summarizes the present situation in more detail.

1.2   Electric Utility Assets

1.2.1   The U.S. Transmission System

The North American transmission system has historically provided relatively open, non-competitive bulk power supply across the continent through voluntary compliance of public and privately owned electric utilities. Deregulation of the electric utility industry has eroded that spirit of cooperation, transforming the transmission and distribution (T&D) network from a delivery vehicle into a competitive tool. For DG technologies, the physics and financing of the U.S. power transmission system is a key element for market access.

Transmission costs represent about 2% of major investor-owned utilities' operating expenses and 12% of plant investment (distribution, 29%; power generation, 55%). Despite its minority share of the total utility budget, the transmission system provides price signals to encourage efficiencies in the power generation market. Transmission prices, if correctly calculated, send signals to add transmission or generation capacity or indicate where to locate future load. Adding transmission capacity to relieve transmission constraints can allow high-cost generation to be replaced by less expensive generation. Also, following trends in the bulk power supply system can identify regions or utilities most susceptible to problems covering native load, where distributed generation technologies can have the most positive impact, and competitive pricing signals for new capacity may be much higher than currently anticipated.

1.2.2   Utility Choices and Deregulation

Despite the common industry understanding that the T&D segments of the former utility structure will remain regulated, utilities addressed more pressing competitive concerns, and investment in infrastructure slowed in the early 1990s. This decision reflects utilities' experiences in the state-level debates over stranded asset calculations. Some regions have abandoned reporting even uncommitted resource additions needed to satisfy reliability criteria, so great was the confusion over cost recovery. This was also indicative of greater reliance on short-term solutions (gas turbine generators, no new transmission) over longer-term ones (large-scale plants, transmission expansion).

As an increasing number of utilities seek to fulfill native load requirements on the open market, another dilemma appears: multiple buyers relying on the same resource pool for future load coverage. Unexpected peaks (e.g., the heat waves of 1998) and outages (California, 2000 and 2001) necessarily affect large areas. This collision between the time frame for deregulation and capacity planning has sparked concern among regulatory agencies. Although

National Electric Reliability Council (NERC) reliability studies show adequate transmission and generation capacity to maintain reliability through 2002, regional disturbances caused by insufficient transfer capability have already occurred in select regions (e.g., Midwest, Northeast, and California). If this is true for the near term, then the years 2002 through 2006, in which NERC reports greater misgivings about the ability of the bulk supply system, may see disruptions across entire control regions. Figure 1.3 shows U.S. capacity margins in 1998 and, absent further investment, projected margins for 2006.

The U.S., Canada, and portions of northern Mexico are all served by four interconnected synchronous grids (interconnections) moving AC current across both the interconnections and intra-grid DC links. Within each interconnection are ten electric reliability regions (Figure 1.2) and over 140 control organizations that were created to schedule electric power exchanges to maintain system reliability; as there are no switches for routing power, this stability derives from operating the AC generators. However, the natural competition fostered by an open market has provoked a re-evaluation of the reliability regions. The FERC is slowly encouraging utilities to join TransCos, either for-profit or non-profit transmission control regions that will enable full and fair access by all.

1.2.3   Transmission Loading Relief (TLR)

Overloaded transmission lines are incrementally shut down to prevent voltage collapse. The evolution of this somewhat automated procedure can be traced to efforts by various utilities seeking solutions to inadvertent or parallel path flows in electricity transmission and the FERC's functional unbundling of transmission and generation in Orders 888 and 889. Prior to these orders, transmission system overloads were typically handled by the affected control areas by first curtailing their wheeling services for third parties and, if that proved inadequate, redispatching generation.

Prior to the system-wide application of NERC TLR procedures, overloads were handled primarily by local procedures, whereas TLRs are regional. TLR relies on multiple control-area coordinators curtailing transmission flows over a much wider area (based on model-generated measures of their impacts on the constrained facilities). The TLR approach is a flow-based approach that curtails transactions based on actual power flows over the transmission system and their estimated impacts on the overloaded facilities. Historically, utilities instituted local curtailments based on contract path flows. According to NERC, this has proven inadequate to deal with the nature and increased volume of transactions on the transmission grid in recent years.

This response system has become a daily obstacle in the evolution of the wholesale/retail power market. It was a significant contributor to the Mid-America Interconnected Network (MAIN) price spikes of June 1998.

FIGURE 1.2

North American Electrical Council Reliability (NERC) regions: Western States Coordinating Council (WSCC), Mid-Continent Area Power Pool (MAPP), Mid-America Interconnected Network (MAIN), Southeastern Electric Reliability Council (SERC), East Central Area Reliability (ECAR), Electrical Reliability Council of Texas (ERCOT), Florida Reliability Coordinating Council (FRCC), Mid-Atlantic Area Council (MAAC), Northeast Power Coordinating Council (NPCC), Southwest Power Pool (SPP).

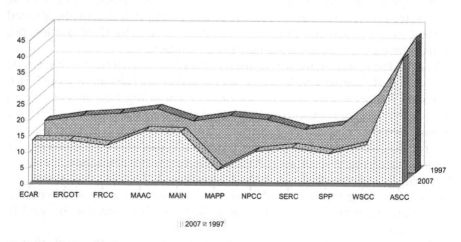

FIGURE 1.3.

Projected U.S. capacity margins by NERC reliability region (Borbely, 1999).

Additional capacity was available from Ontario, Southeastern Electric Reliability Council (SERC), and Mid-Continent Area Power Pool (MAPP), but TLR procedures reduced the import capacity to MAIN. Additionally, PJM (Pennsylvania/New Jersey/Maryland) Interconnection used TLR to curtail all electricity exports during the heat wave.

It is a legal, but potentially devastating, procedure for reliability regions to interrupt contract electricity transfers across their territories if their native load is jeopardized. Dwindling capacity margins are expected to impact the frequency of TLR curtailments.

1.2.4   U.S. Power Generation Assets

In 1996, the U.S. had a total electricity-generating capability of 775,872 megawatts (MW); 91.5% was owned by utilities. The largest portion of utility capability in the country is fueled by coal. The largest plant, Grand Coulee of the Bureau of Reclamation, is a hydroelectric plant on the Columbia River in Washington. The largest utility in the country is the Tennessee Valley Authority (TVA), which provides electricity to seven southeastern states. Although investor-owned utilities account for over three-quarters of U.S. retail electricity sales, both Grand Coulee and TVA are federally owned. The average price of electricity in the U.S. in the year 2000 was 6.9 cents per kilowatt-hour.

The share of U.S. generating capability derived from coal has been steadily falling over the past half-century, and now comprises less than 40%. The non-utility share of capability more than doubled from 1986 to 1996, so that in 1996 nonutilities provided almost 9% of the total. Although the share of utility gas capability increased, the share of net gas-fired generation declined. In 1996, almost one-fifth of electricity was generated at nuclear plants. Figure 1.4 gives a breakout of utility-owned generation by fuel source.

1.2.4.1   *Non Utility-Owned Power Generation*

Several industries rely heavily on their own power sources for protection from outages, reduced dependency on grid-supplied electricity, greater control over power quality, or as a means to control energy costs. Figure 1.5 displays the industries with the greatest concentration of non-utility power generation assets by capacity (MW). Such information hides the tens of thousands of small generators in homes, clinics, and schools that barely register when counted as an aggregate. It is, however, in the market below 1 MW demand that DG will eventually find a foothold.

1.2.5   Double Counting — How Much Is Really Out There?

Capacity resources are sold on a nonfirm (interruptible) basis and reported by purchasers as firm capacity. Power marketers frequently aggregate nonfirm

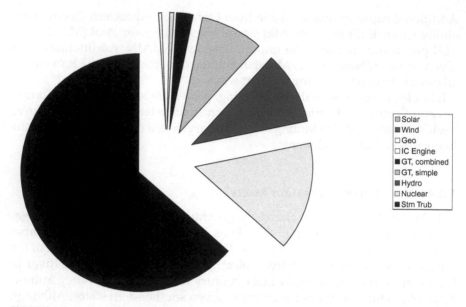

FIGURE 1.4

U.S. utility-owned generation capacity by fuel source (1998); sources are shown from top to bottom in order of the table at right. Solar, wind, and geothermal sources are too small to show at this scale (U.S. DOE Energy Information Administration).

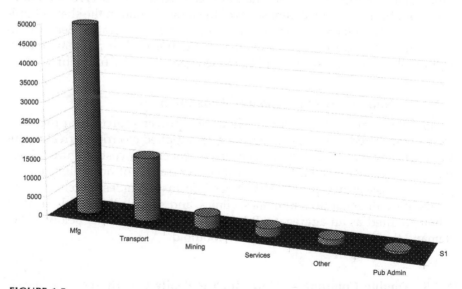

FIGURE 1.5

Non utility-owned power generation capacity by six market sectors; units are MW (U.S. DOE Energy Information Administration).

purchases in their portfolios to allow sales on a firm basis, similar to the way utilities use generation capacity margins to firm up supply. Under current reporting practices, the entity that sells nonfirm capacity to a power marketer would report this as its own installed capacity because, as a nonfirm contract, it really is available to serve the seller's own firm demand. The entity that purchases this illusory firm capacity from the marketer's portfolio also reports it as firm capacity, thus doubling the capacity reported as available. When this practice is combined with utilities' increasing reliance on "unknown" sources for future supply, the reliability implications are compounded.

Identifying the source of an energy supply can be difficult for the control area operator in an environment where transactions may change ownership several times and be subdivided and recombined before reaching a given customer. Without real-time metering, the operator lacks the necessary transparency to identify a supplier's portion of a system's load. It can become impossible to identify situations where energy demand exceeds reserved transmission capacity. System operators must be able to identify points of delivery, intermediary control areas, points of receipt, and levels of firmness so that supply interruptions respect the contractual priorities of each transaction.

A significant portion of all power sales are part of extended daisy chains, involving the repeated retrading of power by market participants who have no intention of ever physically delivering the power. These deals represent multiple resales of the same generation that used to flow directly from vertically integrated utilities to their ultimate customers or to other distribution utilities for resale. Up to 80% of power marketer transactions are financially firm, without any physical transfer of power. The entrance of new marketers, many of whom deal only in financial transactions, has helped to increase the capital size of power markets. But, like printed currency off the gold standard, many of these transactions are not backed by the ability to physically deliver power, exacerbating the volatility of prices under peak conditions.

1.2.6 What Price Power?

One substantial impact of deregulation is the emergence of more accurate locational and temporal pricing of electricity. Regulated utilities were allowed to amortize investments over several decades and dilute price spikes across their entire customer base. While this system denied customers any choice in electricity suppliers, it did reduce their price risk. One mile of a 64kVA distribution line can vary tenfold in construction costs, however, depending upon local requirements (above ground versus overhead, rights-of-way procurement, environmental impacts, relationship to other lines, etc.). Additionally, as Figure 1.6 shows, the distribution grid is designed and built for the highest level of service required, a capacity rarely used.

The generation assets vary widely in the marginal cost of power. Petroleum and coal-fired steam turbines running constantly to supply the baseload may be able to sell into a regional power exchange at $0.02/kWh. But, as additional power is needed (e.g., after 4 p.m. in the residential sector), gas turbines or even combustion engine generator sets may be brought online and used for a much smaller percentage of the temporal load (as few as 200 to 400 hours of operation annually), driving up the incremental cost of power.

New generation assets in a deregulated market will no longer adhere to a 10- or 20-year amortization plan. Additional capacity today will be sited to provide incremental peak power, not baseload. This is a logical response to return-on-investment and risk hedging for private investors. And, as Figure 1.6 shows, a truly deregulated, open market may require much higher price signals before new capacity is built.

One of the ongoing debates in electricity deregulation involves identifying the market price signals for new capacity. If a combustion turbine plant costs $350/kW to construct, with annual operation and maintenance (O&M) costs of 20%, at a capacity factor of 5% (peak shaving, 438 hours/year), the fixed costs for operation would result in:

$$(350 \times .20)/438 = \$0.16/kWh, \text{ or } \$160/MWh$$

If the new capacity were required for a smaller percentage of the peak, e.g., needle peak for 100 hours each year, the ultimate cost of producing electricity would be:

$$(350 \times .20)/100 = \$0.70/kWh \text{ or } \$700/MWh$$

Thus, covering peak demand is far more expensive on the open market than customers have previously come to expect when prices reflect embedded costs to all customers across the utility's load. In an open market, prices would need to rise significantly above current regulated retail prices before sufficient demand could be proven to produce incentives for supply expansion. See Figure 1.7 for a comparison of necessary price durations to build new capacity at $350/kW and $650/kW installed cost, respectively. This is also a powerful argument that prices will ultimately rise, not fall, across the country as deregulation takes effect and existing capacity erodes.

Commodity prices may not fall but may actually increase as new generation is sited only in high-cost areas, eventually raising the cost of electricity to a more even balance between demand and supply, rather than lowering the cost to meet the current lowest-cost supplier.

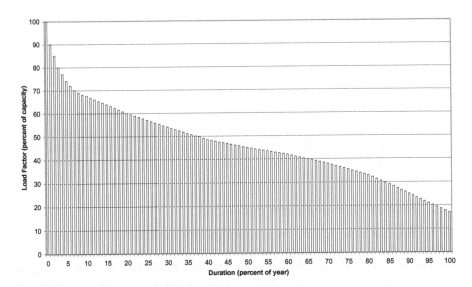

FIGURE 1.6
Temporal load duration for distribution grid.

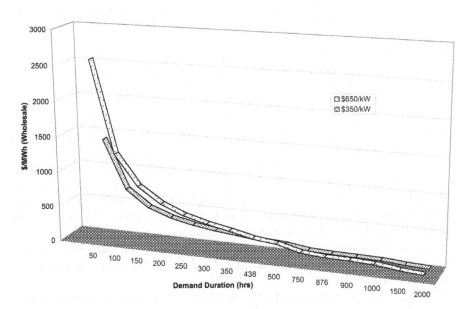

FIGURE 1.7
Price/duration requirements for new capacity investment.

1.3   The Natural Gas Industry

Mike Godec, ICF Consulting

For the coming decade, distributed generation will rely primarily on natural gas as a fuel source. The relationship between natural gas and DG is covered in Chapter 13; this section is intended as an overview to the natural gas industry.

1.3.1   Restructuring and Competition

The natural gas industry historically divided into three segments represented by three different types of companies — producers, pipelines, and local distribution utilities. The producers extracted and produced the gas, sold most of it to the pipeline companies, and held back some percentage for direct sale to large consumers within their production areas. The pipelines transported and sold the gas to distribution company customers and to some large industrial and electric generation customers located along their pipeline routes. The distribution companies then delivered and sold gas to their retail customers of all sizes.

This functional pattern began to change in the early 1990s as gas pipelines were ordered by the FERC to become common carriers and give up their historic merchant function. While the companies maintain their original functions today, the production and distribution companies, along with gas marketers, have assumed the merchant void created when the pipelines were prohibited from buying and selling gas. Now producers sell gas directly to the pipeline companies, past (historic) customers, or gas marketers. Pipelines, with minor exceptions, transport gas from producing areas to designated locations along their lines. They do not resell gas. Distribution companies buy gas from producers and marketers, take deliveries from the pipelines, and make deliveries and sales to their customers. The next step in this restructuring may turn distributors into common carriers for their retail customers who choose to buy gas elsewhere. It is not yet clear how far this latter trend will go, since the public utility regulators for each state must ultimately decide on this change.

Consequently, the fundamental restructuring of the natural gas industry in the last decade (and continuing today) and the emerging restructuring of the electric utility industry are creating an extremely competitive energy marketplace. Both the electric and gas industries are essentially commodity businesses, which promote competition for price and product variety. The rapidly evolving Btu market increases fungibility between energy sources, and financial mechanisms will continue to evolve to reduce the risk of price volatility for those who choose greater predictability of supply costs.

1.3.2 The Future of North American Gas Supplies

Most expert perspectives on the potential of natural gas in North America's energy market are quite bullish. The U.S. will experience tremendous gas demand growth over the next 10 years; total incremental gas demand in the U.S. could increase 10 Tcf (trillion cubic feet) by 2010. Over two-thirds of increased U.S. demand will be in the electric sector, with two-thirds of the remainder in the industrial sector. Capital cost advantage, load matching, environmental considerations, and low gas prices will drive decisions to build a significant amount of new gas-fired power generation capacity in the U.S.

Growth in gas prices will likely be moderate — prices at Henry Hub will increase to $2.10–2.40 per Mcf (thousand cubic feet) on average (1995 dollars) through 2010. A changing mix in gas supplies (as the Rockies and Canada make a larger contribution) will keep overall average U.S. wellhead prices essentially flat, in the $2.00–2.10 per Mcf range. In many areas of the country, basin differentials, on average, will begin to close. This is especially true along those corridors where significant expansions in pipeline capacity are likely. Generally, gas will still tend to flow along most of the traditional corridors, though at substantially higher volumes. In the near term, there may be temporary and seasonal price rises caused by the confluence of higher rates of return demanded by drilling financiers and a shortfall of producing wells at the start of the 21st century (Poruban, 1999).

Overall U.S. production will be able to supply future U.S. demand for gas at reasonable prices. U.S. gas production is forecast to grow substantially by 2010, driven by moderate but stable prices and advancing technology. Improved application of offshore technologies will enable the Gulf of Mexico to continue to be a dominant supply source. While deep-water offshore production will continue to grow rapidly, shallow-water production also remains relatively steady. Unconventional gas production will grow rapidly as production technologies continue to improve. Growth in reserves in discovered fields will allow traditional producing areas to continue to contribute to U.S. supplies.

1.3.3 The Future of Natural Gas Supplies

Unlike crude oil, most natural gas produced around the world is generally consumed relatively near to where it is produced. The limitations associated with restricted access to long distance markets have substantially reduced the commercial viability of gas resources around the world to date. The U.S., the largest consumer of natural gas in the world, has a relatively well developed resource base. Despite this, as discussed above, a substantial resource endowment still remains in the U.S.

In other areas of the world, the natural gas resource base is far less developed than crude oil reserves. This measure of relative youthfulness can be characterized by two measures: the large quantity of gas resources discovered and proven, relative to the amount consumed worldwide versus crude oil, and the relative discovery rates of major gas fields in the world, relative to major oil fields.

Approximately 6700 Tcf of gas resources are believed to exist in the world; this is nearly 100 times current levels of annual consumption. About 75% of the world's gas resources are believed to exist in North America, the former Soviet Union, and the Middle East. However, despite this concentration, major accumulations of natural gas exist in all major areas of the world, allowing natural gas to play a key role as a fuel source in the 21st century.

1.4   The Distributed Generation Technologies

Distributed generation is any small-scale electrical power generation technology that provides electric power at or near the load site; it is either interconnected to the distribution system, directly to the customer's facilities, or both. According to the Distributed Power Coalition of America (DPCA), research indicates that distributed power has the potential to capture up to 20% of all new generating capacity, or 35 Gigawatts (GW), over the next two decades. The Electric Power Research Institute estimates that the DG market could amount to 2.5 to 5 GW/year by 2010. DG technologies include small combustion turbine generators (including microturbines), internal combustion reciprocating engines and generators, photovoltaic panels, and fuel cells. Other technologies including solar thermal conversion, Stirling engines, and biomass conversion are considered DG. In this book, the term DG is limited to units below 10 MW electrical output.

DG can provide a multitude of services to both utilities and consumers, including standby generation, peak shaving capability, peak sharing, baseload generation, or combined heat and power that provide for the thermal and electrical loads of a given site. Less well understood benefits include ancillary services — VAR support, voltage support, network stability, black start, spinning reserve, and others — which may ultimately be of more economic benefit than simple energy for the intended load.

DG technologies can have environmental benefits ranging from truly green power (i.e., photovoltaics) to significant mitigation of one or more pollutants often associated with coal-fired generation. Natural gas-fired DG turbine generators, for example, release less than one-quarter of the emissions of sulfur dioxide (SO_2), less than 1/100th of the nitrogen oxides (NO_X), and 40% less carbon dioxide (CO_2) than many new coal-boiler power plants; these units are clean enough to be sited within a community among residential and commercial establishments (DPCA, 1998).

Electric restructuring has spurred the consideration of DG power because all participants in the energy industry — buyers and sellers alike — must be more responsive to market forces. Central utilities suffer from the burden of significant "stranded costs," which are proposed to be relieved through temporary fixed charges. DG avoids this cost. DG is a priority in parts of the country where the spinning reserve margins are shrinking, where industrial and commercial users and T&D constraints are limiting power flows (DPCA, 1998).

Additional impetus was added to DG efforts during the summer of 1998 due to the heat wave that staggered the U.S. and caused power shortages across the Rust Belt. The shortages and outages were the result of a combination of factors such as climbing electricity demand, the permanent or temporary shutdown of some of the region's nuclear facilities, unusually hot weather, and summer tornadoes that downed a transmission line.

In spite of several notable reasons for DG growth discussed throughout the book, it must be recognized that DG is a disruptive technology and, as was the case with past technologies, may initially offer worse economic or technical performance than traditional approaches. As commercialization continues, however, these new technologies will be characterized by rapid performance improvements and larger market share. But, because the products described below tend to be simpler and smaller than older generations, they may well be less expensive to own and operate, even in the near-term. Table 1.2 provides an overview of feasible present or near-term DG technologies. Each technology summarized below is described fully in separate chapters later in the book.

TABLE 1.2

Summary of Distributed Generation Technologies

	IC Engine	Microturbine	PVs	Fuel Cells
Dispatchability	Yes	Yes	No	Yes
Capacity range	50 kW–5 MW	25 kW–25 MW	1 kW–1 MW	200 kW–2 MW
Efficiency[a]	35%	29–42%	6–19%	40–57%
Capital cost ($/kW)	200–350	450–1000	6,600	3,750–5,000
O&M cost[b] ($/kWh)	0.01	0.005–0.0065	0.001–0.004	0.0017
NO_x (lb/Btu)				
Nat. Gas	0.3	0.10	—	0.003–0.02
Oil	3.7	0.17	—	—
Technology status	Commercial	Commercial in larger sizes	Commercial	Commercial scale demos

[a] Efficiencies of fossil and renewable DG technologies are not directly comparable. The method described in Chapter 8 includes all effects needed to assess energy production.

[b] O&M costs do not include fuel. Capital costs have been adjusted based on quotes.

Source: Distributed Power Coalition of America; Kreider and Associates, LLC (with permission).

1.4.1 Internal Combustion Engines

Reciprocating internal combustion engines (ICEs) are the traditional technology for emergency power all over the world. Operating experience with diesel and Otto cycle units is extensive. The cost of units is the least of any DG technology, but maintenance costs are among the greatest. Furthermore, diesel and gasoline engines produce unacceptable emission levels in air quality maintenance areas of the U.S. Natural gas ICE generators offer a partial solution to the emissions problem but do not solve it entirely. However, the NG-fired IC engine is the key competition to all other DG technologies considered here.

The key barriers to ICE usage include the following:

- Maintenance cost – the highest among the DG technologies due to the large number of moving parts.
- NO_x emissions are highest among the DG technologies (15–20 PPM even for lean burn designs).
- Noise is low frequency and more difficult to control than for other technologies; adequate attenuation is possible.

Attractive ICE features include:

- Capital cost is lowest of the DG approaches.
- Efficiency is good (32 to 36%; LHV basis).
- Thermal or electrical cogeneration is possible in buildings.
- Modularity is excellent, nearly any building related load can be matched well (kW to MW range), part load efficiency is good; the need for this is described later.

1.4.2 Microturbines

A microturbine (MT) is a Brayton cycle engine using atmospheric air and natural gas fuel to produce shaft power. Figure 1.8 shows the essential components of this device. Although a dual shaft approach is shown in the figure, a single-shaft design is also used in which the power produced in the expander is supplied to both the compressor and the load by a single shaft. The dual shaft design offers better control, but at the cost of another rotating part and two more high speed bearings. Electrical power is produced by a permanent magnet generator attached to the output shaft or by way of a gear reducer driving a synchronous generator.

Figure 1.9 is a photograph of a small MT showing most of the key components except for the recuperator. The recuperator is used in most units because about half of the heat supplied to the working fluid can be transferred from the exhaust gas to the combustion air. Without a recuperator the overall efficiency of a MT is 15 to 17%, whereas with an 85% effective recuperator the efficiency can be as high as 33%. MTs without recuperators are basically burners that produce a small amount of electricity with thermal output to be used for cogeneration.

A handful of MT manufacturers have announced products in the U.S. Sizes range from 25 to 150 kW, with double digit power ratings the most common.

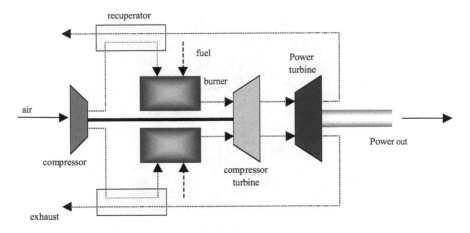

FIGURE 1.8
Schematic diagram of dual shaft microturbine design.

By early 2000, fewer than 1000 MTs had been shipped to U.S. locations. Attractive MT features include low capital cost, good efficiency (30–33%; LHV basis), modest emissions (<10 PPM NO_x quoted by manufacturers), thermal or electrical cogeneration is possible in industry and in buildings, and modularity is excellent (nearly any load can be matched well by multiple units of small to medium capacity).

The key barriers to MT usage include maintenance cost (the exact costs are unknown but are expected to be lower than ICEs due to fewer moving parts) questionable part load efficiency (manufacturer's data vary), limited field experience, use of air bearings is desirable to reduce maintenance but air filtration requirements are stringent, and high frequency noise is produced but is relatively easy to control.

1.4.3 Photovoltaics

Photovoltaic (PV) cells directly convert sufficiently energetic photons in sunlight to electricity. Because sunlight is a diffuse resource, large array areas are needed to produce significant power. However, offsetting this is the zero cost of the fuel itself. Today, there is a PV market worldwide of the order of 100 MW per year. Figure 1.10 shows a typical flat plate PV panel.

Prices for PV arrays have dropped by at least two orders of magnitude in the past three decades but still appear to be too high for many applications in the U.S., where the present utility grid offers an alternative. However, in mountainous areas where the grid does not exist or in developing countries where electricity infrastructure investments may never be made, PVs can produce power more cheaply than the common ICE alternative.

Attractive features of PV systems include emission-free operation, no fossil fuel consumption, low-temperature thermal cogeneration (using building

FIGURE 1.9
10-kW microturbine (courtesy of U.S. Department of Energy, Pacific Northwest National Laboratory).

FIGURE 1.10
Solar PV panel (courtesy of NREL).

integrated modules) possible for space heating, excellent modularity (nearly any building related load can be matched well by multiple units), maintenance negligible, except where batteries are involved, and excellent part load efficiency.

The key barriers to PV usage include: (1) the price of delivered power exceeds other DG resources; subsidies exist in some states that make PV-produced power competitive, and (2) temporal match of power produced to load is imperfect; batteries or other systems are often needed.

1.4.4 Fuel Cells

A fuel cell (FC) is a device in which hydrogen and oxygen combine without combustion to produce electricity in the presence of a catalyst. One design is shown in Figure 1.11. Several competing technologies have been demonstrated and are listed below with their nominal operating temperatures.

- Phosphoric acid (PA) — 300°F
- Proton exchange membrane (PEM) — 200°F
- Molten carbonate (MC) — 1200°F
- Solid oxide (SO) — 1300°F

As indicated in Table 1.2, fuel cells cost too much to be immediately competitive against grid-supplied electricity, but industry experts have indicated that with mass production prices should fall. Installed cost will not always be the deciding factor in choosing a given technology. Where environmental

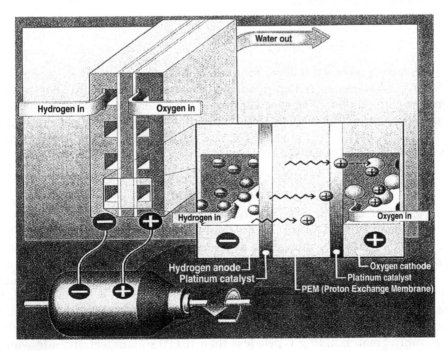

FIGURE 1.11
PEM fuel cell schematic.

regulations are strict, fuel cells offer the only truly clean solution to electricity production outside of the renewables sector. The key barriers to fuel cell usage include cost (predicted cost reductions have not materialized; in fact, one large firm recently announced a 60% price increase), hydrogen fuel (widespread adoption will require a new fuel distribution infrastructure in the U.S. or on-site reforming of natural gas, i.e., methane), maintenance costs are uncertain, and transient response to building load variations is unacceptable for load following for some technologies.

In contrast to these barriers are some very attractive FC features, such as the only byproduct is water — NO_x emissions are very low (< 1PPM), efficiency is good (50–60%, LHV basis), thermal or electrical cogeneration is possible in processes and in buildings, and modularity is excellent — nearly any building related load can be matched well (kW to MW range).

1.5   Matching the Load: Buildings and Industrial Processes

To set the context for distributed electrical generation and cogeneration, the important features of building load profiles in North America are summarized here. Buildings are expected to be one of the key early adoption sectors for distributed generation.

1.5.1   Commercial Buildings

In 1997, there were 4.6 million commercial buildings in the U.S., occupying 58.8 billion square feet of floor space and consuming 126,500 Btu of delivered energy per square foot of space annually. Sixty percent of U.S. commercial buildings range from 5000 to 100,000 square feet. Eighty-two percent range from 1000 to 200,000 square feet. The size class with the largest membership is the 10,000 to 25,000 square foot range.

1.5.1.1   Buildings Disaggregated by Building Type and Floor Space

The use of building space is a key influence on energy consumption and a key determinant of DG plant size. Of the total square footage of commercial office space, 67% is used for mercantile and service purposes, offices, warehouses and storage, or educational facilities. The average square footage for all building types ranges from 1001 to 25,000 square feet. The largest building types, from 20,000 to 25,000 square feet, are lodging and health care facilities. Medium-sized building types from 10,000 to 20,000 square feet are vacant buildings, public order and safety, offices, mercantile and service, and public assembly. Small building types, less than 10,000 square feet, include warehouse and storage facilities, education facilities, and food service and sales. Table 1.4 summarizes the typical sizes of different types.

TABLE 1.3

Size Distribution of U.S. Commercial
Building Space — Commercial Building
Size as of 1995 (Percent of Total Floor Space)

Square Foot Range	Percent
1001 to 5000	10.80
5001 to 10,000	12.80
10,001 to 25,000	18.90
25,001 to 50,000	13.10
50,001 to 100,000	13.60
100,001 to 200,000	11.50
200,001 to 500,000	9.40
Over 500,000	9.00
Total	100.00

TABLE 1.4

Commercial Building Sector Size and Typical Floor Area —
1995 Average and Percent of Commercial Building by Principal
Building Type

Building Type	Floor Space (%)	Average Floor Space/Building (ft²)
Mercantile and Service	22	11260
Office	18	12870
Warehouse/Storage	14	6670
Education	13	1770
Public Assembly	7	12110
Lodging	6	22900
Health Care	4	22220
Food Service	2	4750
Food Sales	1	4690
Public Order and Safety	2	14610
Vacant	9	18480
Other	2	—

1.5.1.2 End Use Consumption by Task

Finally, one must know the end use category — space heating, cooling, water heating, or lighting — in order to assess whether DG is appropriate for a given sector. Space heating and lighting are generally the largest energy loads in commercial office buildings. In 1995, energy consumed for lighting accounted for 31% of commercial energy loads. Space heating consumed 22%, and space cooling consumed 15% of commercial energy loads. On average, water heating is not high at 7%, but this average is largely variable. Health care facilities and lodging are unique in their high water heating loads; however, offices, mercantile and service facilities, and warehouses require minimal hot water.

Another approach to considering the data in the preceding figure is to consider the end uses aggregated over all buildings but further disaggregated over the nine main end uses in commercial buildings. Figure 1.12 shows the data in this way.

1.5.1.3   *Commercial Energy Consumption and Intensity by Principal Building Activity (1995)*

Commercial buildings were distributed unevenly across the categories of most major building characteristics. For example, in 1995, 63% of buildings and 67.1% of floor space were found in four building types: office, mercantile and service, education, and warehouse. Total energy consumption also varied by building type. Three building types — health care, food service, and food sales — had higher energy intensity than the average of 90.5 thousand Btu per square foot for all commercial buildings. Figures 1.13 and 1.14 show the 13 principal building types and their total consumption and intensity.

1.5.1.4   *Energy Consumption by Fuel Type*

Five principal energy types are used in U.S. commercial buildings: (1) natural gas, (2) fuel oil, (3) liquefied petroleum gas (LPG), (4) renewables and other, and (5) on-site electric. Table 1.5 shows the relationship between the end use types shown in Figure 1.13 and the corresponding energy sources. Space heating, lighting, and water heating are the three largest consumers of energy. Natural gas and electricity directly compete in three of the major end uses — space heating, water heating, and cooking. In each of these three, natural gas consumption greatly exceeds electricity consumption.

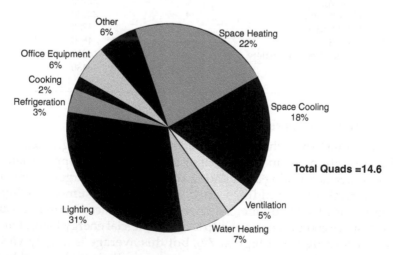

FIGURE 1.12
Commercial building energy end uses aggregated over all building types.

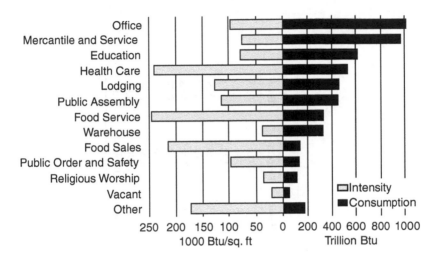

FIGURE 1.13
Energy usage and usage intensity by building type.

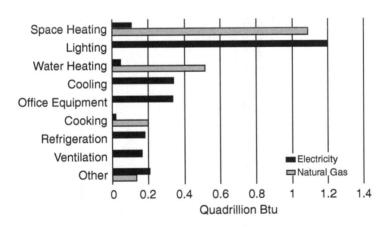

FIGURE 1.14
Gas and electric competition in commercial buildings by task.

1.5.2 The Industrial Sector

The U.S. industrial sector consists of more than three million establishments engaged in manufacturing, agriculture, forestry, fishing, construction, and mining. In 1997, these enterprises occupied 15.5 billion square feet of floor space and used 37% (34.8 quadrillion BTUs) of total U.S. primary energy consumption for their process loads and building conditioning.

After the transportation sector, the manufacturing sector consumes the most energy in the U.S. Of the 37% of primary energy consumption in the industrial sector in 1997, 33% was used for manufacturing purposes and 4%

TABLE 1.5

Fuel Type Usage in Commercial Buildings

End Use	Natural Gas	Fuel Oil (2)	LPG Fuel (3)	Other	Renewable Energy (4)	Site Electric	Site Total	Site Percent	Primary Total
Space Heating	1.58	0.37		0.11		0.16	2.22	29.10	0.53
Space Cooling	0.02					0.34	0.35	4.60	1.08
Ventilation						0.17	0.17	2.20	0.53
Water Heating	0.75	0.07			0.02	0.09	0.93	12.10	0.29
Lighting						1.22	1.22	15.90	3.90
Refrigeration						0.18	0.18	2.40	0.58
Cooking	0.23					0.02	0.25	3.30	0.07
Office Equipment						0.40	0.40	5.30	1.30
Other	0.21	0.04	0.08	0.03	0	0.25	0.61	8.00	0.81
Miscellaneous	0.59	0.12				0.61	1.32	17.20	1.95
Total	3.37	0.6	0.08	0.14	0.02	3.44	7.65	100	11.03

for non-manufacturing purposes. Thus, manufacturing establishments consume the majority of the energy in the industrial sector even though they are far outnumbered by non-manufacturing establishments. Due to lack of information regarding the non-manufacturing sectors and the fact that the majority of energy is consumed in manufacturing, the manufacturing sector is the main focus of this section.

Standard industrial classification (SIC) groups are established according to their primary economic activity. Each major industrial group is assigned a two-digit SIC code. The SIC system divides manufacturing into 20 major industry groups and non-manufacturing into 12 major industry groups. Of the 20 major industry groups in the manufacturing sector, in 1991 six groups accounted for 88% of the consumption of energy for all purposes and 40% of the output value for manufacturing: food and kindred products, paper and allied products, chemical and allied products, petroleum and coal products, stone, clay, and glass products, and primary metals. These six are clearly very energy intensive in their production. Table 1.6 summarizes the key characteristics of the energy-using SIC categories with an overview of each. Table 1.7 shows the floor space inventory by SIC.

Of a total of 15.5 billion square feet of manufacturing space, 17% is used for office space and 83% is used for non-office space. Six groups account for 50% of this space: industrial machinery, food, fabricated metals, primary metals, lumber, and transportation (PNNL, 1997).

Manufacturers use energy in two major ways: to produce heat and power and to generate electricity, and as raw material input to the manufacturing process or for some other purpose.

Three general measures of energy consumption are used by the U.S. Energy Information Administration (EIA). According to its 1991 data, the amount of total site consumption of energy for all industrial purposes was 20.3 quadrillion Btu. About two-thirds (13.9 quadrillion Btu) of this was used to produce heat and power and generate electricity, with about one-third (6.4 quadrillion Btu) consumed as raw material and feedstocks. Energy end uses for industry are similar to those for commercial buildings, although the magnitudes are clearly different. Heating consumes 69% of delivered energy (45% of primary energy usage). Lighting is the second largest end-use with 15% of delivered energy (27% of primary energy usage). Finally, ventilation and cooling account for 8% each.

As with commercial buildings, a variety of fuels are used in industry. Petroleum and natural gas far exceed energy consumption by any other source in the manufacturing sectors. Figure 1.15 displays the fuel mix. DG applications can address electrical and many thermal loads of the industries discussed above. However, consideration of the use of thermal energy, by way of cogeneration, requires one additional piece of information — the process temperature. The process temperature is not easily generalized by the SIC or any other index. Specifics of the industrial process — heat rates and specified temperature levels — are needed.

TABLE 1.6

General Characteristics of Industrial Energy Consumption SIC

Standard Industrial Code	Major Industry Group	Description
High-Energy Consumers		
20	Food and kindred products	The high-energy consumers convert
26	Paper and allied products	raw materials into finished goods
28	Chemicals and allied products	primarily by chemical (not physical)
29	Petroleum and coal products	means. Heat is essential to their
32	Stone, clay, and glass products	production, and steam provides
33	Primary metal industries	much of the heat. Natural gas,
		byproduct, and waste fuels are the
		largest sources of energy for this
		group. All, except food and kindred
		products, are the most energy-
		intensive industries.
High-Value Added Consumers		
34	Fabricated metal products	This group produces high value-
35	Industrial machinery and equipment	added transportation vehicles, industrial machinery, electrical
36	Electronic and other electric equipment	equipment, instruments, and miscellaneous equipment. The
37	Transportation equipment	primary end uses are motor-driven
38	Instruments and related products	physical conversion of materials
39	Miscellaneous manufacturing industries	(cutting, forming, assembly) and heat treating, drying, and bonding. Natural gas is the principal energy source.
Low-Energy Consumers		
21	Tobacco manufactures	This group is the low-energy
22	Textile mill products	consuming sector and represents a
23	Apparel and other textile products	combination of end-use
24	Lumber and wood products	requirements. Motor drive is one of
25	Furniture and fixtures	the key end uses.
27	Printing and publishing	
30	Rubber and miscellaneous plastics	
31	Leather and leather products	

Source: Energy Information Administration, Office of Energy Markets and End Use, Manufacturing Consumption of Energy, 1991, DOE/EIA-0512(91).

1.5.3   Residential Buildings

Although residential buildings are not primary early targets for DG, it is important to consider that sector as smaller generation units come on line. Even now, there are niche applications for DG in the residential sector. The following data summarize residential energy use in the U.S. In 1993, there

TABLE 1.7

Industrial Building Floor Area Distribution (1991)

SIC	Manufacturing Industry	Office Floor Space	Non-Office Floor Space	Total Floor Space
20	Food	203	1207	1410
21	Tobacco	6	51	56
22	Textiles	42	581	623
23	Apparel	73	451	523
24	Lumber	53	1135	1187
25	Furniture	49	521	569
26	Paper	72	827	899
27	Printing	351	477	827
28	Chemical	185	714	899
29	Refining	20	105	125
30	Rubber	97	768	865
31	Leather	9	44	53
32	Stone, Clay	57	808	864
33	Primary Metals	81	1121	1202
34	Fabricated Metals	182	1175	1357
35	Industrial Machinery	337	1149	1485
36	Electronic Equipment	266	629	894
37	Transportation	289	776	1065
38	Instruments	225	170	395
39	Misc. Manufacturing	52	190	242
Total		2641	12,898	15,539

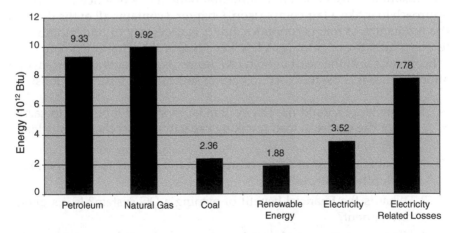

FIGURE 1.15

Industrial consumption by fuel type.

were 76.5 million residential buildings in the U.S. The households consisted of 69% single-family, 25% multi-family, and 6% mobile homes. These buildings consumed 107.8 million Btus of delivered energy (or 187.5 million Btus of primary energy) per household.

TABLE 1.8

U.S. Residential Buildings
Disaggregated by Size — Household
Size in Terms of Heated Floor Space
as of 1993

Sq. Foot Range	Percent
Fewer than 600	7.8
600 to 999	22.6
1000 to 1599	28.8
1600 to 1999	12.8
2000 to 2399	10.0
2400 to 2999	8.5
3000 or more	9.6
Total	100.0

1.6 Economic Considerations

The final judgment regarding the installation of any DG system usually comes down to an economic decision. Accounting for all costs and benefits properly, the DG system owner will decide if the benefits outweigh the costs. Because the costs and benefits are numerous and quite different from one another — for example, capital and energy, operation and maintenance, and insurance costs and environmental benefits — one needs a uniform approach to make a fair comparison. Chapter 8 contains all of the details of the mechanics of microeconomics, but it is necessary to understand a few features from the beginning. A few key ideas will be presented here.

Economics is at the heart of both DG design assessments and system operation because it is the tool used to answer several very basic questions:

1. How large should the DG system be — should it be able to carry all of the electrical load (peaking capacity) or just the average load (base load)?

2. What is the cost of DG-produced electric power? Is it competitive with other sources?

3. What is the financial benefit of owning a DG system? Is it a good investment?

4. Under a certain set of electric and gas rates and electrical demands in a building, is it worthwhile to operate an installed DG system? It may not make sense to make power that can be purchased elsewhere at the specific hour under consideration. One must make this judgment every hour of the year.

TABLE 1.9

Residential End-Use Consumption by Fuel Type and by End Use

End Use	Natural Gas	Fuel Oil[a]	LPG Fuel[b]	Other	Renewable Energy[c]	Site		Percent	Primary[d]	
						Electric	Total		Electric[d]	Total
Space heating[e]	3.58	0.84	0.32	0.15	0.61	0.50	6.00	54.8	1.61	7.10
Space cooling[f]	0.00					0.54	0.54	4.9	1.72	1.72
Water heating[g]	1.27	0.10	0.07		0.01	0.39	1.83	16.8	1.24	2.69
Lighting						0.40	0.40	3.6	1.27	1.27
White goods[h]	0.05					0.78	0.82	7.5	2.49	2.54
Cooking[i]	0.16		0.03			0.23	0.42	3.9	0.74	0.93
Electronics[j]						0.27	0.27	2.5	0.86	0.86
Motors[k]						0.05	0.05	0.5	0.18	0.18
Heating appliances[l]						0.10	0.10	0.9	0.31	0.31
Other[m]	0.09	0.00	0.01				0.10	0.9		0.10
Miscellaneous[n]						0.41	0.41	3.7	1.30	1.30
Total	5.15	0.94	0.43	0.15	0.62	3.66	10.94	100.0	11.73	19.01

a   Indicates 0.94 quads distillate fuel oil.

b   Kerosene (0.09 quad) and coal (0.06 quad) are assumed attributable to space heating.

c   Compound of 0.60 quad wood (space heating), 0.01 quad geothermal (assumed space heating), and 0.01 quad solar (water heating).

d   Site-to-source electricity conversion (due to generation and transmission losses) = 3.21.

e   Fan (0.18 quad) and pump energy use included.

f   Fan energy use included.

g   Includes electric recreational water heating (0.11 quad).

h   Includes (1.26 quad) refrigerators, (0.39 quad) freezers, (0.09 quad) clothes washers, (0.05 quad) natural gas clothes dryers, (0.62 quad) electric clothes dryers, and (0.15 quad) dishwashers. Does not include water heating energy.

i   Includes (0.14 quad) microwaves and other "small" electric cooking appliances.

j   Includes (0.29 quad) color televisions, (0.06 quad) personal computers, and (0.51 quad) other electronics.

k   Includes devices whose energy consumption is driven by motors.

l   Includes appliances such as electric blankets, irons, waterbed heaters, and hairdryers.

m   Includes swimming pool heaters, outdoor grills, and natural gas outdoor lighting.

n   Energy attributable to the residential buildings sector, but not directly to specific end-uses.

Source: EIA, AEO 1999, Dec. 1998, Table A2, p. 113–115, Table A4, p. 118–119, and Table A18, p.135; BTS/Little, A.D., *Electricity Consumption by Small End-Uses in Residential Buildings*, Appendix A for electric end-uses.

All of these questions are answered using the same principles of *microeconomics* in different ways.

Costs of DG systems are two basic types:

- Initial investments — what it takes to acquire a system, e.g., the installed cost of a microturbine, a one-time payment ($)
- Ongoing costs — what it costs to maintain and operate a system, e.g., the annual maintenance contract on the microturbine, an ongoing series of payments ($/yr)

For a complete picture, we need to be able to express both ongoing and one-time costs in a uniform framework. In this book, all costs are reduced to annual cash flows in units of $/yr. Any currency can be used — the key is to reduce all costs to an annual time scale. This is not all that unfamiliar. Everyone knows that the price of a car is often paid in monthly installments. The car's initial cost has been converted to an equivalent monthly payment; annual costs are used in this book because of the fiscal-year basis of most companies.

Likewise, benefits are accrued over the economic lifetime of a system. An example of the benefit of a DG system is the reduced electricity bill of the owner. In very simple terms, the DG investor prefers a system that will produce benefits larger than the costs.

Suppose that a 100 kW DG system operates at full capacity for 8000 of the 8760 hours in a year. If the electricity produced offsets grid power priced at $0.05/kWh, then the annual benefit B of this particular DG system is

$$B = (100 \ kW \times 8000 \ hr/yr) \times \$0.05/kWh = \$40,000/yr$$

The installed cost of this system is also known to be $1200/kW or a total of $120,000. If a loan for this system is paid off in eight years, then the annual payment A (ignoring interest) is

$$A = \$120,000/8 = \$15,000/yr$$

The DG system superficially appears to be beneficial because it saves $25,000 per year. This hasty conclusion is far from correct because a number of key costs have been ignored:

- Cost of money (interest charges at 8% for eight years will add $6000 to the annual cost above)
- Cost of maintenance of equipment (depending on the technology, maintenance could add $7000 to the cost above)
- Cost of fuel (this is very dependent on the local gas cost, but could amount to $3000–$4000/yr)

Therefore, properly accounting for all costs has changed a strongly feasible project to a marginally feasible one.

On the other hand, suppose that the DG system produces exhaust heat that can be used for useful purposes on site. That will reduce the portion of the fuel cost associated with power production by two-thirds. Perhaps a better loan interest rate can be found for the DG system initial purchase. With this combined scenario, the DG economics are much more advantageous. The conclusion is that careful cost and benefit analysis must be conducted for every project. The use of rules of thumb (e.g., if the "spark spread" is $7/MM Btu, then DG is feasible anywhere) is not correct in every case. For example, notice that the three simple cases above had the same spark spread, but three sets of assumptions resulted in three levels of economic feasibility.

Three key economic indices are developed in Chapter 8. The *simple payback period* is the time required to pay back an initial DG investment using the energy sales proceeds to pay off the equipment. For example, the system above saved $40,000 per year with a system cost of $120,000. The payback period *P* is

$$P = \$120{,}000 \,/\, (\$40{,}000/yr) = 3 \; years$$

The payback period is seriously deficient in evaluating projects because of all the shortcomings and ignored costs listed above.

A better index is the *internal rate of return* (IRR), defined as the earning rate that the DG system produces for the system owner. The owner invests in the DG system and earns money on the electricity (and thermal energy) sold much like one would invest in stocks and earn dividends on that investment. The IRR is more difficult to calculate, but the effort is worth the trouble because the IRR is the correct way to rank competing investments.

The third approach used to assess DG costs and benefits is called *life cycle costing* (LCC). This approach can also be used to correctly rank investment options. The costs that are considered in the LCC and IRR methods include:

- Cost of money
- DG system capital cost
- Maintenance
- Insurance
- Operational costs
- Fuel
- Taxes

The benefits that are weighed against the listed costs include the following:

- Electricity payment savings
- Exhaust heat energy savings
- Power quality support benefits
- Other ancillary services that DG can provide
- Environmental benefits

Note that the payback period includes only a few of the listed key parameters. Exactly who owns DG equipment, pays for the energy, and reaps the benefits depends on the ownership scenarios — leased, owned by utility (i.e., energy service company), or owned by building owner.

The final topic that needs to be introduced, with details to follow, is the method for calculating the cost of electricity produced by DG systems. This calculation is simple after the terms listed above have been determined. The cost of power is the number of kilowatt-hours produced divided into the annual cost of owning and operating the DG system. For example, suppose that a DG system produces 100,000 MWh per year using a system that costs $7.2 million per year to own and operate. The cost C_e of electric power produced is found from

$$C_e = \$7,200,000 \, / \, (100,000 \; MWh/yr \times 1000 \; kWh/MWh) = \$0.072 \, / \, kWh$$

1.7   Environmental Issues

Dr. David Shearer, AeroVironment, Inc.

The integration of emerging technologies into existing markets can prove difficult given the conventional technology's low cost, proven performance, and established service networks (National Renewable Energy Laboratory, 1995). DG advocates will need to identify market drivers where the unique characteristics of emerging DG technologies can offer different value propositions relative to conventional approaches. The potential environmental benefits of DG technologies are possible key determinants in securing market niches. The primary environmental drivers, which range in scope from having regional or mesoscale impacts to having only local implications, include:

- Broad policy issues related to electric utility deregulation and the distributed generation market
- Product-specific issues related to emission characteristics and regulatory settings
- Emissions trading issues which can potentially provide customer purchase incentives

1.7.1   Background

It is well known that the traditional electricity industry is a major source of air pollution (Table 1.10). The combustion of fossil fuels, which accounts for 67% of electricity generation, results in the release of a stream of gases into the atmosphere. These gases include several pollutants that pose direct risks to human health and welfare, including sulfur oxides (SO_2), nitrogen oxides (NO_x), particulate matter (PM), volatile organic compounds (VOCs), carbon monoxide (CO), and various heavy metals including lead and mercury. In addition, the combination of VOCs and NO_x in the presence of heat and sunlight forms ozone. Other gases may pose indirect risks. Carbon dioxide (CO_2) may contribute to global warming. Electricity generation accounts for 66, 29, and 35% of the total national emission inventory for SO_2, NO_x, and CO_2 emissions, respectively (Clean Air Network, 1997).

TABLE 1.10

1996 National Estimated Emissions from Fossil Fuel, Steam-Electric Utilities

	CO_2		NO_x		PM_{10}		SO_2		Hg	
	kiloton	%	kiloton	%	kiloton	%	kiloton	%	kiloton	%
All	2,047,368	35	6034	26	282	9	12,604	66	52	33
Coal	1,851,787		5517		258		12,114		51.6	
Oil	56,340		96		5		412		0.2	
Gas	136,689		269		1		21			
Other	2552		151		18		57			

Notes: Kilotons refer to thousands of short tons. "%" refers to percent of all sources of the listed pollutant.
Source: Electric Power Annual, 1996, Energy Information Administration, DOE.

Of the fossil-fired steam generators, coal-fired facilities contribute a disproportionately large share of these airborne contaminants. While coal accounts for about 84% of fossil-fuel fired electricity generated, it accounts for 90% or more of the emission gases (Parker and Blodgett, 1998).

Besides the fuel, the location of a generator can also have important consequences for air pollution impacts (for CO_2, source location is immaterial). Location can be important with respect to local ambient conditions and downwind areas due to long-range transport of air pollutants. For example, with prevailing air movement from west to east, nonattainment of the ozone air pollution standard in the Northeast has directed attention to the concentration of Midwest coal-fired generating facilities as possible sources of NO_x, a precursor to ozone.

After many decades of operating in a regulated market structure, the electric utility industry is facing significant change, from both new generating and transmission technology and shifting policy perspectives with respect to competition and regulation. The policy shift is prompted by the theory that market forces can and should replace the current regulatory structure that provides natural monopolies for electric utilities. The restructuring effort attempts to reduce and alter the role of government in electric regulation by identifying transactions, industry segments, regions, or specific activities that might benefit from fewer regulations. The overall purpose of restructuring is to promote economic efficiency, which presumably will lead to lower overall rates.

Some argue that this singular focus on economic efficiency could come at the expense of other values that the regulatory system traditionally has balanced against economic efficiency (Northeast States for Coordinated Air Use Management, 1998). The environmental concern with respect to restructuring is that the new economic signals implicit in a competitive market could result in increased emissions of undesirable pollutants (National Resources Defense Council, 1997). It is postulated that lower baseload prices would increase electricity demand and, therefore, increase generation and concomitant emissions. In addition, it has been suggested that the restructured market's revaluation of existing facilities relative to the marginal cost of constructing new capacity (along with lower operating costs) would encourage the rehabilitation and full utilization of older, more polluting generating facilities.

The relationship between restructuring electricity generation and environmental consequences is not simple (Begley, 1997). The final environmental outcome will result from dynamic process-balancing decisions which address (1) future electricity demand, including renovation of existing generating capacity or development/deployment of new emerging generating technologies for new construction or existing load locations (i.e., distributed generation), and (2) decisions which either implement existing environmental regulations or catalyze the promulgation of future environmental regulations. Restructuring would influence each of these trends to varying degrees, encouraging some, such as renovating existing capacity, and challenging others (e.g., existing environmental regulations).

1.7.2   Regulatory Setting

Many federal and state laws govern the siting and permitting processes for new energy facilities, including distributed generation technologies. The mix of regulatory authorities results in a complex federal/state process for overseeing the electric utility industry.

1.7.2.1   *Federal Regulations*

1.7.2.1.1   *The Clean Air Act (CAA)*

The Clean Air Act (1963) imposes a complex regulatory structure on air pollution sources (Leonard, 1997). From a historical perspective, the regulatory environment for a major emission source has been largely dependent on two factors: where the facility is located (that is, whether in an area meeting clean air standards or not) and the facility age (new or old source). At the federal level, the CAA and its various amendments (1965, 1967, 1970, 1977, and 1990) provide the critical statute for restrictions on electric generation technologies. Title I establishes national ambient air quality standards (NAAQS) that prescribe the maximum permissible concentration of pollutants allowed in ambient air (Environmental Protection Agency, 1996). Specifically, the act requires the United States Environmental Protection Agency (EPA) to establish standards for six criteria pollutants: CO, NO_x, SO_2, PM, ozone, and lead. Regions of the country where air pollution levels persistently exceed these standards are called non-attainment areas.

In general, the responsibility for reducing air pollution levels has been assigned to the states. Each state is required to promulgate a state implementation plan (SIP) providing for the implementation, maintenance, and enforcement measures necessary to attain the ambient air standards by the deadlines prescribed by the CAA. The EPA has the responsibility of reviewing each state's SIP, and is authorized to direct a state to revise its SIP if necessary. Two elements that an SIP must contain are federal new source performance standards (NSPS) and new source review (NSR) rules. The NSPS specify maximum pollutant emission rates for various processes, including combustion equipment. The EPA has promulgated new source performance standards for SO_2, NO_x, and PM. NSPS are based on the level of control that can be achieved by the best demonstrated technology. NSR rules govern the permitting of new emissions sources and are triggered if a new source emits or has the potential to emit at an annual rate specified by the NSPS. NSR rules distinguish between attainment and nonattainment areas with less stringent prevention of significant deterioration (PSD) rules applying to attainment areas. The trigger for NSR PSD rules is 250 tons/year for any regulated pollutant. Nonattainment areas are differentiated in classes based on severity of ambient pollutant concentrations: marginal, moderate, serious, extreme, and severe (Table 1.11).

TABLE 1.11

New Source Review Thresholds for Nonattainment Areas

Pollutant	Area Designation	Threshold (tpy)[a]
Ozone precursors (NO_x, VOC)	Marginal, moderate	100
	Serious	50
	Severe	25
	Extreme	10
Inhalable particulate matter (PM_{10}) and	Moderate	100
PM_{10} precursors (NO_x, SO_2, VOC)	Serious	70
Carbon Monoxide	Any nonattainment area	100
Nitrogen Oxides		
Sulfur Dioxide		

[a] Short tons per year.
Source: Leonard, R.L., *Air Quality Permitting*, CRC Lewis Publishers, Boca Raton, FL, 1997 (with permission).

In nonattainment areas, in order to construct and operate a new power plant or DG (or to make major modifications to an existing plant), the owner needs to obtain a permit from the state environmental agency if NSR levels are exceeded. The NSR process requires the owner to analyze alternative locations, sizes, production processes, and control techniques, and to demonstrate that the plant benefits outweigh its environmental and social costs. Facilities are also required to have control technology that meets the standard for the lowest achievable emission rate (LAER). The control technology required to meet the LAER is established by each state on a case-by-case basis for each emission source as it is permitted.

Furthermore, the owner of the plant is required to purchase offsets for each criteria pollutant that is in nonattainment. The EPA requires that emission offsets provide a positive air quality benefit to the area. Owners are, therefore, required to obtain more than one offset for each unit of pollutant emitted. The offset ratio depends upon the extent to which the region is in nonattainment. This offset requirement has promoted the establishment and trading of emission reduction credits for NO_x and VOCs among industries in 12 states.

The process for reviewing new facilities is slightly different in attainment areas. Owners are also required to obtain permits to construct and operate new plants (or make major modifications to existing plants) to ensure that new pollution sources do not make the region slip into nonattainment. These PSD permits require a review of the air quality impacts of the proposed facility. New plants are required to install best available control technology (BACT) for all pollutants regulated under the CAA. The control technology required to meet BACT standards is established by each state on a case-by-case basis for each emission source.

Historically, less stringent controls apply to existing electric utility facilities in attainment areas due to grandfathering statutes. This provides clear advantages when competing with new sources that require specific emission controls as specified by NSPS (Biewald et al., 1998). This situation may be changing, as a host of new regulatory initiatives could result in more stringent controls for existing facilities, especially coal-fired facilities. These controls could have a significant influence on the cost of power from coal-fired facilities, making them less attractive in a competitive marketplace.

1.7.2.1.2   National Environmental Policy Act

The National Environmental Policy Act (1969) requires the federal government to consider, at all stages of decision making, any federal action that could significantly affect the quality of the human environment. The NEPA framework includes a requirement for preparation of an environmental impact statement (EIS) if the project is deemed significant based on an initial environmental assessment (EA). If a federal government agency requires the preparation of an EIS, it could influence the decision to install a DG facility.

1.7.2.1.3   Other Federal Regulations

While air contaminant regulations are expected to be the dominant regulatory hurdle for distributed generation technologies, a DG project could be impacted by a suite of other federal laws. These include the Clean Water Act

(1987), Resource, Conservation and Recovery Act (1976), Occupational Safety and Health Act (1970), Toxic Substances Control Act (1976), Endangered Species Act (1973), Coastal Zone Management Act (1972), and Historic Sites Act.

1.7.2.2 State Regulations

1.7.2.2.1 Air Quality

Each state is responsible for implementing programs that conform to the mandates of the federal CAA and associated amendments. This has manifested as either the implementation of federal regulations or the promulgation of more stringent requirements for areas with severe air quality problems (e.g., Los Angeles). In both cases, the responsibility for these actions falls on air pollution control officials at the local level.

The potential for diverse state requirements can lead to inconsistent requirements that pose barriers or opportunities to the restructured electric utility industry. Differing requirements could allow generation companies to choose which state had the least stringent requirements, while the power could be transmitted to the demand location. Conversely, inconsistent or uncertain requirements might be an added incentive for construction of distributed generation capacity that has low air contaminant signatures and is therefore not subject to state permitting regulations.

1.7.2.2.2 Environmental Impacts of Government Decisions

Similar to NEPA requirements, 19 states have enacted laws that provide guidelines for state agencies to incorporate environmental factors into the decision process. Based on whether a DG project would require a state permit, a determination of environmental impacts of the DG project may be necessary.

1.7.2.3 Local Environmental Regulations

Based on the projected emission signatures of distributed generation technologies, it is anticipated that federal regulatory requirements for NSR will not be triggered. Local agencies will therefore be the primary regulatory authority overseeing DG projects relative to environmental drivers. Depending on the attainment status of the local area, permitting thresholds specified by air pollution control districts (APCDs) may limit the operating schedule (i.e., numbers of hours per year) or emission limits of a DG device if a predetermined power rating is exceeded. In addition, control technologies may be required. If the permit thresholds are not triggered, DG technologies will be exempt from local air quality regulations. Given the dynamics of evolving distributed generation technologies and the deregulated electric utility industry, it is likely that permitting requirements will change over the next several years. These changes may broaden the regulatory envelope to include select DG technologies.

1.7.3   Broad Policy Issues

Deregulation of the electricity market coupled with the concomitant development of distributed generation technologies has prompted interest on the part of federal, regional, and state agencies relative to what impacts these technologies or deregulation strategies will have on ambient air quality and environmental policy. There are three salient policy issues: (1) attainment of NAAQS for ozone and particulate matter, (2) global warming, and (3) stratospheric ozone depletion. Each area addresses a different combination of pollutant releases.

1.7.3.1   Ozone and Particulate Matter Attainment

Ozone and particulate matter attainment focuses on the requirement that all geographical areas in the U.S. must not have ambient troposphere ozone and particulate matter concentrations in excess of either federal or state standards (not all states have separate standards). There are over eighty areas in the U.S. that do not meet current or proposed ozone standards, primarily east of the Mississippi (an area called the ozone transport region or OTR) and in the large metropolitan areas of the far west (e.g., Los Angeles, San Francisco, Phoenix, Las Vegas, and Seattle). Particulate matter (PM), with several standards at stake (annual and daily PM_{10} and $PM_{2.5}$ values), also impacts hundreds of rural and metropolitan areas throughout the U.S. While electric utilities or distributed generation technologies do not directly emit ozone, PM_{10}, or $PM_{2.5}$, they do emit NO_x and VOCs which are precursor compounds for the production of ozone, PM_{10}, and $PM_{2.5}$.

It is uncertain what impact a competitive electric market and the integration of DG into the electric utility infrastructure will have on either ozone or particulate matter attainment strategies. This is a complex problem consisting of many broad air quality policy and engineering questions requiring mediation resolution at both state and federal levels. An increase in NO_x emissions could exacerbate the regional transport of ozone from relatively clean upwind areas in the Midwest to severely polluted downwind areas on the northeastern seaboard that do not emit significant concentrations of ozone, PM_{10}, and $PM_{2.5}$ precursor compounds. The impact of transferring electricity loads from hundreds of large utility generating stations (i.e., point sources of air pollutants) to tens of thousands of distributed generation units (i.e., area sources of air pollutants) could reduce peak (i.e., one-hour ozone concentrations) ozone values but increase daily ozone levels (i.e., eight-hour ozone concentrations), thereby threatening compliance with the proposed eight-hour ozone standard. A fundamental theme underlying these policy uncertainties is the resolution of jurisdictional authority between state agencies and the federal government. In the past, the state-regulated utility system meshed reasonably well with the state-implemented air quality controls. As the utility industry becomes more competitive and potentially more regional and as air quality also becomes more regional (regional haze, long-range pollutant transport), state-directed controls on existing sources may prove to be less efficient and effective.

1.7.3.2 Global Warming

Global warming is climate change postulated to be caused by the release of greenhouse gases. While CO_2 emissions are the primary global warming pollutant, other anthropogenic compounds of concern include nitrous oxide and halogenated fluorocarbons. The electric utility industry accounts for approximately 35% of total U.S. CO_2 emission inventory. CO_2 is not currently regulated under the CAA. However, if the U.S. ratifies the Kyoto Protocol (i.e., the international treaty mandating control of global CO_2 emissions), the U.S. would be required to reduce greenhouse gas emissions below 1990 levels. Both conventional utility and select distributed generation technologies will release CO_2 emissions with significant increases predicted in a deregulated electric utility environment (Parker and Blodgett, 1998). The absolute amount is dependent on the efficiency design criteria for the combustion processes. For example, a defined maximum allowable heat rate for distributed energy sources (i.e., minimum efficiency) will lower combustion gas emissions by minimizing rate of fuel use. Maximum allowable heat rate will also reduce use rate of nonrenewable natural capital (e.g., natural gas).

The global warming situation is complicated by the observation that for select combustion technologies (i.e., turbines), CO_2 emissions will be inversely proportional to NO_x emissions. This relationship is critical for understanding the air quality policy implications of distributed generation technologies. Each pollutant has unique air quality implications and, therefore, will require individual policy initiatives. For example, if CO_2 is prioritized for emission reduction strategies, ambient ozone control strategies will require recalibration to offset increases in the NO_x inventory.

1.7.3.3 Stratospheric Ozone Depletion

Stratospheric ozone depletion refers to thinning of the ozone layer located in the upper atmosphere by chlorofluorocarbons (CFCs) and other anthropogenic ozone depleting chemicals (ODCs). The ozone layer shields organisms from solar ultraviolet radiation that can cause skin cancer. It also assists in trapping infrared radiation (i.e., heat), thus maintaining the earth's heat balance. The primary sources of ODCs are (1) refrigerators, (2) heating, ventilation, and air conditioning (HVAC) systems, (3) foam packaging, and (4) certain cleaning solvents. Given the importance of stratospheric ozone depletion as a global environmental problem, several legal instruments have been established which will force the elimination of ODCs. These include the 1990 CAA Amendments and the 1992 Montreal Protocol.

Using a co-generation configuration, several distributed generation technologies (e.g., microturbines) have the potential to include an innovative ODC emission reduction strategy for HVAC systems; the waste heat from kinetic energy pathways can be captured for space heating or absorptive cooling purposes. Absorptive cooling is a well established HVAC strategy that obviates ODCs for cooling. This ODC reduction strategy may have national or state policy implications as (1) a possible "environmental credit" for trading purposes, (2) an incentive to promote distributed generation technologies, or (3) an environmental offset strategy that may aid in the solution of other intractable environmental policy issues associated with distributed generation technologies.

1.7.4 Environmental Attributes of DG Technologies

Of the different air, liquid, and solid emissions that may be associated with DG technologies, air emissions have the strongest influence on a project's viability relative to permitting regulations. NO_x and CO_2 emissions are the critical path emission categories given their magnitude for conventional electric generation technologies. While SO_2 is also an important emission category for tradition electric utilities, SO_2 emissions are expected to be negligible for DG technologies.

The air emission signatures of selected distributed generation technologies and conventional utilities are tabulated in Table 1.12. For the DG devices, these values are first-order proxy based on theoretical calculations or laboratory source testing (Cler and Lenssen, 1997 and National Renewable Energy Laboratory, 1995). The actual air emission waste streams for each distributed generation technology are specific to the end-design features of the DG device and the characteristics of the end-use location.

As shown in Table 1.12, the emission characteristics of DG technologies differ considerably. Fuel cells are the cleanest option, followed by microturbines and reciprocating engines, respectively. Fuel cells are potentially a very low source of air emissions for all pollutant categories. With the exception of CO_2 emissions, microturbines also exhibit low emissions for all classes of pollutants. In contrast to conventional electric generation technologies, DG devices generally provide opportunities for considerable emissions reductions on a per-kWh basis.

1.7.5 Other Environmental Drivers

Conventional air quality regulations place fixed limits on emissions from individual sources. However, compliance may be more economically feasible for some sources than it is for others. Currently, federal and state environmental policy is evolving from prescriptive command and control regulations to descriptive market-based strategies. The goal of market-based pollution control strategies is to optimize the process of emission reductions based on cost and the particular needs of the industrial process in question. This development was triggered by the realization that a single environmental regulation

TABLE 1.12

Emission Characteristics of Electric Generating Technologies

	Pollutant			
Technology	NO_x[b] lb/MM Btu[a]	CO_2 lb/MM Btu[a]	CO lb/MM Btu[a]	SO_2[c] lb/MM Btu[a]
Conventional				
Coal	0.1->2	55.9		0.07–2.55
Natural gas	0.005->1	31.7		0.3
Residual fuel oil	0.05->1	46.8		
DG				
Micro turbine	0.4	119	0.11	0.0006
IC engine (gas)	3.1	110	0.79	0.015
IC engine (diesel)	2.8	150	1.5	0.3
Fuel cell	0.003	ND	ND	0.0204

[a] Conversion of lb/MM Btu to g/kWh uses a Btu to kWh conversion unique to the device engineering specification based on efficiency data.
[b] New source performance standard for NO_x electric utilities: 0.15 lb/MM Btu.
[c] New source performance standard for SO_2 electric utilities: 0.3 lb/MM Btu
[d] Conventional system data from Clean Air Network, *Poisoned Air: How America's Outdated Electric Plants Harm our Health and Environment*, 1997.
[e] DG system data from Cler, G. and Lenssen, N., Distributed generation: markets and technologies in transition, E source 1997, and National Renewable Energy Laboratory, Distributed utility technology cost, performance, and environmental characteristics, NREL/TP-463-7844, 1995.
Source: Cler and Lenssen, 1997, and National Renewable Energy Laboratory, 1995.

template will not be effective in controlling and reducing emissions for a myriad of air pollution sources in a diverse physical and meteorological landscape.

A key element in all market-based pollution control strategies is emissions trading programs. Trading programs consist of institutionalized frameworks where discrete quantities of pollutant releases can be bought or sold by one entity from another entity (Bearden, 1999). A trade can occur under different circumstances which would typically include: (1) a company purchases pollution credits because it emits more contaminants than it is allowed to and it therefore has to make up the difference for accounting purposes, (2) a company sells pollution credits because it emits contaminants at lower levels than it is required to, or (3) a group buys or sells a pollution credit for investment purposes. Trading has the potential to improve air quality in cases where a pollutant disperses over a broad geographic area and the environmental objective is to control total emissions rather than limit local emissions from individual sources.

In the U.S., experience with trading programs began in the 1970s when the EPA used its authority under the CAA to develop more flexible policies that states could pursue to comply with federal air quality standards. Most trading programs have focused on pollutants generated from large stationary sources that accounted for a significant share of total emissions. Several trading programs are either operational today (e.g., the acid rain trading program) or are being currently proposed by federal or state agencies (the Ozone Transport Commission's NO_x trading rule).

Many issues need to be resolved prior to the broad integration of pollution trading markets into federal and state environmental policy. These include: (1) interpollutant emissions trading, (2) inter- versus intrastate emissions trading, (3) trading directionality based on environmental justice concerns, and (4) emission credit reconciliation. Notwithstanding these uncertainties, using DG resources to meet new generation needs will invariably yield both net NO_x emission reductions and NO_x emission credits. NO_x emission credits will therefore be a tangible variable which will need to be addressed in the sale of distributed generation systems. NO_x credits can provide an incentive for customers to purchase microturbine products if it is determined that the NO_x emission reduction credit belongs to the distributed load producer.

1.8   Communications and Controls Technologies

Dr. Sunil Cherian, Sixth Dimension, Inc.

Historically, there was no demand for command and control solutions for distributed resources that had to take into account the diverse interests seen today. However, the emerging competitive energy environment brings this need to the forefront as a necessary ingredient for optimal market operation — it should be possible to buy and sell various value components of distributed resources in an open market. The current debates around distributed generation are a manifestation of various interested parties jockeying for position in the emerging competitive environment or, in many cases, attempting to delay the transition to full competition for self-preservation.

While there are new products to be built and new business models to be developed, there are few, if any, technical barriers to overcome to facilitate an open distributed generation (DG) market. However, the types of technologies that are required for effective operation of such an open market have never been widely utilized in the highly centralized, hierarchical control systems commonly found in utility supervisory control and data acquisition (SCADA) systems.

Utility-owned generation facilities and substations are generally equipped with SCADA systems that enable centralized control of the whole system. These systems are custom designed and put in place for long periods of time. They are not designed for a dynamic DG marketplace where customers with generation assets may be continually added or removed from the system, where power is bought and sold in real time, and where different market participants may have control over different DG value streams.

The entities that will ultimately make an open market for DG work are energy consumers empowered with the tools to make rational market choices about energy services and service providers. Communication and control systems for managing distributed generation have to account for multiple service providers, a dynamic customer base, market-making capabilities, and a fluid asset base where generators can join and leave the network with little impact on the rest of the system.

It is highly likely that a well developed DG marketplace will share more in common with business-to-business e-commerce than with conventional utility SCADA systems. Such systems will include mechanisms for real time price discovery for the various value components of DG and mechanisms for dynamic value chain creation for delivering complete solutions to energy consumers. A fundamental requirement for such a flexible and scalable communication and control system — one that can meet the needs of market-based distributed generation — is distributed intelligence.

1.8.1   Distributed Intelligence

Recently, *EPRI Perspectives on the Future* noted that distributed generation will play out ...

> ... in much the same way the computer industry has evolved. Large mainframe computers have given way to small, geographically dispersed desktop and laptop machines that are interconnected into fully integrated, extremely flexible networks. In our industry, central-station plants will continue to play an important role, of course. But we're increasingly going to need smaller, cleaner, widely distributed generators — combustion turbines, fuel cells, wind turbines, photovoltaic installations — all supported by energy storage technologies. A basic requirement for such a system will be advanced electronic controls: these will be absolutely essential for handling the tremendous traffic of information and power that such complicated interconnection will bring.

On the information side, monolithic, centrally located control systems in charge of system coordination cannot scale economically to meet the demands imposed by such systems. The solution is to design communication and control systems matched to the distributed, multi-participant, and dynamic nature of distributed resource management.

With the emergence of the Internet and the World Wide Web, the ability to communicate has become pervasive down to the smallest of embedded controllers. Currently, most vendors are Web-enabling their products by building thin Web servers into them. They are providing remote monitoring and control interfaces for individual assets. This capability is the first step towards distributed intelligence. Aggregating and coordinating the operation of large numbers of systems from different vendors is the next step. Advances in distributed computing, embedded control, and wide area communications address this need. The final step is the development of a ubiquitous platform for service delivery to end customers. Once again, the Internet and the World Wide Web offer the most widely available and fastest growing platform for service delivery.

This approach follows a bottom-up design strategy — a highly scalable control strategy based on pushing intelligent decision-making capability to every relevant part of the system. This strategy turns the entrenched utility practice of centralized, top-down control on its head. For distributed generation to achieve its full potential, communication and control systems have to break out of the top-down mold and adopt a bottom-up strategy based on distributed system intelligence. Most of the enabling technologies are available today, and the race is on for market-leading solutions in this area.

1.8.2 Enabling Technologies

Solutions for effective control, coordination, and optimization of distributed power generation in an open market environment require the integration of four distinct capabilities: distributed control strategies, distributed computing, pervasive communications, and embedded microprocessor-based control.

1.8.2.1 Distributed Control

Traditionally, distributed control refers to a class of concepts and techniques used to solve complex control problems which may be formulated as a number of smaller interconnected sub-problems. The sub-problems involve some degree of coordination in their solutions. Distributed control systems are appropriate for large-scale systems with hundreds of variables that make centralized control infeasible. A typical application of distributed control is for designing fault tolerant systems where the failure of a single subsystem must not lead to catastrophic failure of the whole system.

Another related area, intelligent control, deals with methods aimed at enhancing the capability and flexibility of controllers from the servo level to the level of process management and process coordination in complex systems. In intelligent control, techniques such as expert systems, neural networks, and genetic algorithms are used along with recent communication, distributed processing, and operations research strategies to extend the performance and range of controller operation. These techniques rely on the accuracy of system models and the reliability and response characteristics of inter-process communications and are firmly rooted in conventional control theory. While there are several viable distributed control strategies, choosing a particular strategy depends on application-specific tradeoffs.

1.8.2.2 Distributed Computing

One of the primary features that distinguishes a software agent from other computational processes is the concept of autonomy. Currently, when a user executes a command on a computer, a single process is activated that executes until it terminates. More sophisticated software applications wait for user input and then carry out some action or sequence of actions and return some result to the user. This is a passive view of computation — the direction or motivation for action comes from the user.

In agent-based distributed computing, this passive view gives way to a self-motivated computational process (software agent). Agent-based distributed computing always seeks to satisfy some internal goals with a minimum of human intervention. An autonomous software agent is viewed as one that can reason and plan a solution strategy once a task is delegated to that agent. The choice of strategy will depend on many factors. An autonomous agent will have the ability to select appropriate strategies for solving ill-posed problems without the user having to provide the decision-making intelligence at each step.

Another important characteristic of software agents is their ability to migrate across networks and carry their data and execution state with them. There are three main ways to access information in a computer network:

- Client software applications (such as web browsers) can access information directly through a fixed set of standard communication protocols.

- Intermediate software between the client application and the information source can provide a layer of abstraction that encapsulates the details of finding and retrieving relevant data.

- Software processes with specific goals can migrate across the network and search remote sites for relevant data.

The last of these three methods alleviates the need to maintain stable communication channels between a client and a server and the need to transport data to the client site before any processing can be carried out on it. Ideally, a mobile agent can perform most of its computation where the data resides and simply bring back the results.

Software agents are well suited for distributed problem solving applications such as those encountered in DG. Distributed problem solving strategies address situations where top-down problem solving turns out to be very difficult. A classic example of distributed problem solving is encountered in restaurants. Each individual who works at the restaurant is in charge of a specific function — seating guests, preparing food, cleaning up, etc. There is no central authority that continuously tells each person what to do next. If each employee handles his or her duties well, the overall objectives of the restaurant — business profitability and customer loyalty — are automatically achieved. The duties and responsibilities of each employee are defined in advance, and his or her activities are regularly monitored to ensure smooth running of the establishment. This type of distributed problem solving is qualitatively different from conventional strategies where a centralized control mechanism issues commands about what each system component should do next.

The success of a centralized problem solving strategy depends on the continuous availability of global information about the state of the entire system to the central decision-making authority. As problem domains become larger, this assumption seldom holds. Besides, the nature of certain problems makes the availability of complete and accurate system information impossible. Unfortunately, most problems in the real world are of this kind. This makes software agents promising candidates for distributed problem solving. From scheduling meetings to running manufacturing processes and air traffic control systems, distributed artificial intelligence researchers are applying agent-based techniques for solving complex real world problems.

1.8.2.3   Embedded Hardware

There are two ways to communicate with DG systems from a remote location. The most common approach is to use a simple data acquisition unit such as an RTU for data collection and communication. Back-end applications periodically poll the remote data acquisition system and send command signals to initiate some action. This approach works well for static systems with a well defined hierarchical control structure. The decision-making intelligence of the system is centrally located, and the remote units simply serve as interfaces to transducers.

The second approach is to push significant decision-making capabilities to the remote units, allowing them to make local decisions based on information acquired from various parts of the system. This requires embedded microprocessor systems with significantly more hardware and software capabilities than simple data acquisition systems. The cost of microprocessors has been dropping fast, and functionality has been doubling every 18 months, making it possible to embed significant processing capabilities into remote hardware units that interface with generators. This makes it possible to design economical software solutions for managing distributed generation consistent with the demands imposed by an open market for distributed resources.

The embedded processor has to serve as the gateway between distributed assets and remote applications used by service providers. It also has to have sufficient intelligence to manage the assets in case remote communications fail or abnormal conditions are detected locally. Several hardware vendors currently offer products to meet this need.

References

Bearden, D., *Air Quality and Emission Trading: An Overview of Issues*, CRS Report 98-563, 1999.
Begley, R., Electric power deregulation: will it mean dirtier air?, *Environ. Sci. Tech.*, October 1997.

Biewald, B., White, D., Woolf, T., Ackerman, F., and Moomaw, W., Grandfathering and Environmental Comparability: An Economic Analysis of Air Emission Regulations and Electricity Market Distortions, Synapse Energy Economics prepared for the National Association of Regulatory Utility Commissioners (NARUC), 1998.

Borbely, A., Distributed generation: the physics of asset management, *Energy Frontiers*, presented at the International Conference on Distributed Generation, May 1999.

Clean Air Network, *Poisoned Air: How America's Outdated Electric Plants Harm Our Health & Environment*, 1997.

Cler, G. and Lenssen, N., Distributed Generation: Markets and Technologies in Transition, E source, DE-1, 1997.

Distributed Power Coalition of America, 1998.

Energy Information Administration, Commercial Building Energy Consumption Survey, 1995.

Environmental Protection Agency, National Air Quality and Emissions Trends Report, 454/R-96-005, 1996.

Hill, L. J., A Primer on Incentive Regulation for Electric Utilities, Oak Ridge National Laboratory, Oak Ridge, TN, ORNL/CON-422, October 1995.

Maes, Pattie, Situated agents can have goals, in *Designing Autonomous Agents: Theory and Practice from Biology to Engineering and Back*, MIT Press, Cambridge, MA, 1990.

National Renewable Energy Laboratory, Distributed Utility Technology cost, performance, and Environmental Characteristics, NREL/TP-463-7844, 1995.

Natural Resources Defense Council, Benchmarking Air Emissions of Electric Utility Generators in the Eastern United States, Public Service Electric and Gas Co. and Pace University, 1997.

Northeast States for Coordinated Air Use Management, *Air Pollution Impacts of Increase Deregulation in the Electric Power Industry: An Initial Analysis*, 1998.

Parker, L. and Blodgett, J., Global Climate Change: Reducing Greenhouse Gases — How Much from What Baseline, CRS Report 98-235, 1998.

Patterson, W., *Transforming Electricity*, Earthscan Publications, London, UK, 1999.

PNNL, An Analysis of Buildings-Related Energy Use in Manufacturing, PNNL-11499, April 1997.

Poruban, S., U.S. wellhead deliverability slide may spike gas prices in late 1999, *Oil Gas J.*, p. 27, August 30, 1999.

2

Combustion Engine Generator Sets

Eric Wong, Herb Whitall, and Paul Dailey

CONTENTS

Among distributed power generation technologies, combustion engines (CEs) are the most mature prime movers. Advantages include comparatively low installed cost, high shaft efficiency, suitability for intermittent (start-stop) operation, high part-load efficiency, and high-temperature exhaust stream for combined heat and power (CHP). Additionally, a sales and technical support structure is already in place, parts are readily available and generally inexpensive, and service technicians (from both the dealer and customer's staff) have experience with maintenance and repair. Figure 2.1 shows a typical CE-based generator set (genset). This chapter details not only traditional internal combustion engine generators, but also developments in external combustion, or Stirling, engines.

FIGURE 2.1
Engine–generator set package (courtesy of Caterpillar, Inc., with permission).

Almost 2600 cogeneration, independent power and small power facilities, most fueled by natural gas, already existed in the United States as of the year 2000. Engine-driven generators account for 46% of the installations but only about 1.5% of the total capacity of 99 GW. Engine generator systems dominate below 1 MW capacity. Stationary reciprocating engines represent 146 GW (5%) of the world's 3000 GW of installed electric generating capacity. In the U.S., engines comprise 52 GW (7%) of 780 GW total installed capacity. In some parts of the world, engines provide a far greater share of generating capacity. For example, in The Netherlands, China, and Indonesia, engines make up more than one-fourth of total installed capacity. In the U.S., there are approximately 300,000 stationary engines. In electric power generation service, there are 76,500 diesel- and gaseous-fueled units greater than 350 kW and 150,000 units below 350 kW.

To date, the largest users of engine-driven generators are gas, electric, and water utilities. About 3100 engine generators are in use or on active standby in the electric utility industry, most at municipal utilities and rural electric cooperatives. The next largest users are manufacturing facilities, hospitals, educational facilities, and office buildings. Sales of prime movers above 1 MW — both engines and turbines — have grown significantly in the past decade. In capacity terms, reciprocating engine sales grew nearly sixfold from 1988 (2 GW) to 1998 (11.5 GW), while combustion turbine sales increased more than threefold. From 1990 to 1998, gas engine sales went from a small fraction of gas turbine sales to outselling gas turbines more than five to one in 1998. Figure 2.2 shows the trend.

Diesel engines are the leading power sources in the 1 to 5 MW size range, mainly because of their low first-cost position. However, gas engines grew

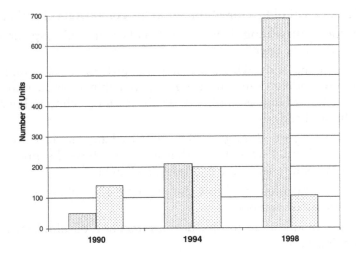

FIGURE 2.2
Gas engine and gas turbine sales trends (courtesy of GRI, with permission).

from 4% of engine sales in 1990 to more than 14% (of a much larger market) in 1998. That trend is likely to continue for two reasons. First, increasingly strict air emissions regulations make diesel engines impractical for continuous duty in many areas. For example, California's South Coast Air Quality Management District, which includes Los Angeles, limits stationary diesels to 200 operating hours per year. Other California air management districts and the entire state of New Jersey impose similar restrictions. Second, gas engine performance has steadily improved for the past 15 years. For example, one manufacturer reports that its gas engine fleet mechanical efficiency (400 kW and larger) increased from 31.9% in 1986 to 34.7% in 1995 — a 10% improvement in ten years. Over the same period, nitrogen oxide (NO_x) emissions decreased 70%, from 14.1 to 4.0 g/bhp-hr. Diesel engines will remain viable for standby service, some peak shaving systems, and other intermittent duty. However, natural gas is already the fuel of choice for engine-driven generation involving long or frequent runs.

2.1 Internal Combustion Engine Design Overview

As shown in Figure 2.3, internal combustion (IC) engines convert heat from combustion of a fuel into rotary motion of a crankshaft which, in turn, drives a generator in a distributed generation (DG) system. IC engines consist of:

- Air filter to filter wear particles from incoming air
- Cylinder block, cylinder liners, and cylinder heads to contain combustion and moving components

- Intake manifold to direct the air to the combustion chamber
- Pistons with sealing rings; the piston fits in and rides up and down in a cylinder liner
- Connecting rods which connect the pistons to the crankshaft; pistons undergo a reciprocating motion within a combustion chamber (the volume is confined by the piston, cylinder liner, and cylinder head)
- Valves or ports through which the air or air/fuel mixture enters the combustion chamber; exhaust valves allow the products of combustion to leave the combustion chamber
- Exhaust manifold to direct the products of combustion from the combustion chamber
- Lubrication system to produce a lubricating oil film between all moving parts
- Cooling system for heat rejection; this may be a system of ducting and fins for air-cooling or an internal system of passages for liquid cooling
- Turbocharger(s) or rotary compressors which may be used to increase the charge and power density (amount of power per cubic centimeter or cubic inch of piston displacement) of the engine

Figure 2.4 shows some of the components of the top end of a compression ignition engine.

Engines are classified as spark ignited or compression ignited. When the piston is farthest from the cylinder head, it is at bottom dead center (BDC). When the piston is at the top of its stroke, it is at top dead center (TDC). The compression ratio is the ratio of the volume confined by the combustion chamber at BDC divided by the volume at TDC. At a compression ratio of about 16:1, the gas is hot enough to self-ignite without a spark plug. A diesel engine operates at around 16:1 and is a compression-ignition engine.

2.1.1   Two-Stroke versus Four-Stroke

All engines go through four cycles: intake, compression, power, and exhaust. Engines can be either two stroke or four stroke. In two-stroke designs, the four cycles are completed during each complete revolution of the crankshaft. Therefore, there is a power stroke during each revolution. In four-stroke designs, the four cycles are completed every two revolutions of the crankshaft; there is a power stroke only every other revolution of the crankshaft. For this reason, two-stroke engines have higher power density than four-stroke engines. However, because the intake and exhaust functions do not achieve completion, the two-stroke engine is not as efficient as a four-stroke engine and has higher emissions.

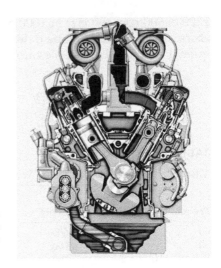

FIGURE 2.3
Crossection of ICE showing from top to bottom: turbochargers, intake manifold, pistons and cylinders, connecting rods, crankshaft, and oil sump (courtesy of Caterpillar, Inc., with permission).

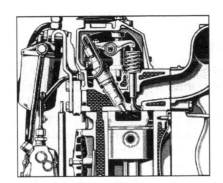

FIGURE 2.4
Induction and combustion chamber detail (courtesy of EGSA, with permission).

2.2.2 Engine Speed

The output power of an engine is the product of its torque and its speed; therefore, higher speed engines have higher power densities. However, piston size and weight limit the speed at which an engine can operate. The speed at which the flame front travels through the compressed gas also limits the speed. Slower speed engines are more efficient and can burn lower grade, less expensive fuels.

Engines are usually directly coupled to the generator and operate at synchronous speeds (Table 2.1). These speeds are defined by the frequency in Hz (cycles per second) of the electric grid in the country involved (usually 60 or 50 Hz). Small engines generally operate at 3600 rpm (60 Hz) or 3000 rpm (50 Hz) with 2-pole generators. Medium size engines operate at 1800 rpm (60 Hz) and 1500 rpm (50 Hz) with 4-pole generators. Large engines operate at

1200 or 900 rpm (60 Hz) or 1000 or 750 rpm (50 Hz) with 6- or 8-pole generators. Engine speed can be determined by solving this equation when frequency and number of poles are known:

$$Frequency = \frac{RPM \times Number\ of\ Poles}{120}$$

TABLE 2.1

Variation of Synchronous Speed with Number of Poles and Frequency

Frequency (Hz)	RPM			
	Two Pole	Four Pole	Six Pole	Eight Pole
60	3600	1800	1200	900
50	3000	1500	1000	750
RPM per Hz	60	30	20	15

2.1.3   Cooling Systems

Cooling systems use either air or liquid. Air-cooled engines are primarily used in small- to medium-sized engines. Fins added to the engine exterior increase surface area for heat loss by radiation and convection. Air-cooled engines tend to have higher ambient noise because of the fan blowing air over the engine. Liquid-cooled engines have internal passages throughout the engine. Figure 2.5 shows the main components of a liquid cooling system with remote radiator and fan.

Normally, there is also a lubricant cooler to keep the lubricating fluid, usually oil, at about 10°C (15°F) above the coolant temperature but below 125°C (255°F). The radiator may be attached to the engine and the radiator fan driven by the engine crankshaft. Alternatively, the radiator may be remotely mounted and the fans driven by electric motors. The latter case usually results in the least noise in the engine room. In total energy installations, the heat is recovered from the oil cooler, engine coolant, and exhaust gases.

2.1.4   Efficiency and Fuels

A venerable rule is that one-third of the energy content of the fuel is useful work, one-third is rejected to the coolant, and one-third is exhausted and radiated from the engine itself. Current technology has improved these old values so that modern, efficient engines approach 40% efficiency. However, this occurs only at the most efficient design point of engine operation. At off-design (i.e., lower) loads and speeds, the efficiency drops. When comparing engine efficiencies (or Btu/kWh ratios), one must make sure that all

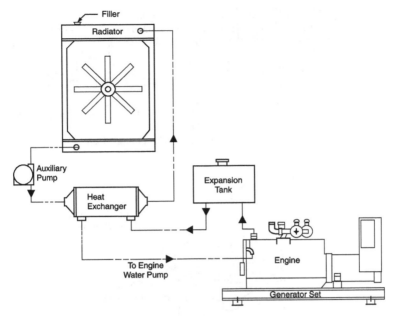

Remote Radiator Installation

FIGURE 2.5
ICE cooling system with remote radiator (courtesy of EGSA, with permission).

accessories and parasitics are accounted for, including fans, alternators, and fuel pumps.

The difference between diesel engines and natural gas engines centers around when the fuel is introduced to the combustion chamber. In diesel engines, the fuel is introduced at high pressure into either the combustion chamber or a pre-chamber after the air in the combustion chamber has been compressed; after compression, the air has reached a temperature at which the fuel droplets will burn. For all engines powered by low-pressure natural gas, the fuel is introduced via a carburetor after the air filter but before the intake manifold, as shown in Figure 2.6.

The fuel and air exist in the combustion chamber before the mixture is compressed. The mixture is ignited with a spark plug. For this reason, natural gas engines operate at about a 10:1 compression ratio, while diesel engines operate at compression ratios ranging from 13:1 to 22:1. Horsepower output, all other things being equal, increases (at a decreasing rate) with compression ratio. Therefore, with two engines of equal displacement and speed, the diesel engine will produce more power than the natural gas engine because of its higher compression ratio. Natural gas engines are less responsive to sudden load changes, and operate better with more constant loads.

Because of its higher power density, a diesel engine will occupy less space than a natural gas engine for the same power output, and a diesel engine will be less expensive than a natural gas engine for the same power output.

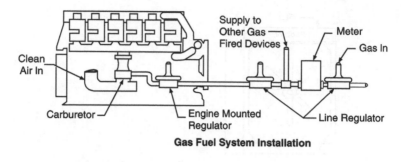

Gas Fuel System Installation

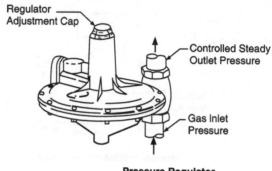

Pressure Regulator

FIGURE 2.6
Natural gas fuel system with pressure regulator detail (courtesy of EGSA, with permission).

However, natural gas as a fuel is usually less expensive than diesel fuel for the same heat content (Btus), and therefore, if the engine is used for a large number of hours per year, the total cost to own and operate the natural gas unit may be lower. Natural gas is not available in all locations, while diesel fuel can be transported anywhere. Even though natural gas engines are less efficient than diesel engines, in total energy installations, where exhaust heat can be recovered and used to heat occupied spaces and/or water, a natural gas engine-based CHP can be more efficient, utilizing up to 90% of the Btu content of the fuel. Diesel engines can also be used for total energy systems with efficiencies in the 85% range. Figure 2.7 compares key operating indicators for spark and compression ignition engines.

2.1.5 Emissions

Natural gas engines have fundamentally lower NO_x and particulate emissions than diesel engines because of the lower peak cylinder pressures in natural gas engines. Lean-burn natural gas engines have particularly low emissions but pay a small penalty in their ability to respond to sudden load changes. With the advent of electronic control of diesel fuel injection systems,

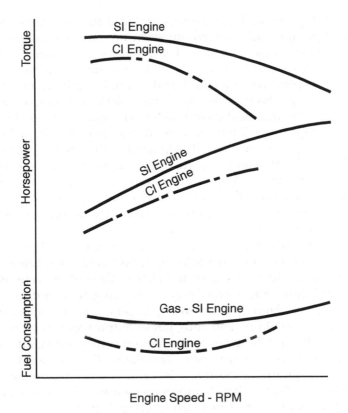

Comparison of SI Engine to CI Engine with
Same cu. in. Displacement

FIGURE 2.7
Comparison of spark ignition and compression ignition engine characteristics (courtesy of
EGSA, with permission).

smoke on transient load changes has been reduced greatly. Electronic control
and very high injection pressures have also reduced all emissions (NO_x, CO,
and particulates) on newly designed engines.

2.2 Past and Current Trends in Engine Development

Natural gas engines have been used for power generation since the mid-
1940s, and have evolved steadily. The earliest gas engines were derived from
diesel blocks and incorporated many of the same components as diesel

engines. Spark plugs and carburetors replaced fuel injectors, magnetos replaced fuel pumps, and lower compression-ratio pistons were substituted to run the engines on gaseous fuels. The first gas engines were set to run at optimum power output without regard to emissions and fuel efficiency. Where fuel was plentiful and could be delivered at little or no cost (often straight from the gas well to the engine fuel inlet), these engines were economical for producing local power for electric generation or for driving pumps and compressors. Their mechanical efficiency was about 25 percent. Most of these engines were naturally aspirated — the absence of turbochargers and charge air aftercoolers made them simple to apply and maintain. However, power levels could not reach above 100 psi brake mean effective pressure, while diesel versions could achieve more than 200 psi. (Brake mean effective pressure, or BMEP, is a measure of engine power output because HP = BMEP × piston area × stroke × speed.)

Since the mid-1980s, gas engine manufacturers have faced growing pressure to increase fuel economy while lowering NO_x emissions. Leaner air/fuel mixtures requiring turbochargers and charge air coolers were used. The leaner fuel mixtures and lower in-cylinder firing temperatures sharply reduced NO_x from about 20 to below 5 g/bhp-hr. Lower cylinder temperatures also meant that BMEP (and, thus, power output) could increase without the damaging effects of hotter exhaust gases on valves and manifolds.

Lean-burn designs, however, have drawbacks. Fuel economy decreases because of turbo pumping losses and less efficient lower-temperature combustion. Additionally, the risk of detonation increases. Detonation, a premature explosive combustion of fuel, can severely damage pistons, liners, and cylinder heads. In response to this, manufacturers began to install control systems, including detonation sensors, that automatically retard ignition timing. Manufacturers also began to install swirl plates and high "squish" pistons to increase turbulence in the cylinders at spark ignition. The result is more complete combustion and greater efficiency. Other advances include solid-state ignition controls to improve timing accuracy and electronic controllers that automatically maintain the optimum air/fuel ratio, adjusting for changes in air temperature and pressure, fuel heating value, and other operating variables.

2.2.1   New Developments in Gas Engine Gensets

Today's gas engine generator sets are used for peak shaving, intermediate load, and base load installations. The use of reciprocating engines for power generation is constrained by the prices of gas and electric utility power. Lower specific capital cost and greater efficiency are the obvious keys to improving gas engines' competitive position. Figure 2.8 shows how specific capital cost and efficiency affect a distributed generation power plant, used primarily for peak shaving at 3200 operating hours per year. The internal rate

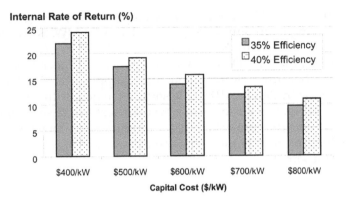

FIGURE 2.8
Internal rate of return for various initial engine costs.

of return, based on an after-tax cash flow analysis, climbs rapidly when installed costs are reduced and efficiency is enhanced.

Recognizing environmental and efficiency needs, engine manufacturers, industry associations, and government agencies have embarked on programs to advance gas engine technology. The next several years should see significant investments in basic research and development dealing with engine air intake, fuel and combustion systems, controls, and safety systems, as well as power generation and interconnection devices.

In 1996, the Gas Research Institute (GRI) launched a series of gas engine development projects to target improvements specifically for the distributed generation market. These efforts included two joint programs with engine manufacturers. The Advanced Reciprocating Gas Engine Technology (TARGET) program, launched in 1997, aims to develop, demonstrate, and commercialize a higher-speed (1800 versus 1200 rpm), high-output, lean-burn, spark-ignited gas engine. From the base unit, a widely used 16-cylinder engine rated at 820 kW, the project goal is to deliver:

- 67% more power output (to 1.35 MW)
- 30% lower engine and generator first cost ($/kW)
- 30% lower maintenance cost ($/kWh)
- 38% engine shaft efficiency (a 6% increase from 36%)
- NO_x emissions of 0.5 g/bhp-hr

The program began in late 1997, and the TARGET engine completed test cell runs in 1998 and 1999.

The second GRI project involved development of a high-output, dual-fuel engine using micropilot prechamber technology. Compression ignition of diesel pilot fuel eliminates spark plugs and ignition system components. The project's primary goal is to boost power output in an existing engine series by 40%, resulting in an output range from 1.1 to 3.3 MW. Shaft efficiency is projected at 38 to 41% with NO_x emissions less than 0.75 g/bhp-hr.

Longer term, the U.S. Department of Energy, with technical assistance from GRI and Southwest Research Institute, supports the Advanced Reciprocating Engine Systems (ARES) consortium, aimed at further substantial advances in gas engine performance over the next five to seven years, starting in 2001. The ARES project goal is to produce a commercially viable gas engine delivering 50% engine shaft efficiency and NO_x emissions of less than 0.1 g/bhp-hr.

2.2.2   Increasing Speed

Engine-driven power generators typically must run at fixed (synchronous) speeds to maintain a constant 50 or 60 Hz output. In the 60 Hz market, common operating speeds are 900, 1200, 1800, and 3600 rpm. Substantial power output increases can be achieved by increasing synchronous operating speed. Table 2.2 shows the dramatic effect on installed cost if speeds are increased. The increase in power output has an exact inverse effect on unit installed cost. Note that a modest increase in cost could be encountered due to higher strength requirements for reciprocating members in higher speed engines. However, at the low speeds listed, the effect is not strong.

TABLE 2.2

Decrease of Capital Cost with Increased
Engine Speed

Speed	Power	Cost
900 → 1200 rpm		
900 rpm	365 kW	$300/kW
1200 rpm	480 kW	$227/kW
1200 → 1800 rpm		
1200 rpm	480 kW	$300/kW
1800 rpm	725 kW	$200/kW

In the TARGET program, the base engine typically operated at 1500 rpm for 50 Hz applications and at 1200 rpm for 60 Hz. Running at 1800 rpm, which is within the engine's original design capability, will enable greater electric power output at the same BMEP.

2.2.3 Increasing BMEP

Higher BMEP is attained primarily by increasing combustion cylinder air pressure. The TARGET program, for example, seeks to boost BMEP by 15% along with increased speed. High-efficiency turbochargers, larger air delivery, and intercooler systems will efficiently provide the higher air flow required. Table 2.3 indicates the effect of BMEP on power output and, therefore, on unit installed cost. This table shows the effect of increasing BMEP from 175 to 250 psi on a 40 L, 1800 rpm engine.

TABLE 2.3

Effect of BMEP on Power Output and Unit Cost

	BMEP (psi)	Power Output (kW)	Specific Cost ($/kW)
Base Case	175	725	300
Step 1	200	830	262
Step 2	225	930	233
Step 3	250	1035	210

Improved lubrication, lower oil temperatures, long-life oil filters, more heat-resistant valve train materials, and spark plug firing end changes will be investigated. The goal is to achieve an annual maintenance schedule based on 2500 operating hours per year. Together, this would reduce maintenance costs by an estimated 10%.

2.3 Utilizing Existing Standby Power Gensets for DG

Standby engine-generator sets in industrial, commercial, and institutional buildings are a significant potential source of distributed generation. These units typically operate only a few hours per year for testing, and during utility power outages, which typically total a few more hours. According to the GRI, standby units may represent up 40 GW of generating capacity in the U.S. Conversion of a substantial share of these idle investments into revenue-producing assets might offer a quick start to DG for buildings in the U.S.

Already, some utilities recruit customers with standby generators for peak load reduction programs, offering payments or rate relief for limited operation during peak periods. Most standby generator sets are installed according to building code requirements and are designed to carry critical electric loads during outages. Facilities such as hospitals, computer centers, and manufacturing plants with critical high-value processes often have substantial standby generation capacity. Diesel engines are the power sources of choice for standby because of their low first cost. At present, emissions regulations are a substantial barrier to the use of these engines for distributed power.

Dual-fuel retrofit technology may pave the way for more usage of existing standby units for distributed power. This approach builds on the inherently high BMEP of diesel engines while incorporating the clean combustion, low fuel price and convenience of natural gas. The shaft efficiency of lean-burn, dual-fuel, or micropilot engines generally ranges from 36 to 40%. The engines usually retain the ability to run on diesel fuel alone in case the gas supply is interrupted.

2.3.1   Engine Control

Contemporary electronic control technology performs six basic functions:

1. Monitoring — sensor technology enables close, precise, automatic monitoring of all major engine systems. Sensors continuously measure starting air pressure, oil pressure, oil temperature, coolant level and temperature, crankcase pressure, fuel temperature, inlet air temperature, engine speed, cylinder temperature, ignition timing, detonation, and other parameters. Control systems can display these values for operator information and record them in memory as part of an ongoing engine history.

2. Protection — controls are programmed to prevent engine damage by shutting down the engine if sensor readings indicate that key parameters are outside acceptable limits. Typical safety shutdown parameters are low oil pressure, high oil or coolant temperature, high exhaust stack temperature, low coolant level, and overspeed.

3. Diagnosis — advanced controls have self-diagnostics that speed troubleshooting and service, enabling repairs to be made quickly before a costly breakdown occurs. Fault LEDs commonly indicate problems with gauges, fluid levels, fuel supply, air intake, exhaust, ignition, and starting systems. Some controls signal problems in the interface with driven equipment. Fault indicators are also generally included for control system components, such as sensors, actuators, modules, and wiring.

4. Sequence Automation — control systems automate engine startup and shutdown, following procedures either built into the system or custom programmed by the user. Startup sequences include prelubrication, cranking, ignition, disengagement of the starting motor cranking, and acceleration to rated speed. Shutdown typically includes a programmed cooldown period.

5. Combustion Control — electronic feedback systems automatically regulate air–fuel ratio, ignition timing, and engine power to

compensate for changes in ambient air temperature, barometric pressure, fuel heating value, engine load, and other operating variables. Some combustion controls are available as add-on modules; others are fully integrated with the engine. Timing control generally includes protection against detonation, a form of uncontrolled, explosive fuel combustion that can severely damage cylinder components. If in-cylinder sensors detect detonation, the control automatically retards the timing. If detonation persists after timing has been retarded to the full extent available, the control triggers a safety shutdown.

6. Remote Capability — engine monitoring and control from off-site is essential to efficient distributed generation on a large scale (see VPP discussion in Chapter 7). Routine operating parameters and fault indicators are displayed at a central control center. More importantly, remote control also enables centralized dispatching of multiple distributed power sources.

2.3.2 Systems Considerations for DG Applications for Combustion Engines

Reciprocating engines are currently a promising technology for distributed generation systems up to 5 MW. The installed cost for engine-generators in distributed power ($350 to $500 for diesel units, $600 to $1000 for gas [$2000]) is about half the cost for central steam power plants. Besides competitive first and life-cycle costs, engines offer high operating flexibility. They perform efficiently in continuous duty or in intermittent service and efficiently accommodate variable loads. Multiple generator sets can be configured so that one or more units cycle with the load while others produce continuous full rated power for optimum performance and fuel economy. Multiple units also provide redundancy for emergencies and enable staggered maintenance intervals, keeping the system on-line while individual units are down for service.

2.3.3 Combined Heat and Power

Traditional fuel-based, large-scale electric power generation is typically about 39% efficient, while separate boilers are about 50% efficient. In either case, the excess heat is simply lost. Engine-driven CHP systems recover the heat from engine exhaust, jacket water, and lubricating oil as described earlier. CHP systems using reciprocating engines range from a few kW to 5 MW electrical output for a single unit. Electrical efficiencies range from 34% in small units to 41% in larger installations. Thermal efficiency is typically 40 to 50%; thus, total efficiency approaches 90%. Figure 2.9 displays the heat sources.

ICE-based CHP is already a healthy industry, as shown in Table 2.4.

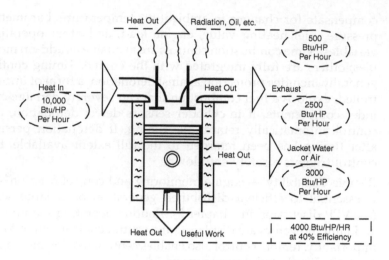

Hypothetical Figures for Engine Heat Rejection

FIGURE 2.9
ICE heat sources for CHP systems.

TABLE 2.4

CHP Sites Using Reciprocating Engines

State	Number of Sites	% of U.S. Market
California	493	42.0
New York	136	11.6
New Jersey	117	10.0
Massachusetts	46	3.9
Illinois	44	3.7
Pennsylvania	44	3.7
Connecticut	43	3.7
Michigan	28	2.4
Texas	23	2.0
Virginia	17	1.4
Florida	16	1.4
Arizona	15	1.3
Top 12 Totals	1022	87.1

2.4   The Utility Interconnection

Today, most engine gensets power facilities also served by a utility. Therefore, power must be transferred from or shared with the utility when the distributed source operates. Interconnection considerations are discussed in detail in Chapter 12. This section mentions a few matters that affect IC engine systems. The four basic mechanisms for transferring or sharing load between the on-site engine-generator and the utility are described as follows.

2.4.1   Open-Transition Transfer Switch

Historically, standby power systems have used conventional double-throw transfer switches for shifting from one power source to another. These systems (sometimes referred to as "break before make") are the simplest arrangements, requiring no interface or coordination with and no protection from the power utility source, since there is never a possibility of interconnection. The obvious drawback to open-transition transfer switches is an inherent momentary power interruption to the load during transfer.

2.4.2   Closed-Transition Transfer Switch

An alternative for loads sensitive to interruptions is a closed-transition transfer switch (sometimes called "make before break"). The closed-transition transfer switch momentarily allows the two sources to operate in parallel, typically for 100 milliseconds or less. Thus, there is no interruption of the load.

2.4.3   Soft Loading Transfer System

When a closed- or open-transition transfer switch transfers load between sources, the load is applied in a block — the load on the switch is connected to the generator in one step. Generator sets are typically rated close to the demand of the load to be handled. When a load approaching the full load rating of the generator set is applied in one step, the generator's frequency will drop. In the worst case, the generator set may not recover, or, at best, loads may be disturbed by the dip. In such cases, a closed-transition transfer switch can be modified to allow soft loading, or ramping of the load from the utility to the engine-generator.

2.4.4   Parallel Operation

In applications involving long run hours, continuous paralleling with the electric utility can be advantageous. Parallel operation is especially beneficial in combined heat and power systems, as CHP equipment is typically sized to carry the building or process heat load and may satisfy only 25 to 50% of the electric power requirement. In a paralleling system, the generator is directly connected to the utility grid. The on-site engine-generator runs continuously at a specified output, while the utility satisfies the balance of the facility load, including load variations.

2.4.5 Maintenance and Service

Some owners of DG power systems secure training for in-house staff from the equipment manufacturer, then assume responsibility for maintenance and service, drawing on the manufacturer only for technical support or major repairs and overhauls. Other owners contract with the manufacturer or its local dealer representative for all or part of a complete service package.

Manufacturer-sponsored maintenance and service programs provide a broad range of options to suit a variety of customer needs. Program specifics vary, but common options include:

- Fluid analysis
- Planned maintenance
- Customer support agreement
- Total maintenance and repair agreement
- Operation and maintenance agreement

2.5 Stirling External Combustion Engines

A Stirling engine is an external combustion heat engine and, therefore, does not require a specific fuel; a Stirling engine-generator can convert any sufficient heat source into useful electrical power. These generator sets are also physically small and very efficient even below 100 W (e). With the added advantages of high reliability, long life, very low noise, and maintenance-free operation, Stirling engines are ideal for distributed generation applications where the generator must be located within a residence or business office, and for cogeneration as shown in Figure 2.10.

2.5.1 Design

Stirling engines operate on a closed thermodynamic cycle where a temperature differential is converted into mechanical and/or electrical power. External heat is supplied at a high temperature to the engine heater head, and thermodynamic waste heat is rejected to ambient temperature. An internal displacer piston physically shuttles the helium working fluid between the hot and cold regions, creating a varying pressure value. That pressure wave causes the power piston to reciprocate. The reciprocating motion can be used to produce shaft power similar to an IC engine or may be used to generate electricity directly using a linear alternator. At no time during the cycle does the working fluid enter or leave the engine, which is hermetically sealed. Therefore, the cycle is defined as closed.

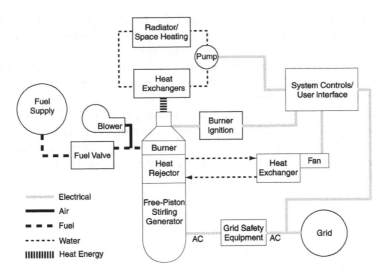

FIGURE 2.10
A grid-connected Stirling micro-cogeneration system.

Several varieties of Stirling engines have been developed by both private and government organizations. The varieties may be grouped into two fundamental categories: kinematic and free-piston. Kinematic engines have a crankshaft and flywheel and may be used in place of internal combustion engines to provide shaft power. The disadvantage of kinematic engines is that the rotating shaft and reciprocating rods must be sealed so that the working fluid does not leave the engine and lubricants do not mix with the working fluid. Due to the difficulty of successfully sealing the shaft, kinematic Stirling engines have limited life and reliability. Free-piston engines, however, have no crankshaft and no seals to maintain. The generator portion, a linear alternator, can be sealed in a pressure vessel along with the engine so that the only items penetrating the pressure vessel are feed-throughs for the electrical output. The two pistons in a free-piston Stirling engine are mounted to allow free axial motion but little or no radial displacement. This is typically accomplished by employing flexural or gas bearings. Other than the mountings, the pistons do not come into contact with any part of the engine, so there are no lubricants needed and no rubbing parts to wear out.

Free-piston Stirling generators can be understood as thermally actuated mass–spring–damper systems. The pistons and moving portion of the alternator provide the masses, their mountings provide the springs, and the magnetic field of the linear alternator provides the damping. Once these values are determined, the engine can be mathematically modeled using linear second order differential equations with known solutions. As with any mass–spring system, a designer may control the natural frequency of the system by altering the mass or spring force. This way, free-piston Stirling generators can be designed to produce AC power at whatever voltage and frequency the application requires. These systems are also load following when attached to the power grid, so if the frequency of the grid changes slightly, the engine will simply change its operating frequency to suit. It is important to note that some load must always be applied to a running free-piston Stirling generator in order to prevent damage to the engine from over-stroke of the pistons. The AC power can be easily converted to DC to charge batteries or operate electronics.

When employed as a remote battery charger (Figure 2.11), a Stirling-powered generator can run continuously and requires only one-quarter the amount of batteries required for a gasoline or diesel system. This configuration could power a remote telecommunications relay, an automated pipeline monitoring station, or an off-grid home.

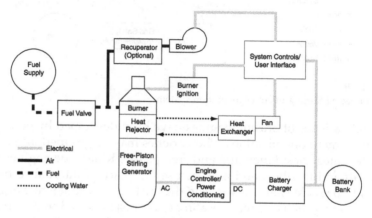

FIGURE 2.11
Schematic of a battery charger using a free-piston Stirling generator.

While there are no fundamental limits on power output, the capacity of most free-piston Stirling engines currently available is under 5 kW. This is partly due to thermodynamic and heat transfer considerations as well as the mechanics of mounting the pistons. The heater head, where the heat energy is supplied to the cycle, for small capacity engines may be fabricated from simple monolithic shapes. In larger capacity engines, those with a 7 kW or greater output, such a monolithic heater head does not provide adequate surface area for the required heat energy input. These typically require complex and more costly tubular heat exchangers to get the required heat energy into the cycle for full power operation. The size of the pistons and their amplitude serve to further complicate the design of larger capacity engines since they require more complex mounting technology. These requirements could increase the cost and size of larger engines substantially. The smaller engines (under 5 kW) are easier to design and build, and may be most cost effective.

The efficiency of a Stirling generator system is affected by a number of variables including fuel type, operating temperatures, and mechanical design of the engine. Currently available Stirling engines have generator efficiencies, which are the ratio of engine heat energy input to electrical power output, ranging from nearly 30% for systems as small as 50 W to around 40% for 3 to 5 kW capacity generators. The generator efficiency is largely determined by the efficiency of the alternator and the effectiveness of the regenerator used for the Stirling cycle. The regenerator is the most important and often most expensive single part of the engine. Practical engines must be designed with cost in mind. Therefore, some efficiency is often sacrificed to lower the production cost of the engine.

The total system efficiency (the ratio of fuel input to electric power output) is largely affected by the system employed to supply heat to the engine. In most cases, gaseous fuel such as propane is combusted in a burner, and the resulting heat is transferred to the engine heater head via convection or radiation. Burner technology is constantly improving. Currently, simple, cost-effective burners are 50% efficient. Recuperative burners can achieve much higher efficiencies but at a greater cost. It should be noted that Stirling generator systems are ideally suited for micro-cogeneration, where the exhaust heat from the burner and waste heat from the engine are used for water and space heating. With appropriate heat recovery techniques, cogeneration systems can approach 98% efficiency since nearly all of the heat energy is used in some way.

2.5.2 Fuels

One of the singular advantages Stirling engines have over internal combustion engines is that they are truly multi-fuel capable. The Stirling cycle requires only a sufficient heat source to operate and does not rely on carefully timed fuel injection and combustion processes as do internal combustion engines. Practical Stirling cycle engines may be operated using propane, natural gas, gasoline, diesel, radioisotopes, solar energy, and even wood or other biomass. The only limitation on fuel source imposed by the engine is that a sufficient amount of heat must be transferred to the cycle at a controllable temperature. There are often minimal changes to fuel systems needed to accommodate different fuel types, but the engine itself requires no new hardware.

The freedom to choose fuel type allows other considerations to be made when selecting fuels. Cost and availability of each fuel are usually the first items considered, and the Stirling generator can operate on whatever fuel is most readily available. However, cost is not the only concern when choosing a fuel. The environmental impact of storing and combusting the fuel must be considered. Low emission burners have been developed for fuels such as gasoline and natural gas, and the technology is being developed for the use of biomass fuels. Some analysis should be performed to make certain that a fuel is compatible with the combustion technique and heat transfer systems used.

2.5.3   Technical Developments and Outstanding Barriers

Stirling cycle engines have developed considerably in recent years. The free-piston configuration has served as an enabling technology for a great deal of energy conversion development. Stirling Technology Company (STC) and Sun Power Inc. are the leading developers of free-piston Stirling cycle engine generators with capacities ranging from 10 W to 5 kW. These generators have demonstrated maintenance-free operating lives far beyond that of gasoline or diesel engine generators. Ongoing endurance testing has proven so far that these generators can run continuously without maintenance for over six years, about 55,000 hours, with no observable degradation in performance. The potential operating life of these generators is well over twenty years. Similar prototype engines are currently available to utilities and other interests for evaluation, while the designs are being refined to lower costs and prepare for mass production. Development work is also under way to produce more reliable and cost-effective balance-of-plant equipment.

Advancement of engine technology is paralleled by advancements in the design and performance of the linear alternators used to convert the piston motion into electricity. Recent work has lead to more efficient and easier to build alternators that drive down cost while enhancing generator performance. Much of the current development work is aimed at bringing the cost of the generators down by making them ready for mass production. A large scale production run of 1 kW generators is planned for 2002, with limited production of 3 kW generators to follow.

Despite the numerous advantages over other technologies, Stirling engines do have some limiting characteristics that must be considered. The primary limitation of cost-effective, free-piston Stirling generators is power generation capacity. As mentioned previously, engines with capacities over 5 to 7 kW require costly and complex heat exchangers and piston mountings. The scale of current generators is similar to most portable generation systems. For this reason, Stirling generators are best used when small amounts of power are required continuously over a long period of time, such as for a remote telecommunications site or in a home or small office building. With their high reliability and silent operation, Stirling generators lend themselves well to being located in such places. The development of Stirling micro-cogeneration systems, where both heat and electricity are supplied at the same time, makes placement in residential and small business sites very practical. It is also possible to operate multiple Stirling generators in parallel to fulfill a larger power requirement.

Stirling engines are also somewhat larger and heavier than IC engines of similar capacity, primarily due to stroke limits of the piston mountings and the need for a safe, hermetically sealed pressure vessel. For this reason, it is best to consider Stirling generators for applications where the system will not need to be moved frequently by hand. However, lighter-weight engines are being developed as space probe power systems, and the technology is certain to find its way into commercial generators.

Stirling generators can provide tangible alternatives to small IC engines and add a whole new level to power distribution. Information about current developments in Stirling generators and supporting technology can be obtained by contacting STC and other companies who manufacture Stirling engines and generators.

2.5.4 Controls and Communications — Dispatchability

The characteristics of Stirling generators allow them to be located inside homes, businesses, and similar environments where IC generators would be intolerable because of their high noise and maintenance levels. The most popular application of Stirling generators is in micro-cogeneration systems. In these systems, the generator is coupled to a household boiler (a water heater for hydronic space heating), so electricity and hot water are produced simultaneously from the same fuel. Stirling micro-cogeneration systems are gaining popularity in Northern Europe, where they are connected to the grid and use natural gas. The micro-cogeneration system is treated much like any other home heating system as far as the consumer is concerned. The systems can be configured to supply electricity only when heat is called for, or they can be programmed to dump unwanted heat when only electricity is needed.

Being connected to the grid is an excellent way to operate Stirling generators. The grid controls the generator frequency and voltage, so very little is needed for the connection other than the requisite safety equipment. Connecting a free-piston Stirling generator to the grid is a fairly simple operation and requires minimal hardware for the European grid. Utilities in the U.S. are only now becoming exposed to numerous consumer level, grid-coupled generators, and must evaluate implementing them with the U.S. grid.

Stirling generators also operate very well off-grid as battery chargers. The generator can run continuously, charging a battery bank, to handle peak loads that may be greater than the engine's capacity. Many photovoltaic (PV) systems are installed in this fashion but require a large number of batteries, since they can only provide charging power when the sun is shining. A Stirling generator can run day and night, thereby requiring a minimum amount of batteries. Since the generator doesn't need to be shut down, the required amount of batteries is far less than even gasoline and diesel generator systems need. When an engine is run off-grid, system electronics can automatically activate the generator when battery voltage drops below a preset threshold and shut it down again once the battery bank is fully charged. AC power can be easily obtained from a battery bank by using an inverter. Generators in this type of system can easily power off-grid homes, remote monitoring equipment, or communication relays.

Activation of Stirling generators can be handled in a variety of ways, depending on the application. Remote activation by a utility is certainly possible, but the relatively small capacity of the Stirling generator would be insignificant on a utility scale unless the installed capacity was very large.

When connected to the grid, it is far more practical to turn the generator on and off as needed or activate it as loads within the household reach prescribed levels. This way, power can be produced when it is most needed or most cost effective. Off-grid systems are best run continuously or shut down only when the battery bank is fully charged. This minimizes the size and cost of the battery bank and allows the system to more easily handle loads that exceed the engine capacity.

2.5.5   Utility Interfacing

As previously noted, Stirling generators can be used for both on- and off-grid applications. It is left to the utility companies and their customers to decide what is most appropriate. Stirling generators are reliable enough that a utility company may consider leasing a unit to a remote off-grid customer, thus avoiding the costs of building power lines or larger generators on site. An off-grid Stirling micro-cogeneration system is a very reliable and cost-effective solution for providing heat and power in a remote location. Stirling engines require no maintenance, so there is basically nothing the end user has to do to it once the system is installed. The utility can also choose the fuel used by the generator and make certain it is available to the customer. Reliable, maintenance-free, silent operation and a long operating life are some important advantages Stirling generators offer the utility that cannot be found in conventional small-scale generators. Utility companies in areas where peak demand is very high may encourage their residential customers to invest in grid-connected micro-cogeneration systems. Utilities need to determine the best policies for how the systems are set up and how much power should be produced. It may be desirable for the generator to produce all of the power needed for the home in which it is installed, or it may be more convenient to produce power only when heat is called for. Stirling generators and cogeneration systems are very versatile and can be adapted to fit the needs of the utility and consumer alike.

2.5.6   Costs

Free-piston Stirling cycle engines and generators are currently available on a prototype basis for development and technology evaluation programs. Work is under way to have several sizes of generators mass produced. Full scale production of STC's 1 kW generator is planned for the year 2002, with limited production of the 3 kW generator to follow.

Early commercial engines will have higher capital costs than IC generators, but lower than initial costs for PV and thermoelectric systems. Despite the higher start-up costs, however, Stirling generators can be more cost-effective than even IC engine generators on a life cycle basis. The higher

efficiency of the Stirling engine and lack of significant maintenance requirements bring the operating costs down to levels below most other distributed generation technologies.

Additional Reading

Onsite Power Generation: A Reference Book, 3rd ed., Electrical Generating Systems
    Association, 1998.
Walker, G. and Senft, J.R., *Free Piston Stirling Engines*, Springer-Verlag, Berlin, 1998.
West, C.D., *Principles and Applications of Stirling Engines*, Van Nostrand Reinhold
    Company, New York, 1986

3

Combustion Turbines

Richard Brent

CONTENTS

Natural gas-fired combustion turbines are the most widely adopted prime movers for new power generation worldwide, based on the aggregated power rating; in the year 2000, over 4000 units were sold or ordered. The benefits of gas turbines in power generation are fivefold: (1) comparatively low installation cost per MW output, (2) increasing availability of natural gas for low fixed-price contracts, (3) explosion of demand for peaking capacity in a deregulated energy marketplace combined with (4) the higher electrical efficiencies of aeroderivative turbines, and (5) the ability to site and install units from 1.7 to 40 MW (and larger) in weeks to months, not years.

Basic gas turbine (GT) technology is mature; performance improvements are incremental. Component efficiency is probably the most important performance factor for GTs. A slight increase in efficiency for one component can have significant impacts on the net system efficiency. Incremental improvements to GT components have increased large system conversion efficiency from 25% (LHV basis) in the 1950s to 35–38% in current models.

3.1 Basic Cycle

Gas turbines consist of a compressor, combustor, and turbine-generator assembly that converts the rotational energy into electrical power output. The standard conditions for the ambient air flowing through the GT are assumed to be at 59°F (15°C), 14.7 psia (1.013 bar), and 60% relative humidity. A simple-cycle single-shaft GT is shown in Figure 3.1. As can be seen, single-shaft turbines are configured in one continuous shaft, and, therefore, all stages of the turbine operate at the same speed. These types of units are typically used for generator-driven applications where significant speed variation is not required.

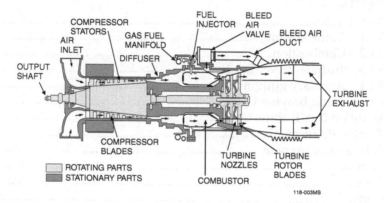

FIGURE 3.1
Gas turbine cross section.

 The low pressure or power turbine rotor can be mechanically separated from the high-pressure turbine and compressor rotor. This feature allows the power turbine to be operated at a wide range of speeds and makes it ideally suited for variable-speed applications. All of the work developed by the power turbine is available to drive the load equipment since the high-pressure turbine exclusively drives the compressor. The starting requirements for the load train are also reduced since the load equipment is mechanically separate from the high-pressure turbine. The simple-cycle GTs are a mature technology based on the thermodynamics of the Brayton (or Joule) cycle with the following paths:

 Path 1–2: Compression of atmospheric air

 Path 2–3: Heating compressed air via fuel combustion

 Path 3–4: Expansion of heated air–fuel mixture through a turbine, rotating the blades

 Path 4–1: Discharging exhaust gases back to the atmosphere

For electric power applications, the nominal rating is measured at the output terminals of the electric generator to include gearing and generator losses. It does not take into account inlet filter or exhaust silencer losses or auxiliary running loads. Natural gas fuel can give a 2 to 3% higher output and a 1 to 2.2% improvement in heat rate over the same machine burning No. 2 distillate oil. Inlet filter and exhaust losses can be equivalent to around a 2% penalty, while auxiliary running losses add about a 0.6% penalty in available power output and heat rate. If gears are used for speed reduction, gearing losses can be as high as 1.5% depending on the specific design and gear ratios, while the electric generator loss is usually about 2% of the GT shaft power output.

Lower heating value (LHV) efficiencies for this design (compressor, combustor, turbine) range from 18 to 35%. The energy losses are consumed by the compressor and other auxiliaries as discussed above. Most of the work produced in the turbine is used to run the compressor, and the rest is used to run auxiliary equipment and produce power.

3.1.1 Compression

Although several compressor designs are available, most GTs utilize multi-stage axial designs, which produce a higher compression ratio than centrifugal designs. Figure 3.2 shows typical compressor efficiency for two staging options. GTs with high pressure ratios can use an intercooler to cool the air between stages of compression, allowing more fuel to be combusted and generating more power. The limiting factor on fuel input is the temperature of the hot gas created and the resultant impact on the metallurgy of the first-stage nozzle and turbine blades. With advances in materials, however, this limiting factor is gradually being reduced.

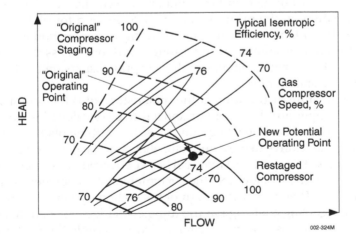

FIGURE 3.2
Gas compressor efficiency.

3.1.2   Combustion

Three alternate combustor design configurations — silo-type, can-annular, and annular — are employed in power GTs. The can-annular design drew on the early experimental combustor can experience of the jet turbine, even though jets eventually selected an annular combustor. Can-annular design can be shipped integral to the unit, allowing for internal temperature profiles. GTs have the capability of burning a variety of fuels including natural gas, blast furnace gas, coal gas, distillate fuel, residual fuel, etc. However, burning fuels that have potentially corrosive elements requires that fuel contaminant specifications be set. In addition, more stringent emissions regulations have limited the application of these alternative fuels. Water and steam injection approaches provided an interim solution to reducing emissions, paving the way for dry, low NO_x approaches. Some manufacturers are active in developing advanced dry, low emission systems for their turbines utilizing one or more of three combustion design approaches: pre-mixed/lean combustion, rich–lean, or catalytic combustion.

The main components of a typical combustion system are the torch ignition system, a dual-fuel injection system, and an annular combustor. Ignition within the main combustor is achieved with a jet of hot gas produced by an auxiliary torch that uses turbine air and fuel. This torch is capable of operation on either liquid or gaseous fuel and uses a low-energy spark plug. When liquid fuel is used, it is atomized with high-pressure air from an external source. The high temperature of the exit gas stream from the torch provides a source of very high energy for the main combustor light-up and thus provides reliable ignition over the range of ambient conditions in which the turbine operates.

Combustion efficiency is affected by evaporation rates for liquids and chemical reaction rates for both gaseous and liquid fuels. Most industrial GTs achieve high reaction rate and evaporation rate efficiencies. A stable, turbulent flow pattern in the primary zone is also required for high combustion efficiency. Important parameters that determine combustion efficiency are combustor volume, combustor operating conditions (mass flow rate, pressure, and primary zone temperature), fuel spray dispersion by droplet size, and fuel evaporation rates.

The temperature distribution at the combustor exit is the combustion system's most important characteristic and has to be developed to optimize the life of the downstream nozzle guide vanes and turbine blades. Because the vanes are fixed relative to the combustor, they must be designed to accept the hottest localized temperature that will be experienced at the combustor exit; however, the turbine blades feel the circumferentially-averaged temperature at any radius. Therefore, the design of the stationary vanes represents the most severe cooling challenge in the turbine, and it is essential to be able to define a non-dimensional parameter which represents the peak measured exhaust temperature.

In a well-developed, practical combustion system, the pattern factor is invariably reduced by approximately 12% during operation using No. 2 diesel when compared to natural gas. However, the radial temperature distribution factor remains essentially the same when either of the two fuels is used. Pattern factor optimization is carried out using natural gas fuel because there is no deterioration of the parameter when the GT is switched to liquid fuel operation.

3.1.3   Turbine Power Production

Gases enter the turbine at 1800 to 2200°F, under continuous full load via a nozzle assembly that restricts, accelerates, and directs the flow of gas into the turbine wheel. As the superheated air–fuel mixture is sprayed into the rotor, the gas expansion process continues as it passes the turbine blades, creating the rotational force. Figure 3.3 shows a typical turbine performance map.

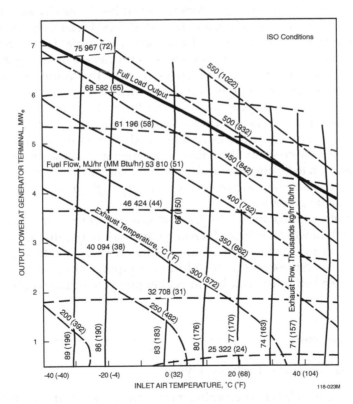

FIGURE 3.3
Performance map for industrial gas turbines (courtesy of Solar Turbines, Inc., with permission).

3.1.4 Ancillary Equipment

A turnkey turbine generator installation will also require a controls package, fuel supply system, electrical system, and attendant power-switching and safety protection features such as grounding, circuit breakers, and transfer switches. The balance-of-plant equipment and labor required to install a turbine generator set can generally be estimated at 30% of the total turnkey cost per kilowatt output installed.

3.2   Recuperated Brayton Cycle

Turbine efficiency can be enhanced with recuperation. Figure 3.4 shows the recuperator location in the basic Brayton (GT) cycle and how the recuperator uses the hot exhaust to preheat compressed air before it enters the combustor. The main obstacle to the use of recuperated turbines is the size and cost of the recuperator. Several manufacturers have developed relatively compact and highly effective recuperators, and the market potential for microturbines (which generally must have recuperators) is further encouraging this development. The recuperated turbine offers an attractive route for the user of smaller gas turbines with higher efficiency and better part load characteristics than the combined cycle machine, but the cost is high. Recuperation is less suitable for high pressure-ratio machines, however, which already have higher simple-cycle efficiencies and in which system maximum operating temperature may be set by the temperature tolerance of the recuperator. For this application to be effective, the turbine exhaust temperature must exceed the temperature at the compressor exit. In very high pressure ratio designs, this may not be possible.

Recuperators do contribute to pressure loss (lower pressure ratio entering the turbine), however. A 90% efficient recuperator may cause a 2% pressure loss on the air side and up to a 4% loss on the exhaust gas side, with an additional 1% loss estimated for the additional piping. Thus, the final system would have a better heat rate but lower net power (electrical) output.

3.3   Modified Gas Turbine Cycle

The combined-cycle GT is becoming increasingly popular due to its high efficiency. The exhaust air–fuel mixture exchanges energy with water in the boiler to produce steam for the steam turbine. The steam enters the steam turbine and expands to produce shaft work, which is converted into additional electric energy in the generator. Finally, the outlet flow from the turbine is

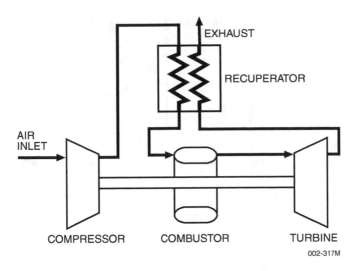

AIR
INLET

EXHAUST

RECUPERATOR

COMPRESSOR   COMBUSTOR   TURBINE
002-317M

FIGURE 3.4
Recuperated GT cycle.

condensed and returned to the boiler. However, GT installations below 10 MW are generally not combined-cycle, due to the scaling inefficiencies of the steam turbine. Figure 3.5 shows the combined cycle schematically as well as how GT exhaust heat is used to produce steam used in a "bottoming" Rankine power cycle.

There is scope for a range of small, high-performance steam turbines for combined cycle duty, but the major obstacle to their use is the complexity involved. Highly efficient steam turbines require condensers, vacuum pumps, cooling water, and water treatment plants before they can operate. The amount of supervision required is disproportionately high compared with the GT components, but the situation may not be unacceptable if the user has sufficient on-site technical staff (e.g., a large manufacturer or processing plant).

Steam turbines may also be used in a backpressure mode, taking high quality steam from a heat recovery boiler on the gas turbine exhaust to generate power exhausting low-pressure process steam for industrial use or space heating. The possibilities for utilizing GT exhaust heat range from producing steam alone (for process use) to producing electric power. In the first case, 80% or more of the heat in the primary fuel may be utilized, and in the second case perhaps only 45% will be utilized, but the output will be all electrical power with a high market value.

Alternatives for improving the efficiency of an open simple-cycle include the following:

- Intercooling the air after compression
- Reheating the exhaust gases after combustion
- Recovering part of the energy lost to exhaust gases

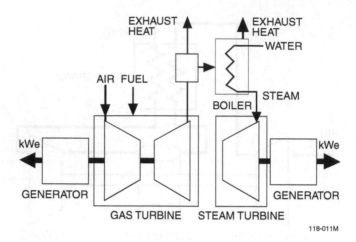

FIGURE 3.5
Schematic diagram of combined cycle.

A GT with a heat exchanger recaptures some of the energy in the exhaust gas, preheating the air entering the combustor. This cycle is typically used with low pressure-ratio turbines.

3.4 Turbine Performance

Factors influencing the performance of GTs include operating temperatures at the compressor and turbine inlets, pressure ratio, and aerodynamic efficiencies of the compressor and turbine sections. Aside from the compressor inlet temperature, all of these tend to be size related. Thus, large GTs — above 30 MW output — may achieve efficiencies up to 40% without exhaust heat recuperation, while typical efficiencies in the 5 MW range may be closer to 35%, falling to as low as 15 to 17% in the unrecuperated microturbine range (25 to 100 kW). The reasons for this variation are summarized below.

The turbine inlet temperature (TIT) together with the pressure ratio (PR) define the amount of energy that can be extracted from the hot gas leaving the combustor; hence, they control the power output of the machine and also its efficiency. In large turbines, it is feasible to use TIT levels that are substantially above the temperatures that can be tolerated by metal components such as turbine blades because blades can be cooled by air introduced into internal passages via the blade roots. This becomes impractical in small blades with thin sections due to manufacturing difficulties and the risk of partial or complete blockage of the cooling passages by dust or oxidation products in the coolant. One alternative is to use a more refractory material; ceramics are currently being developed and tested. Figure 3.6 summarizes the evolution of materials for GT applications.

Pressure loss must be minimized because of its deleterious effect on cycle performance; however, some pressure loss is necessary to promote the turbulence and fuel/air mixing required for efficient combustion. Apart from these requirements, the system should provide stable and smooth combustion, rapid and reliable ignition, freedom from carbon deposits and minimum smoke, oxides of nitrogen (NO_x), carbon monoxide (CO), unburned hydrocarbons (UHC), and sulfur oxides (SO_x). The combustion system has to be capable of fulfilling the above requirements in most cases not only at the design point, but also over a wide range of part-load conditions. The suitability of the use of non-standard fuels must be determined with due regard to meeting these requirements.

The pressure ratio that can be achieved in a compressor is proportional to the square of the rotor tip speed. Tip speeds in the order of 800 feet per second can be generated quite easily on large rotors, but microturbines require rotational speeds up to100,000 rpm. Mechanical complexity usually rules out the use of multiple compressor stages in very small machines. Multi-stage axial flow compressors are made for industrial compressors when a high pressure ratio is desired, but expense may become an issue. The same effect can be achieved more economically with a series of radial-stage compressors.

The aerodynamic efficiency of turbo machines is limited by gas friction turbulence losses at blade tips and the hub attachment and gas–air leakage between the rotor and its casing. The compressor is even more sensitive to these issues. All of these are more difficult to control as the components become smaller and the penalties become especially severe at very small sizes. Above 1 MW output, compressor efficiency ranges from 85 to 90%.

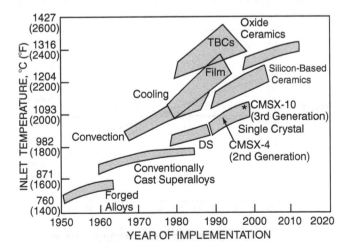

FIGURE 3.6
Evolution of cooling and materials in GT combustors.

Since the GT is an ambient-air breathing turbine, its performance will be changed by anything affecting the mass flow of air to the compressor, most obviously changes from the reference conditions of 59°F and 14.7 psia. Correction for altitude or barometric pressure is simple because less-dense air reduces mass flow and output proportionately; heat rate and other cycle parameters are not affected. A typical temperature correction curve is presented in Figure 3.7. Similarly, humid air, being less dense than dry air, will also have an effect on output and heat rate. This humidity effect has taken on more significance recently because of the increasing size of gas turbines and the utilization of humidity to bias water and steam injection for NO_x control.

Inserting air filtration, silencing, evaporative coolers, chillers, and exhaust heat recovery devices in the inlet and exhaust systems, respectively, causes pressure drops in the system. The effects of these pressure drops are unique to each turbine design. Fuel type will also impact gas turbine performance. Gaseous fuels with heating values lower than natural gas have a significant impact on performance. As the heating value (Btu/lb) drops, the mass flow of fuel must increase to provide the necessary heat input (Btu/hr). At the same time, more air is required to combust the lower heating value fuel. Therefore, there are several associated side effects that must be considered in the GT designs that are likely to burn lower heating value fuels.

Generally, it is not possible to control the factors affecting the GT performance, since most are determined by site location and plant configuration. In the event that additional output is needed, there are several possibilities which may be considered to enhance performance. These include technologies to achieve inlet air cooling, steam and water injection, and peak performance ratings. All turbomachinery experiences loss of performance with time, and for GTs, performance losses can be classified as recoverable or nonrecoverable losses. Recoverable loss is usually associated with compressor

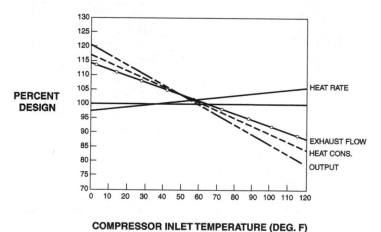

FIGURE 3.7
Example temperature correction curve.

fouling and can be rectified by water washing. Non-recoverable loss is due primarily to increased turbine and compressor clearances and changes in surface finish and airfoil contour. Power output deterioration at the 25,000 hour operating point of a typical turbine could vary from 3 to 5% and heat rate within 1% of the "new and clean" situation. These mechanisms of component efficiency losses can only be recovered through replacement of affected parts at recommended inspection intervals. It is often extremely difficult to quantify the specific performance degradation, but one generalization that holds true is the fact that turbines located in dry, hot climates will degrade less than those in humid climates.

3.5 Future Developments

Recent technical developments in GT technology have concentrated largely on addressing the needs of a prospective mass market for small high performance turbochargers for automobiles and, at the other end of the scale, for achieving higher efficiencies in base-loaded control electric power generators. This former effort has relied extensively on the use of ceramic materials to allow high operating temperatures in uncooled turbines, while the latter effort has concentrated on more advanced metals and cooling systems with an emphasis on combined cycle efficiencies.

By reducing the requirement for cooling air, the use of ceramics in the hot gas path of a turbine improves both output and efficiency because air extracted for cooling does not contribute to power generation. Ceramic components are also relatively free from hot gas attack and do not distort so that aerodynamic efficiency is retained over extended service. However, ceramics are inherently brittle and are subject to failure due to stress generated by temperature gradients. Over the long term, ceramics are subject to morphological changes and, in some cases, to oxidation and chemical attack. While these effects are relatively minor over the life of a GT installation (and a small fraction of that time is spent at full load), they become very important under industrial power turbine conditions which call for outputs of 20,000 kW or more, much of it at full power rating.

While there is no firm limit to the size of turbine that can benefit from ceramic technology, the problems of making large ceramic components and their cost are fundamental to these materials. Another factor is the technical competition from metals. Aircraft turbines and their industrial aeroderivatives have demonstrated the use of internally-cooled turbulent blades and other components, and techniques for casting even small metal blades with cooling passages are well established. Such blades can operate with gas temperatures higher than those that can be tolerated by today's ceramics, although the cooling process involves performance penalties. On large turbines, steam cooling has been used to minimize these penalties, and this approach may filter down to turbines in the industrial or distributed generation sizes; but the possibility of restrictions developing in cooling passages will inevitably call for caution in adopting internal cooling in small components. There still remain other avenues for efficiency improvement. Many of these fall in the category of design detail — they include seals, including self-compensating designs that can deal with problems posed by thermal expansion, and aerodynamic refinements.

3.6   Controls

Controls and safety devices represent a fairly high proportion of the total cost of a gas turbine that is coupled directly or through gearing to an alternator running at synchronous speed. The electrical connection to a power supply grid will require control of speed and voltage. GT generator systems require the following, typically provided by programmable logic controllers (PLC):

- Startup and shutdown sequencing and protection
- Vibration monitoring
- Fuel or steam governors
- Surge control
- Alarm annunciation
- Fire and gas monitoring
- DC conversion/rectification

Several basic starting systems are available for GTs:

- Electric motor supplied from batteries
- Compressed air or gas system
- Small reciprocating engine (fired with diesel liquid or gas fuels)

In larger turbine systems, it is common to start the turbine by using the generator as a synchronous starting motor to crank the turbine up to its self-sustaining speed.

3.7   Costs

The capital cost of combustion turbines is ultimately to be determined by manufacturers' quotes prepared in response to engineers' specifications.

However, for planning purposes, past projects can serve as a guide. For example, the Gas Turbine World Handbook lists budget prices for sample systems which can be adjusted for site using the Means Mechanical System Cost reports. The former publication also provides a comprehensive list of the key performance characteristics of all turbines by major manufacturers. Figure 3.8 shows the average costs of turbines as a function of size. The prices quoted are as of 1995. Techniques described in Chapter 8 can be used to adjust costs to the present dollar value.

3.8 Fuels

A standard liquid fuel system for industrial GTs typically accepts liquid fuels ranging between kerosene and diesel fuel (JP-5, kerosene, No. 1 diesel, Grade 1 and 2 fuel oils, and No. 2 diesel fuel). An alternative liquid fuel system for industrial GTs using high vapor pressure and low viscosity fuels, such as natural gas liquids (NGL) and liquefied petroleum gas (LPG), gasoline, and naphthas, is typically used in pipeline applications where a high-pressure fuel supply is available. A throttle valve rather than a bypass valve is used to control the fuel flow pressure and maintain a higher pressure drop

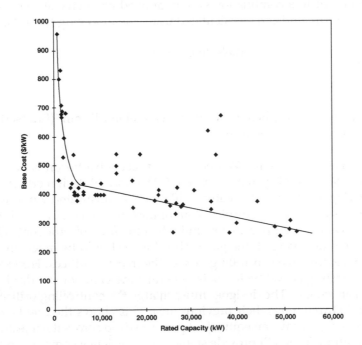

FIGURE 3.8
Average costs of GTs (courtesy of EPRI, with permission).

system. In this system, the fuel orifices are located adjacent to the fuel injectors so that high vapor pressure fuels can be kept at high pressure until the point of actual fuel injection, thus avoiding two-phase flow in any part of the fuel injection system.

The development of dual fuel injectors with dry, low-NO_x combustors by some manufacturers has been a major factor in combustion system versatility. Requirements arise in industrial GTs for burning natural gas as the standard fuel, but with provision for standby operation burning liquid fuel. It has become increasingly more important for GT combustion systems to become more versatile in the use of different fuels. Most industrial gas turbines are designed to operate on both standard natural gas and liquid fuel distillates. With minor modifications to the fuel control system, conventional combustion systems can operate on a wide range of fuels, including NGL, LPG, gaseous fuels rich in hydrogen, and gaseous fuels with a medium heating value, such as landfill or bio-derived gases.

The LHV of a gas is used to classify individual fuels into several distinct classes. These classes require different handling and control systems and, for more radical fuels, redesigned combustion systems.

Gaseous fuels are normally classified by using the Wobbe Index, a standard that accounts for variation in fuel gas density and heating value. The Wobbe Index is used to indicate the changes required to the fuel system so that fuels with different heating values can be accommodated. This index relates relative heat input to a combustion system of fixed geometry at a constant fuel supply pressure and can be calculated using the following formula:

$$Wobbe\ Index = LHV/S.G.$$

where:

LHV = lower heating value of the fuel in MJ/nm^3 (Btu/scf)
$S.G.$ = fuel specific gravity

If two fuels have the same Wobbe Index, direct substitution is possible and no change to the fuel system is required. The normal design criterion is that gases having a Wobbe Index within ±10% can be substituted without making adjustments to the fuel control system or injector orifices. This volume ratio is a significant design parameter, and when the fuel injector controlling orifices have to be changed, the gas Wobbe Index should be inversely proportional to the effective controlling area of the injector orifices. For example, a typical landfill gas Wobbe Index is one-third the value of standard pipeline quality natural gas. The designer must enlarge the controlling orifices on the injectors to three times their previous area. This allows the fuel flow rate of the landfill gas to have an equivalent pressure drop across the injector at full-load condition. This will provide stable, high efficiency combustion with the desired turbine inlet temperature distribution for long combustor and blade life. As fuel heating values decrease below standard levels, the torch igniter and the combustion system may require standard natural gas or liquid fuel for start-up or shutdown, as well as possible restrictions on turbine transient load operation.

Standard fuel gas systems can handle gas with lower heating values down to about 23.6 MJ/nm³ (600 Btu/scf) through minor modifications to fuel injector orifices and control system components. Alternate fuel gas systems use multiple fuel control components, manifolds (or single large manifold), and fuel injectors in parallel to further extend the handling of fuel heating value change to 11.8 MJ/nm³ (300 Btu/scf).

In order to provide greater flexibility for alternative gaseous fuels, a dual fuel system is generally recommended. A standard dual fuel system consists of a standard gas fuel and liquid fuel system and requires the use of dual fuel injectors. Start-up and shutdown with a standard backup fuel of consistent heat content, such as natural gas or distillate liquid fuel, is more reliable and has the added advantage of separating the GT start-up operation from the plant's fuel generating operation. Although it is desirable to have natural gas or distillate liquid fuel for start-up, it is not always required. In fact, for land-fill gas applications, the GT typically starts and operates on the landfill gas.

If an alternative gaseous fuel is used for GT start-up, the extent of modification to fuel handling, control, and injection components to provide a dual gas fuel system is a function of the difference in Wobbe Indices between the two gases. A dual gas fuel system involving large variations in heating value, such as 19.7 MJ/nm³ (500 Btu/scf), medium Btu gas, and a 35.4 MJ/nm³ (900 Btu/scf) high Btu gas, requires two different gas manifolds and sets of

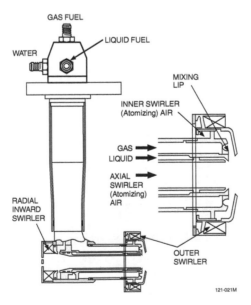

FIGURE 3.9
Typical dual fuel injector (courtesy of Solar Turbines, Inc., with permission).

metering orifices. These are needed to maintain gas injector pressure drop necessary for fuel distribution during start-up and stable operation under load. Figure 3.9 depicts a dual fuel metering system.

Additional Reading

R.S. Means Company (2001). *Means Mechanical Cost Data 2001 Book*, Kingston, MA.

Solar turbines, Inc., *Fuel Flexibility in Industrial Gas Turbines*, 1998.

Product and Technology Development Strategy, Solar Turbines, Inc., 1998.

Kluka, J.A. and Wilson, D.G., Low-leakage modular regenerators for gas turbine turbines, *Trans. ASME*, 120, 358, 1998.

Horlock, J.H., Aero-Turbine derivative gas turbines for power generation: thermodynamic and economic perspectives, *J. Turbineering Gas Turbines Power*, 119, 119, 1997.

4

Photovoltaic Systems

Yogi Goswami and Jan F. Kreider

CONTENTS

0-8493-0074-6/01/$0.00+$1.50
© 2001 by CRC Press LLC

Photovoltaic (PV) systems involve the direct conversion of sunlight into electricity with no intervening heat engine. PV devices are solid state; therefore, they are rugged and simple in design and require very little maintenance. A key advantage of PV systems is that they can be constructed as either grid-connected or stand-alone to produce outputs from microwatts to megawatts. They have been used as the power sources for calculators, watches, water pumping, remote buildings, communications, satellites and space vehicles, as well as megawatt-scale power plants. Because they are lightweight, modular, and do not require a gaseous or liquid fuel supply, PVs fit a niche that is unavailable to other DG technologies. For an overview of solar energy resources, technologies, and design approaches, the reader is referred to Goswami et al. (2000). Figure 4.1 shows a residential PV system.

FIGURE 4.1
PV array located adjacent to residential building.

Figure 4.2 shows the historical trend of PV shipments since the early 1970s when the only significant consumer was the U.S. Space Program. By contrast, all shipments in 1998 exceeded 150 MW,* up 21% from 1997. The U.S. is the world's largest producer, followed closely by Japan. Europe ranks third with significant production levels in Australia, India, China, and Taiwan.

For most of the 1990s, PV modules for buildings were used in off-grid modes. However, as Figure 4.3 indicates, grid-connected sales caught up with off-grid sales in 1997 because of various incentive programs that made PV power more competitive with conventional power. Among these were net metering and state payments based on kW installation levels. However, without subsidies, PV power remains two to five times as expensive as grid power, where grid power exists. Where there is no grid, as is the case in most of the world, PV power is the cheapest electricity source when operating and maintenance costs are considered.

Finally, with the drop of PV prices as shown in Figure 4.4, installation volume grew rapidly. In 1998, for the first time, PV prices dropped below $4/watt (not including the balance of system needed to control and convert the DC power into AC power, for example).

Historically, the photoelectric effect was first noted by Becquerel in 1839 when light was incident on an electrode in an electrolyte solution. Adams and

---

* PV systems are rated by their output under standard sunny conditions. Their average output is considerably less than this peak rating. The average rating is used to determine the value of energy produced, not the peak rating, which is often the basis of costs.

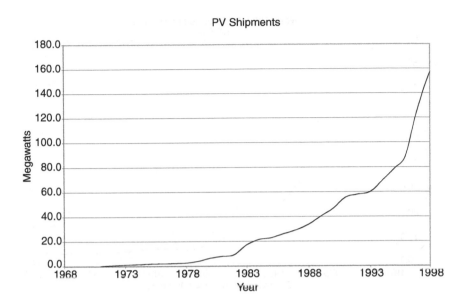

FIGURE 4.2
World shipments of PV modules 1971 to 1998 (courtesy of Paul Maycock).

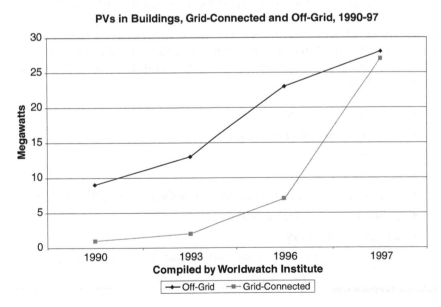

FIGURE 4.3
PV applications for buildings, grid-connected (lower line), and off-grid (upper line) for 1990 to 1998 (courtesy of Paul Maycock).

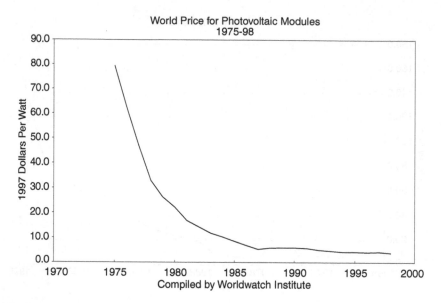

FIGURE 4.4
Average factory prices of PV modules ($1997/watt) for 1975–1998 (courtesy of Paul Maycock).

Day first observed the effect in solids in 1877 while working with selenium. Early work was done with selenium and copper oxide by pioneers such as Schottkey, Lange, and Grandahl. In 1954, researchers at RCA and Bell Laboratories reported achieving efficiencies of about 6% by using devices made of p and n types of semiconductors. The space race between the U.S. and the Soviet Union resulted in dramatic improvements in the photovoltaic devices.

This chapter discusses features of PV systems, devoting a section to each in the following order:

- Semiconductor types
- System efficiency and design
- Technical barriers
- Controls
- Conventional generation capacity displacement by PV systems
- Utility interconnection issues
- Economic summary

4.1 Semiconductor Types

A basic understanding of atomic structure is quite helpful in understanding the behavior of semiconductors and their use as PV energy conversion devices. Any fundamental book on physics or chemistry generally gives adequate background for basic understanding. For any atom, the electrons arrange themselves in orbital shells around the nucleus so as to result in the minimum amount of energy. In elements that have electrons in multiple shells, the innermost electrons have the minimum energy and therefore require the maximum amount of externally imparted energy to overcome the attraction of the nucleus and become free. Electrons in the outermost band of subshells are the only ones that participate in the interaction of an atom with its neighboring atoms. If these electrons are very loosely attached to the atom, they may attach themselves to a neighboring atom to give that atom a negative charge, leaving the original atom as a positive charged ion. The positive and negative charged ions become attached by the force of attraction of the charges, thus forming *ionic bonds*. If the electrons in the outermost band do not fill the band completely but are not loosely attached either, they arrange themselves such that neighboring atoms can share them to make the outermost bands full. The bonds thus formed between the neighboring atoms are called *covalent bonds*.

Since electrons in the outermost band of an atom determine how an atom will react or join with a neighboring atom, the outermost band is called the valence band. Some electrons in the valence band may be so energetic that they jump into an even higher band and are so far removed from the nucleus that a small amount of impressed force would cause them to move away from the atom. Such electrons are responsible for the conduction of heat and electricity, and that band is called a conduction band. The difference in the energy of an electron in the valence band and the innermost subshell of the conduction band is called the band gap, or the forbidden gap.

Materials whose valence bands are full have very high band gaps (> 3 eV). Such materials are called *insulators*. Materials that have relatively empty valence bands, on the other hand, and may have some electrons in the conduction band are good *conductors*. Metals fall in this category. Materials with valence bands partly filled have intermediate band gaps (≤ 3 eV). Such materials are called *semiconductors*. Pure semiconductors are called *intrinsic semiconductors*, while semiconductors doped with very small amounts of impurities are called *extrinsic semiconductors*. If the dopant material has more electrons in the valence band than the semiconductor, the doped material is called an *n*-type semiconductor. Such a material seems to have excess electrons available for conduction, even though the material is electronically neutral.

Silicon, for example, has four electrons in the valence band. Atoms of pure silicon arrange themselves in such a way that each atom shares two electrons with each neighboring atom with covalent bands to form a stable structure. If phosphorous, which has five valence electrons (one more than silicon), is introduced as an impurity in silicon, the doped material seems to have excess electrons, even though it is electrically neutral. Such a doped material is called *n*-type silicon. If, on the other hand, silicon is doped with boron, which has three valence electrons (one less than silicon), there seems to be a positive hole (missing electron) in the structure, even though the doped material is electrically neutral. Such material is called *p*-type silicon. *n*- and *p*-type semiconductors make it easier for the electrons and holes, respectively, to move in the semiconductors.

4.1.1 *p–n* Junction

As explained earlier, an *n*-type material has some impurity atoms with more electrons than the rest of the semiconductor atoms. If those excess electrons are removed, the impurity atoms will fit more uniformly in the structure formed by the main semiconductor atoms; however, the atoms will be left with positive charges. On the other hand, a *p*-type material has some impurity atoms with fewer electrons than the rest of the semiconductor atoms. Therefore, these atoms seem to have holes that could accommodate excess electrons, even though the atoms are electrically neutral. If additional electrons could be brought to fill the holes, the impurity atoms would fit more uniformly in the structure formed by the main semiconductor atoms, however, the atoms would be negatively charged.

The above scenario occurs at the junction when *p*- and *n*-type materials are joined together, as shown in Figure 4.5. As soon as the two materials are joined, excess electrons from the *n* layer jump to fill the holes in the *p* layer. Therefore, close to the junction, the material has positive charges on the *n* side and negative charges on the *p* side. The *p–n* junction behaves like a diode. This diode character of a *p–n* junction is utilized in PV cells as explained below.

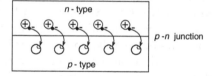

FIGURE 4.5
Schematic diagram of *p–n* junction in silicon.

4.1.2 The Photovoltaic Effect

When a photon of light is absorbed by a valence electron of an atom, the energy of the electron is increased by the amount of energy of the photon. If the energy of the photon is equal to or more than the band gap of the semiconductor, the electron with excess energy will jump into the conduction band where it can move freely. If, however, the photon energy is less than the band gap, the electron will not have sufficient energy to jump into the conduction band. In this case, the excess energy of the electrons is converted to excess kinetic energy of the electrons, which manifests at increased temperatures. If the absorbed photon has more energy than the band gap, the excess energy over the band gap simply increases the kinetic energy of the electron. Note that one photon can free up only one electron, even if the photon energy is greater than the band gap. This is a key limiting factor to PV cell efficiency.

Figure 4.6 shows a schematic diagram of a PV device and its load circuit. As free electrons are generated in the *n* layer by the action of photons, they can either pass through an external circuit, recombine with positive holes in the lateral direction, or move toward the *p* layer. The negative charges in the *p* layer at the *p-n* junction restrict their movement in that direction. If the *n* layer is made extremely thin, the movement of the electrons and, therefore, the probability of recombination within the *n* layer is greatly reduced unless the external circuit is open. If the external circuit is open, the electrons generated by the action of photons eventually recombine with the holes, resulting in an increase in the temperature of the device.

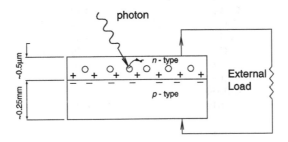

FIGURE 4.6
Simple circuit showing PV cell and resistive load.

4.1.3   Materials Overview

In a typical crystalline silicon cell, the *n* layer is about 0.5 µm thick and the *p* layer is about 0.25 mm thick. The energy contained in a photon E_p is given by

$$E_p = h\nu$$

where *h* is the Planck's constant (6.625×10^{-34} J-sec) and ν is the frequency, which is related to the wavelength and the speed of light *c* by

$$\nu = c/\lambda$$

Therefore,

$$E_p = hc/\lambda$$

For silicon, which has a band gap of 1.11 eV, this expression shows that photons of solar radiation of wavelength 1.12 µm or shorter are useful in creating electron–hole pairs and, therefore, an electric current.

Table 4.1 lists some candidate semiconductor materials for PV cells along with their band gaps. Substances shown in bold type are promising PV materials.

TABLE 4.1

Energy Gap for Some Candidate
Materials for Photovoltaic Cells

Material	Bandgap (eV)
Si	1.11
SiC	2.60
$CdAs_2$	1.00
CdTe	1.44
CdSe	1.74
CdS	2.42
$CdSnO_4$	2.90
GaAs	1.40
GaP	2.24
Cu_2S	1.80
CuO	2.00
Cu_2Se	1.40
$CuInS_2$	1.50
$CuInSe_2$	1.01
$CuInTe_2$	0.90
InP	1.27
In_2Te_3	1.20
In_2O_3	2.80
Zn_3P_2	1.60
ZnTe	2.20
ZnSe	2.60
AlP	2.43
AlSb	1.63
As_2Se_3	1.60
Sb_2Se_3	1.20
Ge	0.67
Se	1.60

Figure 4.7 shows a comparison of the maximum energy conversion efficiency of cells for some of the listed materials. This figure shows that the optimum band gap for terrestrial solar cells is around 1.5 eV.

4.1.4 Manufacture of Solar Cells and Panels

Manufacture of crystalline silicon solar cells is an outgrowth of the manufacturing methods used for microprocessors. A major difference is that the silicon used in microprocessors must be ultra pure; PV cells do not require this. Consequently, a large source of feedstock for silicon solar cells has been the waste material from the microelectronics industry. Solar cells are also manufactured as polycrystalline and thin films. This section briefly describes some of the common methods of manufacture of silicon solar cells.

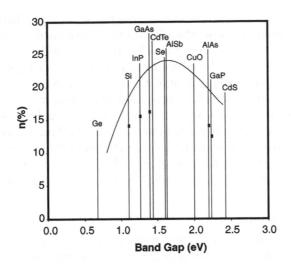

FIGURE 4.7
PV cell efficiency for several practical materials.

4.1.4.1 Single Crystal and Polycrystalline Silicon

Single crystal silicon cells are produced by a series of processes: (1) growing crystalline ingots of *p*-silicon, (2) slicing wafers from the ingots, (3) polishing and cleaning the surface, (4) doping with *n* material to form the *p–n* junction, (5) deposition of electrical contacts, (6) application of antireflection coating, and (7) encapsulation. The Czochralski process is the most common method of growing single crystal ingots. A seed crystal is dipped in molten silicon doped with a *p*-material (boron) and drawn upward under tightly controlled conditions of linear and rotational speed and temperature. This process produces cylindrical ingots of typically 10 cm diameter and 1 m length. Polycrystalline ingots are produced by casting silicon in a mold of preferred shape (rectangular). Molten silicon is cooled slowly in a mold along one direction in order to orient the crystal structures and grain boundaries in a preferred direction. In order to achieve efficiencies of greater than 10%, grain sizes greater than 0.5 mm are needed and the grain boundaries must be oriented perpendicular to the wafer. Ingots as large as 400 cm × 40 cm × 40 cm can be produced by this method. Ingots are sliced into wafers by internal diameter (ID) saws or multiwire saws impregnated with diamond abrasive particles. Both of these methods result in high wastage of valuable crystalline silicon. Alternative methods that reduce wastage are those that grow polycrystalline thin films.

A *p–n* junction is formed in the cell by diffusing a small amount of *n* material (phosphorous) into the top layer of a silicon wafer. The most common method is diffusion of phosphorous in the vapor phase. In that case, the back side of the wafer must be covered to prevent the diffusion of vapors from that side. Electrical contacts are attached to the top surface of silicon crystals in a grid pattern to cover no more than 10% of the cell surface, and a solid metallic sheet is attached to the back surface. The front electrode grid is made either by vacuum metal vapor deposition through a mask or by screen printing. Finally, titanium dioxide (TiO_2) and tantalum pentoxide (Ta_2O_5) are deposited on the cell surface to reduce reflection from more than 30% of untreated silicon to less than 3%. Anti-reflective (AR) coatings are deposited by vacuum vapor deposition, sputtering, or chemical spraying. Finally, the cell is encapsulated in a transparent material to protect it from the environment. Encapsulants usually consist of a layer of either polyvinyl butyryl (PVB) or ethylene vinyl acetate (EVA) and a top layer of low iron glass.

4.1.4.2   Amorphous Silicon

Amorphous silicon (a-Si) cells are made as thin films of a-Si:H alloy doped with phosphorous and boron to make n and p layers, respectively. The cells are manufactured by depositing a thin layer of a-Si on a substrate (glass, metal, or plastic) using glow discharge, sputtering, or CVD methods. The most common method is by an RF glow discharge decomposition of silane (SiH_4) on a substrate heated to a temperature of 200 to 300°C. To produce p-silicon, diborane (B_2H_6) vapor is introduced with the silane vapor. Similarly, phosphene (PH_3) is used to produce n-silicon. The cell consists of an n-layer, an intermediate undoped a-Si layer, and a p-layer on a substrate. The cell thickness is about 1 μm. The manufacturing process can be automated to produce rolls of solar cells from rolls of substrate.

Figure 4.8 shows an example of roll-to-roll, a-Si cell manufacturing equipment using a plasma CVD method. This machine can be used to make multijunction or tandem cells by introducing the appropriate materials at different points in the machine.

4.2   PV System Efficiency and Design

Figure 4.9 shows a typical current voltage curve for a solar cell. The power output is the product of the load current I_L and voltage V_L.

$$P_L = I_L \times V_L$$

The power output exhibits a maximum as shown. To maximize power output, the cell must operate at this condition no matter what the environmental conditions. This so-called maximum power point tracking is readily accomplished by electronic controllers that adjust the operating voltage in real time to the maximum power point voltage.

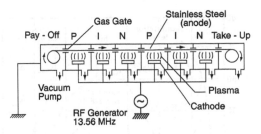

FIGURE 4.8
Schematic diagram of continuous *a*-Si manufacturing process.

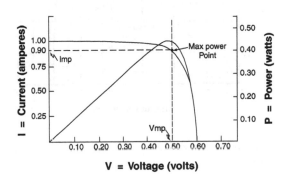

FIGURE 4.9
Typical current and power characteristics of a solar cell. Power axis to the right has units of W.

Figure 4.10 indicates the effect of incident solar radiation intensity and the load resistance on the performance of a silicon cell at a fixed temperature. Cell temperature also affects cell performance in such a way that the voltage and, thus, the power output decrease linearly with increasing temperature. Therefore, PV cells operate best when the cells are cool and the solar irradiance is high. It is worth noting that the maximum power point is at essentially the same voltage irrespective of the array irradiance. PV panels and arrays have similar performance characteristics.

4.2.1 Efficiency of Solar Cells

As indicated above, PV cells behave in this fashion:

1. Power output increases linearly with solar flux.
2. Power output increases linearly with decreasing cell temperature.
3. Power output increases linearly with cell size (follows from 1 above).

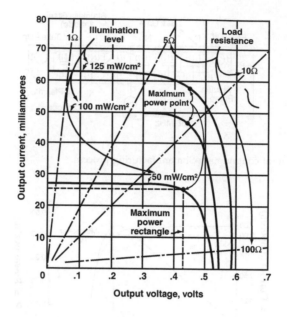

FIGURE 4.10

Example current–voltage characteristics of a silicon cell showing the effects of solar irradiance level and load resistance.

It is easy to show that efficiency of a PV cell or array is given by an equation of this type:

$$\eta = \eta_{ref}\,(1 + \beta[T_{ref} - T_{cell}])$$

The reference efficiency is given by the manufacturer and is usually 12 to 18%. The temperature coefficient of efficiency β is 0.004 C^{-1}. If one knows the average solar radiation, I, the average cell, panel, or array output (in units of watts) can be found from

$$E = \eta\, A\, I$$

The same approach can be used to find instantaneous output if the instantaneous solar radiation and cell temperature are used in the two expressions above.

4.2.2 Basic Design for Photovoltaic Systems

Grid-connected systems do not present any specific challenge to the experienced electrical system designer. The PV system is designed with priority given to local loads with any surplus either sold to the utility or stored on-site. Interconnection safeties and code factors must be accommodated in the design. Off-grid systems require additional attention because no grid backup exists to mask design oversights. PV power may be ideal for a remote application requiring a few watts to hundreds of kW of electrical power. Even where a conventional electrical grid is available, some applications needing uninterruptible or emergency standby power can use PV power.

Examples of off-grid PV applications include water pumping for potable water supply and irrigation, power for stand alone homes, street lighting, battery charging, telephone and radio communication relay stations, weather stations, etc. Additional examples include electrical utility switching stations, peak electrical utility power where environmental quality is a concern, data acquisition systems, and critical load such as ventilation fans, vaccine refrigeration, etc.

The design of a PV system is based on some basic considerations that distinguish one system from another:

- Which is more important, the daily energy output or the peak?
- Is a back-up energy source needed and/or available?
- Is the PV system the priority system or is another DG system the "lead" with PV power used for peaking?
- Is energy storage present?
- Is the PV-produced power needed as AC or DC, and at what voltage?

There are four basic steps in the design of a PV system: (1) estimation of electrical load and diurnal profile, (2) estimation of available solar radiation, (3) design of PV system, including area of PV panels, sizing of other components, wiring design documents, and (4) specifications document. Each of these steps is described by Taylor and Kreider (1998).

4.3 Technical Developments and Barriers

From preceding sections it is apparent that PV systems have several advantages and several challenges. Their most attractive features include modularity, easy maintainability, low weight, and environmental benignness. In contrast, PVs must overcome several barriers, including (1) significant area requirement due to the diffuse nature of the solar resource, (2) higher installed cost basis than other DG approaches, (3) intermittency of power supply without battery storage; low load factor, (4) modest efficiency, and (5) lack of finalized interconnection standards for grid applications.

Although module costs have already decreased significantly (see Figure 4.4), further cost reductions are likely. However, the *cost of energy* delivered by PV systems is unlikely to compete with energy produced by other DG technologies without incentives. However, when the capacity credit of PV systems and their power availability during peak load periods are considered, the economics are much more favorable. The broader financial context in which PV power is marketed will be discussed in the final chapter of this book. When the financial dispatch of electricity matures, the true value of PV power will become established, including its environmental benefits. During the late 1990s, peaking power was sold at high multiples of the average cost. Under these conditions, PV power is niche competitive in grid-connected applications.

4.4   PV System Controls and Diagnostics

Because PV systems do not require moving parts, except for the occasionally used electromechanical sun tracking device, their operational control can be fully electronic. The control features needed and included in all modern controllers are

- Maximum power point tracking
- Inverter operation and overload safety cutout
- Battery charge control
- Grid interconnection switching (for grid connected systems)
- Diagnostics

The first two functions are often combined into one unit in smaller systems. Unfortunately, system control designers have often sized the combined unit incorrectly by simply using the PV array manufacturer's nominal nameplate data and forgetting that higher power outputs and, most importantly, higher open-circuit voltages can be experienced on cold, sunny days. The economic viability of a PV system is immediately degraded when an inverter fails and needs to be replaced — the cost of an inverter is about $1000 per kW.

Battery charge and discharge control is necessary because long battery lifetimes are necessary for acceptable system economic performance. Both levels and charge and discharge rates must be controlled. Charge levels must be controlled to avoid overcharging and excess hydrogen evolution and physical damage to battery plates. Charge levels below 50–60% of the nominal level are to be avoided with lead acid batteries because lifetimes are reduced significantly when depth of discharge levels are exceeded. Likewise, excessive charge and discharge rates reduce battery lifetime. Charge controllers are readily available for PV systems.

Safety aspects of grid interconnection standards adopted by the PV industry require that local DG-produced power be disconnected rapidly and automatically in the event of a grid failure for whatever reason. The details of how these disconnects function are not important for this book. It is sufficient to say that complete and utterly reliable disconnect control hardware and software must be present in every grid-connected PV system. Disconnect standards are discussed in Section 4.6. Modern control systems archive data that are useful for panel and array diagnostics.

4.5 PV System Capacity Credit

PV systems differ from other DG systems in that they are not dispatchable unless equipped with storage. For that reason, they have been viewed as generator fuel saving. However, recent work has shown that there is significant PV capacity credit for grid penetrations of 20% or less. Using the standard Institute of Electrical and Electronic Engineers (IEEE) load duration curves, Rahim and Kreider (1992) showed, using an hourly simulation, that the equivalent load carrying capacity (ELCC) of a PV plant connected to the IEEE-standard grid was approximately equal to the average electrical output of the PV plant. Recall the earlier key distinction between the peak and average PV system outputs. As a general rule, in sunny climates the average output is about half the peak output under standard rating conditions. This rule of thumb obviously does not apply to overcast or hot climates because both of these factors diminish plant output when compared to the output in a sunny, cooler climate.

The ELCC is defined as the amount of standard fossil generation capacity that can be avoided by connecting PV systems to the grid. Therefore, the net installed cost of a PV system is its initial capital cost reduced by the cost of the avoided conventional plant. Since conventional plants cost about $600 to 1400/kW, depending on the size and type, the net PV system cost (expressed in $/kW) is reduced by this amount multiplied by the average to peak plant output ratio noted above to be about 50% in sunny locations. Therefore, a PV plant that has a gross installed cost of $5000/kW would have a net installed cost of $4300/kW if its installation resulted in avoided capacity installation. The prevailing mode for PV system economic analysis rarely addresses capacity displacement. However, with the new financial markets in electric power, this issue becomes moot since the several types of generation will each ultimately need to charge their full costs.

4.6 The Utility Interface

The interface between distributed power sources and electric power systems is the point at which new standards for interconnection will apply. Although at the present time standards are still evolving, the key issues as they apply to PV systems can be summarized. Since the PV industry has taken the lead in the new DG technologies (microturbines, fuel cells, and PV), it is likely that these three, at least, will end up with similar standards.

Interconnection standards result in safe and reliable PV systems. A key feature is the disconnect between the PV source and the grid system if the grid experiences an outage. This prevents powering the utility system from the DG system, thereby protecting line personnel. The independent system operator (ISO) also requires PV operators to operate within frequency and voltage limits. A key driver for new standards is that the existing grid was not designed for DG. Interconnection standards address these main features:

- Coordination with distribution system equipment, such as upstream voltage regulators and overcurrent protection devices
- Transformer connections and grounding
- Monitoring, data telemetry, and utility remote control
- Testing and verification of interconnection relays, switchgear, and distributed generation control equipment

4.6.1 Example Interface Standard — State of Texas

The state of Texas was the first to establish an interconnection standard for DG. Although a national consensus standard will ultimately be adopted by most of the states, this early standard draft indicates what considerations are involved. The following is a condensation of the Texas standard with key points that will be present in any interconnection standard. Note that a standard is required because heretofore only a national guide for the interconnection of emergency power generators existed, issued in the 1980s by the IEEE. In the U.S., only standards can be referenced in building codes, for example; guides cannot be referenced. At the time this book was written, the IEEE was midway through its DG standards writing activity under working group 1547.

Selections from the Texas Draft Standard (with editing by the author)*

The size of DG units considered is generally ten megawatts or less. DG (the generator and any associated interface equipment) operating in parallel with the distribution utility will be required to operate and maintain equipment such that there is no adverse effect on other customers, or on the utility's ability to maintain voltage and frequency in compliance with P.U.C. SUBST. R. 25.51, relating to power quality.

---

* These extracts are for illustrative purposes only and are taken from a standard draft. The final standard was not complete at the time of publication of this book. The reader is referred to the final standard or to the IEEE standard when it is completed (scheduled for 2001).

Standardization of interconnection criteria should recognize that many features to protect safety and meet reliability needs can be, and are, integrated within electrical conversion units. These electric conversion units can comply with independent testing requirements, can be tested on the production line, and can then be installed in a streamlined fashion at multiple sites. Protection functions and the interrupting device should have some way of being individually function-tested during commissioning and on a periodic basis.

These guidelines are not intended for generation that would interface directly with secondary network systems (anything below 2.4 kilovolts) due to the special characteristics of these systems. These guidelines are not intended for generation that would interface directly with the transmission system (at or above 60 kilovolts). Transmission system interconnection is governed by P.U.C. SUBST. R. 23.67 and 23.70 and any successor regulations. (See Project No. 18703, Proposed Changes to P.U.C. Subst. R. §23.67 and §23.70 relating to transmission access and pricing.)

4.6.1.1   Safety Standards

Safety standards are important to continue the safe operation of distribution systems. Changes in other requirements should not unduly compromise safety standards. Safe operation is a concern of the host utility, the owners of distributed generation, and the operators of these facilities. Safety standards implemented by most electric utilities have a long history. Safety must not be unduly compromised in order to implement distributed generation in the State of Texas. However, in order to facilitate installation of distributed generation, flexibility in implementing the present standards of the electric utilities is needed. This flexibility should be focused on allowing installation of products and devices that meet the intent of the safety standards, although they may not meet a particular utility's historical practices and standards.

A device shall be in put in place for the purpose of isolating the source of generation from the utility system. This device should have the ability to physically disconnect the distributed generation source from the utility system. The device should provide for a visible disconnect to provide easy and sure confirmation of the switch status to a utility troubleshooting patrolman. The device shall be capable of interrupting full load current and shall be lockable in the open position. This device shall be accessible by the host utility on a 24-hour basis. The device may be a disconnect switch, a draw-out breaker, fuse block, or other commonly used means of physical isolation. These devices must be able to be controlled on-site and may have the capability for remote control.

The generation system must meet all applicable governmental standards: national, state, and local construction and safety standards. These do not include standard utility policies and practices. The local distribution utility may direct the generator to disconnect completely from the system due to abnormal system conditions, such as system outages, emergencies, and equipment maintenance and repairs on the utility system.

4.6.1.2   System Stability Requirements

System stability requirements are focused on keeping a reliable operating power system. This includes the need for voltage and voltampere-reactive (VAR) support as well as that for over- and underfrequency protection.

If the source of generation is not able to inherently provide the required levels of power factor, between 90% lagging and 95% leading as measured at the customer meter, some means of improving the power factor may be required, such as power factor correction capacitors. The means of power factor correction should be at the discretion of the owner/operator of the distributed generation.

For generators that will operate in parallel to the utility power system, the total harmonic distortion (THD) voltage must not exceed 5% as measured at the generator's terminals when the generator is not paralleled with the utility system. The THD at the point of common coupling needs to conform to IEEE 519 –1992 and is limited to 5% by P.U.C. SUBST. R. 25.51, relating to Power Quality. (See §25.51(c) relating to harmonics.)

4.6.1.3   Protection Requirements

Protection systems are required to protect the local distribution utility's distribution system and service. The protection of the generation system's equipment is the responsibility of the owner/operator of the distributed generation. This guideline does not purport to provide appropriate protection for either the generator or distribution system, but identifies minimal areas of protection that are needed to provide safe and reliable operation. The local distribution utility will not assume any responsibility for protection of the generator or of any other portion of the generation system's equipment for any operating conditions. The generation system owner/operator is fully responsible for protecting distributed generation equipment in such a manner that faults or other disturbances on the utility system do not cause damage to the generation system equipment. The generation system must be protected in such a way that ensures the separation of the generation source from the faulted area of the electrical distribution system. The protection system shall be designed so that the generator shall not energize a de-energized circuit owned by the utility. The owner/operator of the generation system is solely responsible for properly synchronizing the distribution generator with the utility. Protection requirements differ with the technology.

The owner/operator of the distributed generation shall provide a "standard" one-line diagram depicting the electrical system under normal and contingency operations. The one line diagram should include, at a minimum, all major electrical equipment that is pertinent for understanding the normal and contingency operation of the generation system including generators, switches, circuit breakers, fuses, protective relays, and instrument transformers. The diagram should include transformer connection configuration (i.e., Delta, wye, and grounding) where a transformer is required.

A document displaying protection threshold and coordination studies shall also be provided to the host utility on request. At the host utility's option, it may specify settings for protective relays in the interconnection scheme.

For packaged generating equipment where the protective devices are part of a manufactured assembly that has been certified (including necessary relay settings) for use by the utility, each subsequent installation of the manufactured assembly shall be deemed certified for use by the utility.

Many utility distribution systems use breaker-reclosing schemes in order to provide a high degree of reliability to the distribution system. The schemes may use instantaneous reclosing, timed reclosing, or some combination of the two. To safeguard against any misoperation where reclosing is used, isolation within 8–10 cycles may be required. In other cases, isolation will be allowed to occur over a longer period to avoid excessive operations and/or sympathetic trips for faults on feeders. In most cases, the utility will not be able to remove reclosing schemes.

Basic functionality would require that the generator be equipped with adequate protection to trip the unit off-line during abnormal system conditions. For units greater than two megawatts in parallel with the utility, this functionality should include but not be limited to single phasing of utility supply, system faults, equipment failures, abnormal voltage, or frequency.

Some common, basic protective schemes required at the interconnection interface might include the following (numbers in parentheses represent device numbers used to perform the stated function):

- Undervoltage (27) and overvoltage (59) protection
- Frequency (81) protection for over- and underfrequency sensing
- Synchronizing relay (25)
- Phase and ground over-current relays (50/51 and 50/51N)
- Ground overvoltage (59N) for delta transformer connection interfaces
- Phase directional overcurrent relays (67) (if not selling to utility)
- Devices required to initiate transfer tripping utility transfer trip protection

Protection may be obtainable through components of integrated products (hardware and/or software) that meet independent testing facility standards and production line testing. The relays listed here may not always be the best solution. New technologies may achieve the required outcome more efficiently.

4.6.1.4   Switchgear Requirements

This section is concerned with isolation requirements and paralleling of systems. Isolation switches are addressed above in Section 4.6.1.1.1.

The owner/operator of the distributed generation will have interrupting devices sized to meet all applicable codes along with voltage protection and frequency protection. Where a disconnect switch is used, it shall be able to provide visible confirmation. Circuit breakers or other interrupting devices at the interface between generator and utility must be capable of interrupting maximum available fault current. For larger units, a redundant circuit breaker will be required for installations that can generate more power than they consume. This requirement is to cover the contention of a hung breaker. This may not be necessary if a utility grade or high quality circuit breaker is used.

4.6.1.5   Metering Requirements

Energy and other components of electric power are easily addressed through modern electronic meters. For qualifying facilities with 100 kilowatts or less capacity, P.U.C. SUBST. R. 23.66(f)(3) and (4) apply for metering. If metering at the generator is required in such applications, metering that is part of the generator control package will be considered sufficient if it meets all the measurement criteria that would be required by a separate stand alone meter.

4.6.1.6   Generation Control

The control of generation should remain with the owner/operator of the distributed generation. The generation control shall take into account coordination with the utility including system protection and operational control concerns.

4.6.1.7   Testing and Record Keeping

Testing of protection systems shall be limited to records of compliance with standard acceptance procedures as defined by the manufacturer of the protective devices and by industry standards and practices. These records shall include testing at the time of commercial operation and periodic testing thereafter. The utility reserves the right to witness testing of switchgear.

A log of generator operations shall be kept. At a minimum, the log shall include the date, generator time on, generator time off, and megawatt and megavar output.

The host utility has the option to initially qualify a potential generator site as a viable source of capacity and energy prior to signing a contract for resources. In order to ensure the reliable operation of this generation, it has the right to evaluate maintenance records, operating personnel, etc., as part of the contracting process.

4.6.1.8 Review of Interconnection Proposals

A number of products that have not been previously used by electric utilities may satisfy these interconnection guidelines. Because a timely review process is important in any implementation of distributed generation, it is recommended that a review process be installed where the electric utility has four calendar weeks to respond to any request for distributed generation and interconnection approval. Good faith omissions, which can be relatively easy to deal with, ought not to unnecessarily delay approval of the project. If the electric utility rejects any implementation, it should give full reason and explanation. Upon request by the distributed generation project developer, the Commission will review this information and provide some guidance from a regulatory perspective within one month. Any projects so rejected can be proposed at any time in the future with deficiencies corrected. This timeline is based upon the customer providing all of the necessary information up-front. If information were missing, the timeline would be delayed by an amount reflecting the wait for information.

Proposal review shall be confidential, and information therein shall not be used for any competing purpose. When reviewing particular proposals or equipment, the utility shall take into account the intent of safety standards and the ability for the implementation to meet that intent of the host utility's safety and reliability practices and procedures. The review should also take into account the approvals of equipment by other utilities across the United States. It will not be acceptable to reject an implementation due to not meeting a particular utility's standards without addressing the implementation's negative effects on safety and reliability and its conformance to utility standards and practices.

In order to facilitate the review process, the following data shall be supplied as a minimum. Equipment specifications of major equipment including the generator and protection systems should include the following:

- One-line diagram
- System protection data as described above
- Location on utility system
- Commissioning date
- Test data
- Synchronizing method
- Maintenance schedules

For certain small generators (less than 300 kilowatts) there may be a once-only evaluation by an independent laboratory of the technology against the standards. Once the technology is approved, it shall be an acceptable distributed generation technology that requires no further utility review.

4.7   PV System Costs

Factory costs of PV panels are shown in Figure 4.4. However, these are average figures. Specific cases will differ somewhat depending on vendor, size of purchase, and current incentives at the marketer level.

4.7.1   System Costs

Installed prices are quite different from factory costs because many other costs are involved. The balance-of-system (BOS) costs increase the noted factory costs by 30 to 50%. Included are:

- Control equipment — maximum power point trackers, inverters, battery charge controllers (major BOS cost)
- PV array support structures (major BOS cost)
- PV array to field wiring (major BOS cost)
- PV panel to array wiring
- Disconnects (major BOS cost)
- Panel protection diodes
- Battery storage if present (major BOS cost)
- Installation labor and fees (major BOS cost)
- Insurance
- Data acquisition system and sensors (moderate BOS cost)

Because of cost variation among systems in various locations, it is essential to develop a cost estimate early in the design process by consultation with local suppliers. Because of the dynamic nature of the market, costs for similar systems may not be a good reference basis. Of course, the major cost items noted above deserve the most accurate estimation.

4.7.2   Cost of PV Power

There is often some confusion regarding the manner in which the PV industry quotes prices. The standard number is in $/W under specific conditions. The example below indicates how one can use this number to compare PV energy costs with those from other sources on a $/kWh basis used universally by all other DG technologies.

Example

A PV system including BOS has a net cost of \$.550/kW and is located in Denver, Colorado. Its efficiency averages 15% in that climate. The average solar radiation in Denver on a fixed surface is about 500 W/m², whereas the panel is rated at 850 W/m². Standard rating conditions are documented in manufacturers' literature.

To convert the given initial cost to the annualized cost of the system one multiplies the initial cost by the *fixed charge rate (FCR)* described in Chapter 8. A typical FCR value is 10%.

Therefore, the cost of this plant per unit area is

$$Cost\ per\ area = 850\ W/m^2 \times 0.15 \times 5.50\ \$/W = \$700/m^2$$

Since 10% of this cost is paid per year due to the given FCR value, the annual cost is

$$PV\ cost\ per\ year = 0.10 \times \$700 = \$70/m^2\ per\ year$$

Finally, the energy produced per m² per year is

$$Energy\ per\ year\ per\ m^2 = 0.15 \times 500\ W/m^2 \times 4380\ hr/yr$$
$$= 300\ kWh/m^2\ per\ year$$

Then the unit energy cost is just the ratio of the annual system cost to the annual energy output, or

$$Unit\ Cost\ of\ PV\ energy = \$70/300\ kWh = \$0.23/kWh$$

California makes a \$3/W payment to purchasers of some PV systems. If that incentive were applied to this problem, the system cost would be \$32/m²/yr and the cost of energy would be \$0.10/kWh.

References

Goswami, Y., Kreith, F., and Kreider, J., *Principles of Solar Engineering*, Taylor and Francis, New York, 2000.

Taylor and Kreider (1998), *Principles of Solar Energy Applications*, University of Colorado.

Rahim and Kreider (1992), *The Capacity Credits PV Systems*, JCEM Internal Report, University of Colorado.

5

Microturbines

Colin Rodgers, James Watts, Dan Thoren,
Ken Nichols, and Richard Brent

CONTENTS

Microturbines may offer one of the best short-term distributed power production options because of their simplicity and because no major technological breakthroughs are required for their deployment. Low emissions also characterize modern microturbines. This chapter describes three promising turbine technologies: single-shaft gas-fired turbines, dual-shaft gas-fired turbines, and Rankine cycle engines. The differentiators between combustion turbines, described in Chapter 3, and microturbines are four:

1. Size: less than 200 kW net shaft power output (this number is somewhat fuzzy, and various authors will include engines above 200 kW or restrict the size to less than 100 kW)
2. Simple cycle: single-stage compressor and single-stage turbine
3. Pressure ratio: 3:1 to 4:1 instead of 13:1 to 15:1
4. Rotor: short drive shaft with generator on one end with a bearing in the middle

5.1   Single-Shaft Gas Microturbines

The classic open Brayton cycle described in Chapter 3 is also the basis of gas-fired microturbine (MT) engines. The reader is referred to that chapter for cycle basics.

5.1.1   Overview

Several single-shaft MTs have been developed recently by Capstone, Elliott, and Honeywell with ratings between about 20–150 kW. Some published specifications for these four MTs are listed in Table 5.1. A number of MT flow-path configurations are depicted in Figure 5.1, the most compact of which is the wrap-around recuperator with an annular combustor, as used in the Capstone MT. Configuration choice is dependent upon the application. For

TABLE 5.1

Example Microturbine Specifications

Manufacturer	N krpm	Power kWe	Efficiency % (LHV basis)	Recuperated
Capstone	30	96	28	Yes
Elliott	116	45	17	No
Honeywell	75	75	30	Yes

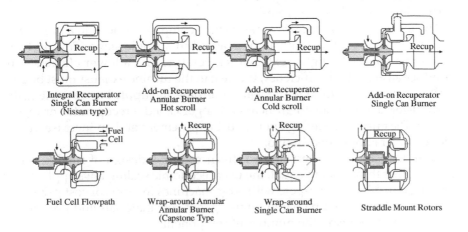

FIGURE 5.1
Various microturbine flowpath designs.

example, if both simple and recuperated cycle MT variants are to be marketed, it could be advantageous to use the add-on recuperator type, with possible provision for recuperator bypass and coupling with fuel cells. Bypassing also enables more heat to be available for cogeneration purposes.

Optimum MT rotational speeds at typical power ratings are between 60 to 100 krpm with compressor and turbine tip diameters of the order of a few inches, similar to small turbochargers. The state-of-the-art of small turbocharger turbomachinery has markedly improved in the last decade with the introduction of advanced computational fluid dynamics (CFD) design methodology and the routine use of composite materials and ceramic bearings.

The major aerodynamic difference between the small gas turbine and the turbocharger is the turbine design. The hurdles to a viable MT are not turbomachinery technology as much as other factors, such as:

- Cost concerns — overall $/kWe and recuperator costs
- Emissions
- Natural gas injection methods and their safety
- Shaft dynamics and bearing design
- Recuperator reliability, effectiveness, and cost

Cost concerns depend heavily on the generator power conditioning and control systems to be discussed later.

A major design feature of small gas turbines is the use of radial flow compressors and expanders. Use of radial flow compressors makes it possible to achieve a high pressure ratio, typically 3 or 4:1 in a single stage compared with about six stages in an axial flow compressor. For higher pressure ratios, more than one radial stage would be used, introducing considerable complexity. Thus, most gas turbines below 100 kW output and all machines in the 100 kW class use single-stage radial compressors. Most of them also use radial flow expanders or turbines.

The inherent characteristics of radial flow compressors lead to a preference for low pressure ratios that can be reached in a single stage. This, in turn, leads to a relatively higher exhaust temperature and a lower efficiency. To produce an acceptable efficiency, the heat in the turbine exhaust must be partially recovered and used to preheat the turbine air supply before it enters the combustor, using an air-to-air heat exchanger called a recuperator or regenerator. The effect is a savings in the fuel requirement and an increase in efficiency by a factor of nearly two.

Simplicity and cost exert major influences in the design of small gas turbines. Other things being equal, the usual economic scaling laws apply, so for small gas turbines to compete with large machines in terms of cost per kW of output, cost must first be pared by using low-cost materials and production methods and reducing the number of components.

The electrical connection to the grid or to a stand-alone load requires control of speed and voltage. MTs usually employ permanent magnet variable-speed alternators generating very high frequency alternating current (AC) which must be first rectified and then converted to AC to match the required supply frequency.

5.1.2   Design Characteristics

The MT prime mover is a simple Brayton cycle (see Chapter 3 for a complete description) with or without recuperation. MTs are operated at much lower pressure ratios (3 to 4) than larger gas turbines (10 to 15). In a recuperated system, pressure ratio is in direct proportion to temperature spread between inlet and exhaust. This allows heat (from exhaust) to be introduced to the recuperator, increasing net cycle efficiency to as much as 30%. Unrecuperated MTs average 17% net efficiency.

5.1.2.1   Combustor Overeiw

Scaling techniques for the design of mini combustors are less defined, due, in part, to the effects of (1) surface area/volume changes with size, (2) increased effects of wall quenching, (3) low fuel flows necessitating a small number of injectors and orifice sizing, and (4) increased effect of leakage gaps on pattern factor.

As a consequence, there is reluctance to directly apply scaling from larger combustors, and alternative design solutions have been considered, for example, by Rodgers (1974), where a single injector rotating cup fuel atomizer was successfully used to enhance cold starting with very low fuel flows.

The key combustor sizing parameter is defined as the heat release rate (HRR):

$$HRR = Fuel\ flow \times LHV/(Primary\ volume \times Pressure\ Ratio)$$

Typical HRRs for small single can combustors range from 6 to 10 million $kJ/m^3/bar$; lower HRRs provide increased residence time and are conducive to reducing CO emissions. Relatively high HRRs can be obtained with catalytic combustors (CC), but they require the addition of some form of preburner (to 430°C) plus additional downstream volume for combustion completion. As a consequence, overall combustor volumes are similar to conventional fuel injection burners. Ultra-lean burn CCs with compressor inlet injection require expensive catalysts such as platinum or palladium.

Emissions and carbon dioxide release are rapidly becoming the dominant criteria in the design of small MTs for hybrid electric vehicles, to the point that the whole engine design may be focused upon the combustor environment and operation. The MT flowpaths shown in Figure 5.1 include a reverse flowpath focused on access and flow uniformity into a single can combustor. The preferred fuel to minimize emissions is clearly natural gas.

5.1.2.2 Natural Gas Fueling

Natural gas (NG) is the fuel of choice for small business and domestic MTs but requires compression from essentially ambient pipeline pressures to levels exceeding MT compressor delivery pressure. The compressor outlet pressure is nominally three to four atmospheres. Adiabatic efficiencies approaching 40% have been recently measured (Rodgers, 1989) and required approximately 6% of the engine output power. Positive displacement, rather than dynamic, compressors more efficiently handle the very small, low specific gravity flow of natural gas. The selection of MT cycle pressure ratio requires consideration of the MT gas supply equipment.

Gas injection at the engine compressor inlet, in combination with burning a dilute air–gas mixture, is being researched. At the time of the writing of this book, at least three small companies are experimenting with this alternative to the gas compressor dilemma. Note that the air–gas mixture results in a 5% lighter gas and will slightly reduce overall compressor pressure ratio. Durability and safety aspects are prime concerns exceeding those of achieving efficient gas compression. Additional alternatives to NG compression include some form of lower pressure ratio (2.0) cycle, or the concept of discrete gas injection at the compressor inlet, and possible migration to a rich primary zone for stable combustion (Rodgers, 1991) plus the inverted Brayton cycle with atmospheric combustion. A conceptual inverted MT is shown in Figure 5.2.

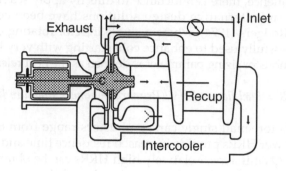

FIGURE 5.2
Inverted Brayton cycle schematic diagram.

During the late 1970s, AiResearch built and tested components of a 35 kW semi-closed, inverted Brayton cycle gas turbine (Friedman, 1977). The unit was designed for 27% thermal efficiency (ETATH) running at 90 krpm and 815°C. The inverted cycle was selected to allow for natural gas injection and combustion at atmospheric pressure and lower speed turbomachinery components.

Pitfalls that the inverted cycle presents include (1) both recuperator and intercooler effectiveness need to exceed 90%, (2) larger overall system size, (3) elevated compressor inlet temperature, (4) subatmospheric inboard seal leakage, and (5) 70% increase in package weight and cost. Emissions might be reduced with partial compressor exit flow recirculation.

5.1.2.3   *Recuperators*

Prime surface metallic recuperators that recover exhaust heat for combustion air preheating are most common, although ceramics hold considerable promise (McDonald, 2000), since seal deterioration on non-metallic regenerators can precipitate inordinately high leakage. Since recuperators are basically air-to-air heat exchangers, rather significant surface areas are needed because air to surface heat transfer coefficients are modest at best.

Both recuperator weight and costs typical of prime surface configurations with moderate pressure drops of 4 to 5% cycle pressure ratio are related to the recuperator effectiveness parameter (ε). Recuperator weight and cost are proportional to key drivers as follows:

$$Weight \propto airflow \times \varepsilon / (1 - \varepsilon)$$

and

$$Cost \propto airflow \times \varepsilon / (1 - \varepsilon)$$

Recuperator effectiveness values exceeding 0.90, therefore, have very large weight and high cost. These can only be justified when thermal efficiency and the value of power produced dominates operating costs.

A more detailed survey of low cost MT recuperators is addressed by McDonald (2000). Clean burning combustion is the key to both low emissions and highly durable recuperator designs, even though gas bypass may be used for surface periodic cleaning. Recuperator bypass can also be used for higher exhaust heat generation for cogeneration systems, but this essentially confines the MT flowpath to the add-on recuperator design.

5.1.2.4 *Bearings*

Of paramount importance in proposed applications is cold-starting the MT generator at cold weather conditions, nominally –40°C. Bearing tests have been satisfactorily conducted to assess the capability of accomplishing cold-starting with both turbocharger type floating sleeve bearings and for air bearings. Air thrust bearings are normally located behind the impeller and may require increased axial spacing between the compressor and turbine. The primary advantage of air bearings is that neither an oil system nor cooling is required.

Magnetic bearings have been considered for larger aerospace turbogenerators where adequate axial space exists to position both the magnetic and catcher bearings, and low cost is not a driver. Rotor assembly balance is a critical item for small turbomachines, and parts reduction is particularly beneficial in effecting balance sensitivity. Monorotor designs are, therefore, often favored. Magnetic bearings are not considered for MT applications.

5.1.2.5 *Generator*

Improvements in permanent magnet materials have resulted in lighter and more efficient permanent magnet generators (PMGs) than wound field generators. The field excitation is provided by permanent magnets that are capable of operation at temperatures up to 260°C. Integrating a high speed PMG with a small gas turbine presents challenges to the designer, such as high speed dynamics and balance, magnet retainment and temperature limitations, choice of cooling system design and evaluation of parasitic losses, maintenance and repair of components, voltage regulation and excitation shut off with internal fault, and frequency conversion to AC power. Overall system DC to AC efficiencies of 95% are possible with small gas turbine-driven PMGs. Power conversion to grid-standard, commercial AC incurs additional losses, due to both inversion and dissipation of heat generation.

An attractive feature of the high speed PMG is the ability to provide a high starting or light-off speed for the turbine, avoiding the need for a separate starting motor and dedicated start fuel injector, thus simplifying the fuel control system. Although high PMG tip speeds may be preferred to reduce rotor length with a stiffer shaft, generator efficiency decreases due to higher windage losses. PMG power capability P is linked to rotor speed and volume by the following relationship:

$$P = \eta\, N\, L\, D^2\, ESS/Constant$$

The terms L and D are shown in Figure 5.3, and the equation has been used to prepared the plot shown for a rotor length to diameter ratio of 4.0. *ESS* is the electromagnetic shear stress, η is the efficiency, and N is the rotational speed. The rotating shaft is shown crosshatched, and the fixed PMG field is shown immediately outboard of the rotor. The trade-offs that can be made between rotational speed, PMG tip speed, and diameter for the relevant power output range are shown. These three parameters have a major influence upon the aerodynamic and structural design of the complete MT. Generator cooling and heat rejection are major considerations and may incur parasitic power losses equivalent to 5% of the MT output. Cooling approaches using integral fan, air-oil mist, and suction from the compressor inlet have been developed.

5.1.3   Single-Shaft MT Cost Considerations

The prime consideration in any MT design is this: can the developed product successfully compete in the open market and against other generation technologies against the cost of utility power supply? Chapter 8 considers all

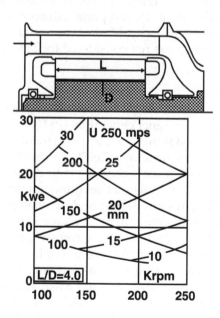

FIGURE 5.3
Generator performance map (power kWe as a function of rotational speed, Krpm). The generator section drawing shows the rotor (crosshatched) and the fixed stator. The lines of constant gap D slope upward to the right while lines of constant tip speed U (meters per second) slope downward to the right.

factors affecting these matters in detail. Even without a detailed analysis, it is clear that the non-recuperated cycle MT configuration would have difficulty competing on an operating cost basis unless coupled with some form of waste heat recovery. Thus, market opportunities for non-recuperated machines will probably be confined to standby power generation.

Variable MT costs include the heat engine assembly itself, the recuperator, and the generator. On the other hand, MT engine accessory and control costs tend to remain nearly constant, i.e., independent of size. Engine control costs also do not follow scalar relationships, since control dynamic relationships (apart from inertial effects) are relatively independent of size.

Typical microturbine system cost percentages are of the order:

Powerhead	25%
Recuperator	30%
Electronics	25%
Generator	5%
Accessories	5%
Package	10%

5.1.4 Single-Shaft MT Cycle Analysis

Thermodynamic performance of open cycle Brayton recuperated cycle small gas turbines is discussed in standard thermodynamics textbooks and by Rodgers (1993, 1997). However, it is important to understand the basics because a standard cycle analysis is the first step in analyzing the effects of all key parameters on MT performance. The analysis can be set up in a spreadsheet where parametric effects can be readily calculated.

MT efficiency and electrical (and thermal) output are basically functions of peak cycle temperature (turbine inlet temperature, TIT), recuperator inlet temperature (i.e., turbine exhaust gas temperature, EGT), compressor pressure ratio, and component efficiencies and size effects (recuperator effectiveness, turbine isentropic efficiency, compressor isentropic efficiency). The TIT is essentially determined by the limits of turbine rotor alloy stress rupture and low cycle fatigue strengths, duty cycle, and rotor cooling options. Likewise, the recuperator inlet temperature, i.e., EGT, is also determined by recuperator matrix material life limitations. The pressure ratio is dictated by the compressor type and material. Pressure ratios higher than 3.0 are desirable to reduce recuperator and combustor volumes. Component efficiencies are related to rotor sizes, aerodynamic excellence, and clearance gaps. Blade thickness and throat area tolerances concern efficiency, casting produciblity, and blade erosion.

The computational routine linking the engine thermodynamic cycle analysis, via specific speed and compressor work factor plus turbine velocity ratio, to component physical size and cost is described by Rodgers (1997). Several conclusions from such calculations are important to note because they apply more generally to any microturbine in the range of 2.5 to 100 kW:

- Pressure ratios above 3.5 show no improvement in efficiency (ETATH) as a consequence of lower compressor and turbine efficiencies and tip speed limits.
- Peak thermal efficiencies are 27.0 and 22.5% for small 2.5 and 5.0 kWe microturbines, respectively, at sea level at 27°C (80°F); for larger turbines, efficiencies approach 30% (LHV basis).
- Increasing recuperator effectiveness from 85 to 90% would increase ETATH by 1.9%, but recuperator size and price would increase by 60%.
- ETATH is relatively flat near optimum rotational speed; therefore, one would choose lower rather than higher speeds near the optimum, thus leaving a margin for future uprating.
- Power-to-weight and cost-to-power ratios versus speed are also relatively flat.

The final design, of course, entails significant further analysis and synthesis including considerations of stresses, thermal aspects, dynamics, and myriad other factors beyond the scope of this book. A cross section of a compact MT is shown in Figure 5.4 with a catalytic combustor, pre-burner, and replaceable recuperator cartridge. Provision for recuperator bypass would be feasible if additional steam capacity were required.

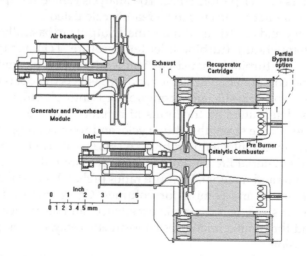

FIGURE 5.4
Example MT design drawing; dimensions depend on the output capacity.

5.1.4.1 Part-Load Performance

Since electrical loads in buildings and industrial processes vary with time, under certain dispatch control scenarios, MT output may need to be modulated. The details of this are described in Chapter 7. Operation at other than full power is commonplace. Therefore, the designer, economic analyst, and manufacturer need to understand part load performance of MTs. This section summarizes that topic.

The part-load performance of open cycle gas turbines is treated by Stone and Eberhardt (1962). They concluded that all conventional fixed geometry open-cycle engines within the normal range of pressure ratio and TITs have similar percentage rates of increase of specific fuel consumption (SFC, fuel used per kW output) with decreasing load, with the exception of moderately low pressure ratio cycles with a high degree of exhaust recuperation and with the ability to maintain a high TIT at part load. The single-shaft, high speed MT with variable speed operation and power conditioning lends itself, therefore, to improved part-load fuel economy, especially with intrinsically increased recuperator effectiveness at lower part load airflows, provided that an adequate compressor surge margin exists.

Figure 5.5 shows the generic part-load characteristics of a typical recuperated MT operating at either constant speed and variable TIT, or variable speed and constant recuperator inlet temperature. The variable speed mode improves part-load performance but requires a control system able to sense load and optimize speed. Exposure to high cycle fatigue failure is increased. Generally, constant speed operation is preferred for both improved life and frequency regulation, which is of concern since it affects the driven electrical equipment.

5.2 Twin-Shaft Gas Microturbines

5.2.1 Configuration

Two-shaft MTs follow an industrial equipment design philosophy similar to that used for chillers, boilers, or furnaces. They are built to meet utility-grade reliability and durability standards while producing electricity at least as efficiently as central generation and distribution technologies currently in use. Two-shaft MTs are designed exclusively for rugged, industrial-quality stationary applications; they fit right in on the plant floor or utility room and include no design compromises inherited from vehicle or aerospace ancestries.

Like single-shaft MT engines, two-shaft designs typically employ metallic radial turbomachinery components. They use "ruggedized" turbocharger components featuring pressurized lube-oil systems consistent with industrial best practice. They operate at relatively low pressure ratios in the 3:1 range using one stage of compression and two turbine stages (Figure 5.6). The first turbine (the gasifier turbine) drives the compressor and the second free-power turbine drives the load generator.

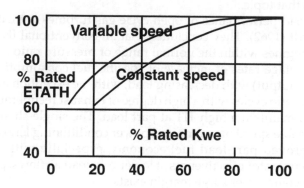

FIGURE 5.5
Generic part load efficiency curves for MTs; the efficiency is shown as a function of the fraction of full load power along the abscissa for two control approaches.

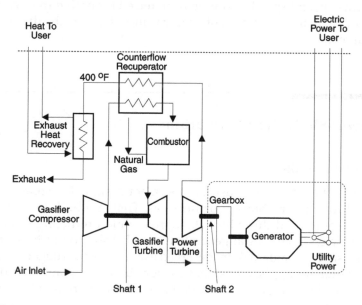

FIGURE 5.6
Typical two-shaft microturbine cycle diagram.

5.2.2 Why Two Turbines?

Designing the MT engine to employ two turbines offers several advantages:

- Long engine life — The two-turbine configuration reduces stress by splitting work output between the turbines. Turbine life is extended even further by the relatively low pressure ratio and TITs employed by the engine.

- Direct mechanical output — Any rotating mechanical component can be driven by the power turbine. Thus, the two-shaft MT engine can be used in a wide variety of applications. For example, two-shaft MT systems have been built and tested to directly drive the refrigerant compressor of a vapor-cycle chiller or a screw compressor that would typically be used in ammonia cycles for refrigeration systems.

- Design point flexibility — The two-shaft engine configuration allows the designer greater flexibility in choosing a design point for the power turbine. For example, the power turbine can rotate at significantly lower speeds to better match a particular load requirement. Thus, conventional, reliable, low-cost induction and synchronous generators can readily be connected to the power turbine through a reasonably sized gearbox. Since the power turbine rotates independently, it can accommodate variable-speed applications such as refrigeration systems where the turbine drives a screw compressor.

- Shaft mechanical design — Shaft design issues related to the power turbine and the load component (compressor impeller, generator, etc.) are independent of those associated with the gasifier/turbine shaft. The complexity of each of these simple shafts is, thus, much less than the shaft design required of single-shaft engines. The latter must account for the rotor dynamic, loading, and sealing complexities associated with placing all rotating components (including the generator) on one shaft.

- Component configuration complexity — An independent power turbine allows greater freedom in laying out rotating components. Requiring multiple components to be crowded onto a single shaft often forces poor design compromises. For example, the air flowing into the compressor is often used first to cool the high-speed alternator. Unfortunately, gas turbine engines quickly lose efficiency and power with rising air inlet temperature, and overall performance suffers accordingly.

- Mechanical safety issues — The lower rotating speed of the power components reduces the danger level of catastrophic rotating failure in the system. The higher the rotating speed, the more deadly the "shrapnel."

- Favorable torque/speed characteristics — Since, to the first order, the mass flow through the engine is not affected by power turbine conditions, the power turbine actually delivers more torque as it slows, the opposite characteristic of the turbine in single shaft designs. This improves the engine's ability to handle load changes and maintain operational stability.

Not every advantage lies with the two-shaft approach, of course. Initial capital cost is still an important consideration even when calculating the full life cost of an MT to a facility. In addition, the cost of a design does not necessarily scale with the number of components. However, MT designers must be very careful about cost because of the competitive pressures of the generation market. Unlike single-shaft designs, a two-shaft MT cannot be started by driving the turbine temporarily by the generator (now acting as a motor). Therefore, the designer must build into the system a starting mechanism to bring the gasifier turbine up to an initial operating speed.

5.2.3   Applications

The two-shaft microturbine can be used in a variety of applications because of the flexibility inherent in its direct mechanical drive capability. Figure 5.7 shows the possible combinations of shaft power/thermal power applications. The most commonly discussed application for MTs today is for cogeneration wherein the turbine drives an electric generator. As shown in Figure 5.6, an additional heat exchanger built into the exhaust stream is typically used to heat water for either space heating or domestic hot water service. In a cogeneration application that can effectively use the waste heat, overall system efficiency can be quite high (80% or more). The system can be designed to provide a variable heat recovery output to accommodate the

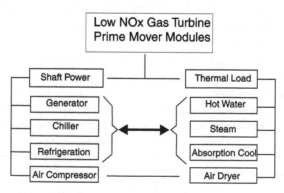

FIGURE 5.7
Application matrix for a two-shaft microturbine.

specific heat requirements of the facility over time. This includes running the MT at full power output with a zero heat load if necessary.

In another two-shaft MT application, the power turbine is directly connected to a centrifugal compressor to drive a vapor-cycle chiller system (Figure 5.8). The MT simply replaces the electric motor/compressor unit commonly used in chillers. Using a conventional refrigerant such as HFC134a, the rest of the chiller system (condenser, expansion valve/economizer, evaporator, etc.) remains the same. When packaged as a chiller, the two-shaft MT systems can be sized to deliver between 30 and 400 refrigeration ton (RT) of chilled water. Typical integrated part-load value (IPLV) coefficients of performances (COP) approach 2.0 (based on HHV) under ARI conditions. In the past decade, natural gas-fueled systems such as these have become less expensive to operate than electric motor-driven chillers, due in part to rising electricity costs and comparatively low gas prices in the summer.

The two-shaft design is also well suited for driving positive displacement screw compressors (Figure 5.9). Again substituting the power turbine for the electric motor, the system can be designed to work with many different types of fluids in refrigeration and gas boosting applications. The operating characteristics of the engine make it best suited for full-load, continuous-duty, industrial refrigeration which, depending on the cycle design, can be configured to provide cooling down to –100°F (–73°C).

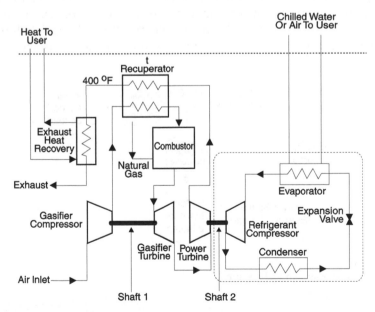

FIGURE 5.8
Microturbine chiller application cycle diagram.

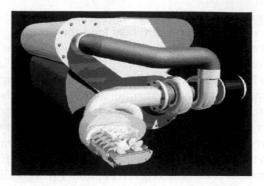

FIGURE 5.9
Microturbine engine/screw compressor assembly.

5.3   Power Electronics and Controls

Microturbines produce high frequency AC power. Development of the power electronics and control systems (PECS) required to convert this unregulated power into a usable form (e.g., 60 Hz, 480 VAC) has proven to be a significant but achievable hurdle in the commercial deployment of MTs. Figure 5.10 depicts a modern DG power system. Although the details depend on the specific power generation technology, the same basic principles apply.

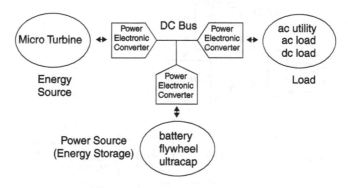

FIGURE 5.10
Modern MT power system architecture.

The unregulated power from the energy source must be processed (or converted) before it is consumed. In the case of an MT, the power produced by the generator is variable both in terms of voltage and frequency and is, therefore, not directly usable. The PECs that perform this power processing and regulation function are often specific to the generation technology. For example, in the case of an MT, the first stage of power electronics may contain a diode rectifier that converts the high frequency AC power produced by the generator into DC. After the power produced by the energy source has been converted to DC, it can easily be combined with power from the element at the bottom of Figure 5.10, labeled "Power Source" (Energy Storage). Energy storage is an important aspect of many DG power systems, often required for supplying starting power to the system as well as for supplying rapid transients demanded by the load. This power source is coupled to the DC bus through another power electronic converter, the details of which are dependent on the energy storage technology employed. For example, if an electrochemical battery is used, a DC-to-DC converter is used to couple the energy from the battery to the DC bus. In the case of a flywheel, an AC-to-DC converter would be used.

Finally, a PEC is required to convert electricity into a form consumable by the load. Once again, the details of this converter are dependent on the specifics of the application. For example, in the case of an electric utility (or stand-alone AC load), the power electronics would take the form of a DC-to-AC inverter. In the case of a hybrid electric vehicle (HEV), the power electronics would take the form of a DC-to-DC converter. Other power electronic elements are not shown in this simple diagram but are essential for realization of a practical DG power system. For example, motor controllers and power supplies are commonly employed in ancillary systems.

5.3.1 Power Electronics Technology

Modern power electronics is primarily based on the technique of switch-mode energy conversion: high power transistors are rapidly switched on and off in order to control the flow of electric power (this is in contrast to a transistor operating in the linear region). At any instant of time, the transistor is either fully on (the voltage across the device is zero), or it is fully off (the current passing through the transistor is zero); therefore, the losses in the transistor are minimized. In reality, it is impossible to create a perfect transistor that has no losses, but significant improvements continue to be made. These high power transistors can be configured into various topologies to form the desired power conversion function (e.g., DC-to-DC, AC-to-DC, or DC-to-AC). Pulse width modulation (PWM) is then normally used to vary the on and off times of the transistors in order to synthesize the desired output voltage and frequency.

For power levels up to approximately 500 kW, the insulated gate bipolar transistor (IGBT) is the most common power semiconductor device used today. Switching frequencies up to 20 kHz can be obtained using these devices. Because the transistors are controlled to be only on or off , the waveforms produced by the converters consist of pulses with very high harmonic content. High power filters are necessary to remove the high frequency harmonics and allow only the useful fundamental component of energy to pass. In order to minimize the losses in these filters, reactive components are used (inductors and capacitors). Filter design is a very complex and challenging aspect of power electronic systems.

Modern power electronic converters can achieve efficiencies greater than 96% for a single power conversion stage (including the filter losses). Even with these very high efficiencies, a significant amount of power is dissipated in the converters. For this reason, thermal management is a critical aspect of power electronics design. Thermal management systems can use both liquid and forced air-cooling techniques, with air-cooling being the most desirable from a cost and reliability standpoint.

Figure 5.11 shows an oscillogram from an MT system showing the current produced by the high-speed generator (Channel 1), along with the resultant current simultaneously supplied to the load (Channel 2), in this case an electric utility. This figure indicates the importance of the power electronics function in an advanced DG power system.

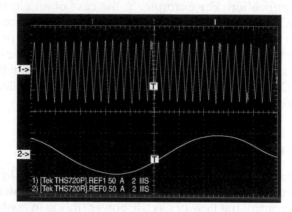

FIGURE 5.11
Oscillogram from an operating microturbine system. (Courtesy of Capstone Turbine Corporation, with permission.)

5.3.2 Digital Control Technology

In a typical DG power system, three aspects or layers of control must be considered. The first and most demanding is the "micro" level of control required to manage the switching of the high power transistors in the power electronic converters. This task requires very high-speed real time control, with sampling rates on the order of microseconds.

The second aspect is the "macro" level of control that is required to manage the system at large, i.e., control the power flow among the energy source, energy storage element, and load. The requirements of this particular task are dependent on the specific generation technology, but sampling rates on the order of milliseconds are typically required. State-of-the-art DG power systems use digital controls to perform these functions. The advantages of digital electronics over analog electronics are well understood, including accuracy, flexibility, and repeatability. In addition, the DG industry is taking advantage of the example set by the computer industry: digital data processing results in higher reliability with increasingly lower cost. However, real time digital control systems are extremely complex, and a significant amount of time and resources go into developing the control software for an advanced DG power system.

Figure 5.12 illustrates the basic principle of a digital control system. After removal of unwanted spectral content, an analog-to-digital converter (ADC) is used to sample the physical analog signal of interest. A high-speed digital signal processor (DSP) is used to perform advanced control algorithms using these signals and produce the desired control action. This processing must be accomplished in real time, meaning that the time it takes from the instant the data is sampled to the instant the output is produced must be small compared to the dynamics of the process being controlled. For example, in the control of a DC-to-AC converter, the voltage and current produced by the converter are sampled, and the DSP subsequently computes the required on and off times of the various high power transistors in order to control the energy conversion process. In the case of controlling an MT engine, the exhaust gas temperature may be sampled, and the DSP computes the required fuel flow in order to control the engine.

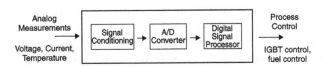

FIGURE 5.12
Digital signal processing and control.

The third and final aspect of control in a DG system is that of communications to external equipment and the outside world. Advanced DG systems provide a variety of digital communications interfaces so that the system can be remotely monitored and controlled. This capability is required in order to aggregate significant numbers of DG systems into larger power generation systems.

A crucial and often overlooked aspect of DG power system design is that the micro level control over the power electronics, the macro level control over the energy source, and the communications function must all be considered together. This aspect makes the system design very difficult and requires close interaction and cooperation between the designers of the power electronics and the designers of the energy source. When this is done well, a very high performance and highly integrated product results.

5.3.3   Applications

Figure 5.13 shows a complete MT generator system. The MT engine can be seen located above the PECS module. Both components of the system are similar in terms of size and weight, and both components are equally important to the mission of generating clean, reliable power. The PECS in a modern DG system allows it to operate in a variety of modes, providing solutions for applications that otherwise would not be practical for small heat and power generation systems. A few example applications are described below.

FIGURE 5.13
Microturbine generator system, showing the engine (on top), and the power electronics and controls (below). (Courtesy of Capstone Turbine Corporation, with permission.)

5.3.3.1   Grid-Connect Operation

The PECS can be configured to operate in a grid-connect (GC) mode of operation. In that case, the system follows the voltage and frequency from the grid and behaves like a controlled current source. Grid-connect applications include load following and peak shaving. One of the key aspects of a GC system is that the synchronization, protective relay, and anti-islanding functions required to reliably and safely interconnect with the grid can be integrated directly into the PECS. This capability is a major advantage of PECS in DG systems, and eliminates the need for very expensive and cumbersome external equipment required in conventional generation technologies.

Another aspect of PECS in DG is the ability to provide power quality functions without any external equipment. For example, the PECS can be controlled to supply reactive power and is capable of performing voltage regulation and power factor correction.

Many performance characteristics of the electric utility grid cannot be easily duplicated by DG technologies. For example, the ability to supply large fault currents and rapid power demands (both real and reactive) is rather limited in DG systems. As opposed to being viewed as a potential replacement for the grid, a DG system with advanced PECS should be considered as an enabling technology allowing improvements to the reliability and power quality of the grid.

5.3.3.2   Stand-Alone Operation

The PECS can be configured to function in a stand-alone (SA) mode of operation. In that case, the system behaves as an independent voltage source and supplies the current demanded by the load (in terms of both magnitude and phase). There are many advantages to utilizing power electronics in an SA generator. Similar to an adjustable speed drive (ASD), the PECS is capable of providing variable voltage and variable frequency power to the load. For example, if an MT system were operating a process control motor, it would be capable of varying the speed to optimize the process. With this capability, a single MT system can replace the combination of a reciprocating engine-generator along with an ASD used to control a motor load.

5.3.3.3 Dual Mode Operation

A PECS system that is capable of functioning in both a GC and SA mode of operation can also be designed to automatically switch between these two modes. This type of functionality is extremely useful in a wide variety of applications, and is commonly referred to as dual mode operation.

5.3.3.4 Multiple Unit Operation

The PECS can be configured to operate in parallel with other DG systems in order to form a larger power generation system. This capability can be built directly into the system and does not require the use of any external synchronizing equipment. Redundancy can be provided such that if one unit shuts down, the others will continue to operate.

5.3.3.5 Flexible Fuel Operation

A sophisticated PECS design can allow a DG technology to operate on a wide range of fuels. It is the flexibility and adaptability enabled by digital control software that allows this to be possible without making significant changes to the system hardware. For example, an MT system can be designed to operate on fuels with Btu content ranging from as high as propane [~2500 Btu/SCF], to pipeline quality natural gas [~1000 Btu/SCF], to digester gas [~600 Btu/SCF], all the way down to landfill gas [~ 350–500 Btu/SCF]. Liquid fuels such as diesel, gasoline, and kerosene can also be readily accommodated with only minor changes required to the MT fuel system.

5.4 Microturbine Performance Improvements

5.4.1 Turbomachinery Performance

The efficiency of turbomachinery components such as radial compressor and turbine stages continues to improve with time as fluid dynamic designers employ new design methods and technologies. For example, the blading of a modern radial compressor stage is typically designed using a set of straight-line elements. These lines are arranged so as to create three-dimensional blade shapes called ruled surfaces. This approach can create fairly complex surfaces and, if designed correctly, produces relatively efficient stage designs. But new techniques are becoming available, including three-dimensional CFD analysis capability, which allow the designer to define more complex blades shapes (called sculpted surfaces). This offers opportunities for even greater stage performance.

Figure 5.14 illustrates the current state of the art in compressor performance using current design techniques. MTs use radial compressor stages that generally fall into the specific speed range of 0.7 to 1.0. As shown, stage efficiencies of 87 to 89% could potentially be achieved for certain types of compressors in this specific speed range. However, compressors typically used in MTs are more likely to only reach efficiency level percentages in the low 80s. Part of the reason for this lies in the relatively small size (typically a few inches in diameter) of the compressors used in MTs. Although state-of-the-art efficiencies might be possible with large diameter designs, several factors significantly limit the efficiency MT-scale compressors can reach. For example, Reynolds number effects impose aerodynamic limits, and clearance ratios are relatively large due to practical limits in the bearings. In addition, the tight cost constraints imposed on MT in order to meet market price targets dictate the use of high-volume/low-cost manufacturing techniques. For example, compressor impellers are cast rather than precision machined to reduce cost. However, cast parts exhibit larger tip clearances, looser dimensional tolerances, and other characteristics that significantly reduce compressor performance. Therefore, there are practical limits to improving turbomachinery performance. Only relatively small overall system gains can

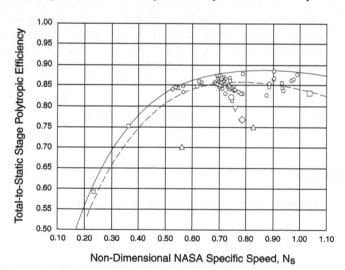

FIGURE 5.14
Compressor efficiency improvements.

be achieved as long as the turbomachinery components are well matched to the operating requirements of the MT in the first place.

5.4.2 Engine Operating Conditions

One obvious way to raise system efficiency is to operate gas turbines with more aggressive operating conditions, i.e., higher pressure ratios and/or greater TITs. For example, the designer can choose to operate at higher TITs or to increase pressure ratios in the system. As Figure 5.15 illustrates, substantial system efficiency improvements are possible with either of these approaches for a typical simple-cycle engine in the MT size range.

However, the strict cost constraints require use of metal components without advanced features found in more expensive gas turbine engines (e.g., air-cooled blading). As a result, rising pressures and temperatures quickly reach limits that exact a steep price in engine component life. High temperatures are especially problematic due to the creep-life and stress constraints of the materials used in MTs. As a point of reference, TITs in current MT designs vary between 1600 and 1850°F [870 and 1000°C]). At this temperature level, the strengths of the high-temperature materials typically used in MTs are very sensitive to temperature and quickly lose strength (and, thus, reduce engine life) with higher temperature.

Increasing system pressure ratio, of course, raises stresses in the system. When coupled with the relatively high temperatures mentioned earlier, it is easy to see why these two aggressive conditions can so dramatically shorten engine life. One of the more important advantages of the two-shaft MT

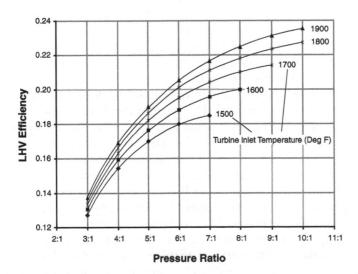

FIGURE 5.15
Pressure ratio and TIT effects for nonrecuperated MTs.

design is that these life-limiting stresses are essentially split between the two turbines. The effect is so favorable that the designer can even use somewhat higher TITs and still preserve long engine life. One other disadvantage of operating at higher pressure ratios is the resulting need for higher fuel delivery pressures because higher output pressures from the fuel gas booster (required in at least half of the U.S. natural gas service area) are required, thereby consuming more parasitic power.

5.4.3   Heat Recovery

Presently, it appears that the most promising way to raise engine efficiency is to recover heat by using a regenerator or recuperator as part of the engine cycle. Therefore, a recuperated cycle has been adopted by nearly all MT manufacturers, although the degree of heat recovery varies considerably. Figure 5.16 shows how recuperation improves the performance of a typical simple-cycle gas turbine engine. Note the dramatic effect in the range of low pressure ratios employed by microturbines (3:1 and higher).

The performance impact on the balance among TIT, pressure ratio, and recuperator effectiveness is shown in Figures 5.17 and 5.18. Note the effect of differing levels of recuperator effectiveness. An engine design with an effectiveness of only 85% would require a TIT of 1800°F and a pressure ratio of 4:1 to reach an engine efficiency of 33% lower heating value (LHV). However, a 1600°F TIT at a pressure ratio of about 3:1 will reach this level only if a 91% effective recuperator is used. An engine running under the latter conditions will enjoy a significantly longer life and reduced maintenance costs.

More effectiveness, however means more surface area and, thus, more weight and volume. Fortunately, the cogeneration, chiller, refrigeration, air compression, and other market opportunities for today's MTs are stationary applications. Within reasonable limits, it does not matter how heavy a system is. Therefore, an MT designed specifically for stationary applications is free to incorporate a very effective recuperator.

Figure 5.19 shows an example of two-shaft MT design that incorporates a high-effectiveness recuperator. The recuperator is contained in the bright metal enclosure in the upper portion of the machine that also provides inlet air and engine exhaust ducting service. The waste heat exchanger for cogeneration applications is also located within this enclosure.

Like other MT components, recuperator life and cost are critical. Recuperators experience large thermal gradients and load swings in the course of normal operation. Inlet temperatures are also fairly high, but are considerably lower than TITs. Increased TIT and pressure ratios do not cause recuperator inlet temperature to increase by much because the expansion ratio of the turbine increases as the ratio of turbine inlet and outlet temperatures increase. Therefore, metals will still be applicable for recuperators even if future turbine components are switched to ceramics to withstand greater operating temperatures.

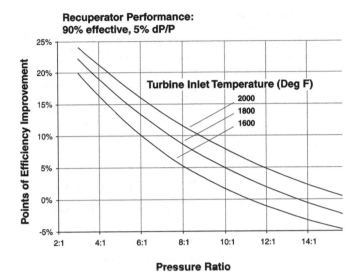

FIGURE 5.16
Recuperator performance improvement.

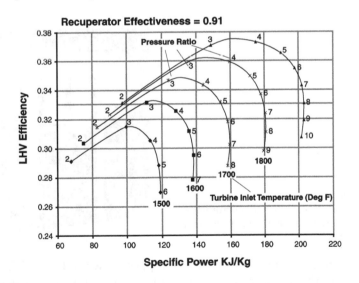

FIGURE 5.17
Efficiency map (91% effective recuperator).

5.5   Rankine Cycle Microturbines

The Rankine cycle is most commonly known from its application in large
steam-driven power plants. In its simplest form, this closed cycle consists of

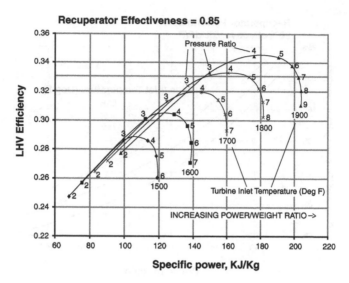

FIGURE 5.18
Efficiency map (85% effective recuperator).

FIGURE 5.19
Highly recuperated, two-shaft microturbine. (Courtesy of Ingersoll-Rand Energy Systems, with permission.)

a boiler, turbine, condenser, and feed pump, as shown in Figure 5.20. Heat is transferred to the Rankine cycle working fluid in the boiler, producing saturated or superheated vapor for the turbine. As the vapor is expanded through the turbine, shaft power is extracted. The low density fluid then passes through the condenser, where it is cooled and converted to liquid which is then pressurized by the feed pump to supply liquid to the boiler, thus completing the cycle.

While the Rankine cycle is best known from steam power plants, it has been used successfully in geothermal and solar binary power systems and waste heat power systems. Figure 5.21 shows a small, high-speed turbine–alternator–pump unit used in a solar-powered Rankine system. In this application, the turbine,

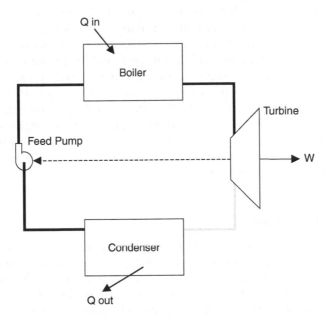

FIGURE 5.20
Rankine cycle schematic diagram.

FIGURE 5.21
Integrated turbine-alternator-pump for solar Rankine cycle.

feed pump, and power generator were all mounted on the same shaft. The feed pump supplied working fluid through the alternator cooling passages, thus preheating the feed water before moving to the boiler.

When combined with other engines, the Rankine cycle engine can increase overall system efficiency by utilizing the exhaust heat stream typically discarded from the primary engine to produce power with no additional fuel expense. Studies show the output of a recuperated gas turbine can be increased 15 to 30%, and the output of diesel and gasoline engines can be increased 12 to 25% by exhaust gas heated Rankine cycle engines (Angelino and Moroni, 1973; Morgan and David, 1974). Thus, Rankine cycle engines can be used to improve the operating cost basis of these primary engines in the distributed generation market.

5.5.1   Working Fluids

Water has been the dominant fluid for use in high temperature Rankine cycle systems because of its large latent heat of vaporization, good transport properties, availability, and ease of use. However, in low temperature heat source Rankine cycles such as heat recovery applications, the large latent heat of vaporization of water results in low boiling temperature and low cycle efficiency. Organic fluids are superior to water in these applications. A wide variety of organic (carbon-atom based) fluids have been used in low temperature heat recovery applications. Refrigerants are typically used up to 400°F. Hydrocarbons such as toluene are used at higher temperatures. Fluid characteristics such as thermal decomposition limits, flammability, cost, and availability must all be considered when designing the low temperature organic Rankine cycle.

5.5.2   Rankine Cycle Engine Performance

Heat source and heat sink temperatures have a strong effect on Rankine cycle efficiency. As shown in Figure 5.22, efficiency for organic Rankine cycle engines ranges from 7–10% for systems using heat source temperatures near 200°F to 25–30% for 600 to 1200°F sources. For comparison, efficient, fossil fueled, power plant Rankine cycles operating at high power levels and high temperatures exhibit cycle efficiencies in the 35 to 40% range. Using the cycle efficiency information in Figure 5.22, the potential power output of a low temperature, organic Rankine cycle engine can be estimated. Figure 5.23 shows predicted power as a function of heat source temperature and amount of available heat.

5.5.3   Rankine Cycle Engine Cost

The installed costs of waste heat recovery Rankine cycle systems producing electrical power vary over a significant range. This variation is caused primarily by output power levels, production quantity, temperature range available, and heat exchanger requirements. In DG applications, where the Rankine cycle engine is combined with primary engines to produce less than 250 kW total, installed costs of $300 to $600/kW are estimated for production quantity Rankine cycle engines.

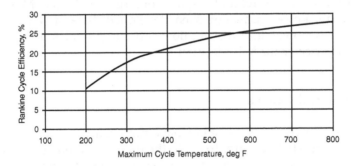

FIGURE 5.22
Simple cycle efficiency as a function of temperature.

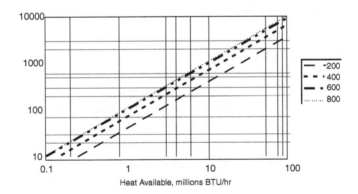

FIGURE 5.23
Potential Rankine cycle power output (kW) as a function of heat rate.

5.6 Challenges

MT manufacturers have demonstrated that efficiency goals can be met and no insurmountable barriers to mass production remain. Several cost studies indicate that with sufficient volume, cost goals can also be met. Sufficient natural gas is available for a significant MT rollout without a new piping infrastructure. However, challenges to MT adoption remain and are summarized below.

5.6.1 Operational Challenges

- Very few MTs (hundreds at most) have been operated for a sufficiently long period to establish a credible field performance data base.
- Methods for control and dispatch of large numbers of MTs selling surplus power have not been developed. Sale of MT-produced surplus power into distribution systems provides a strong economic driver for MT adoption.

5.6.2 Technical Challenges

- Final designs of MT systems have not been verified by sufficient bench and field testing. To avoid the notorious premature adoption pitfalls of solar, packaged cogeneration system and heat pump industries in the U.S., the MT industry must deliver only tested designs. Otherwise, the industry will be poisoned by false claims.
- Bearing durability for high speed turbines in MT applications with cold start has not been established.

5.6.3 Institutional Challenges

- Standards for interconnection to the grid do not exist. Until they do, expensive case-by-case applications need to be made for each connection of an MT to the grid.
- Building codes do not recognize microturbines.
- Incumbent utilities generally know little about microturbines, their operation, control, and role in the future generation mix.

5.7 Nomenclature

EGT	turbine exhaust gas temperature
ETATH	thermal efficiency
ESS	electromagnetic shear stress
HRR	heat release rate
HP	horsepower
L	length
LHV	lower heating value

N	rotational speed
PMG	permanent magnetic generator
SFC	specific fuel consumption
T	temperature
TIT	turbine inlet temperature
U	tip speed
Wa	Airflow
Wt	weight
ε	effectiveness
η	efficiency

Acknowledgments

Colin Rodgers acknowledges the incentive for the single-shaft assessment study from Tony Davies of Bowman Power Systems, as well as the unabated encouragement from his long time friend Colin McDonald.

References

Angelino, G. and Moroni, V., Perspectives for waste heat recovery by means of organic fluid cycles, ASME Paper 72-WA/Pwr-2, 1973.

Craig, P., The Capstone turbogenerator as an alternate power source, SAE 970202, 1997.

Friedman, D., Light commercial Brayton/Rankine conditioning system, IECEC 779030, 1977.

McDonald, C. F., Low cost recuperator concept for microturbine applications, *ASME Trans.*, 2000, submitted.

Morgan, D. T. and David, J. P., High efficiency gas turbine/organic Rankine cycle combined power plant, ASME Paper 74-GT-35, 1974.

Rodgers, C., Performance development history — 10 kW turboalternator, SAE 740849, 1974.

Rodgers, C., Experiments with a low specific speed partial emission centrifugal compressor, ASME Paper 89-GT-1, 1989.

Rodgers, C., Low cost fuel system for a gas turbine, U.S. Patent 5,048,298, 1991.

Rodgers, C., Small (10–200 kW) turbogenerator design considerations, ASME Transactions, 1993.

Rodgers, C., Thermo economics of small 50 kW microturbine, ASME Paper 97-GT-260, 1997.

Stone, A. and Eberhardt, J. P., Part load fuel consumption problem of open cycle gas turbines, ASME Paper 589A, 1962.

6

Fuel Cells

Jacob Brouwer

CONTENTS

Fuel cells are electrochemical devices that convert the chemical energy of a fuel directly to usable energy — electricity and heat — without combustion. This is quite different from most electric generating devices (e.g., steam turbines, gas turbines, and reciprocating engines) which first convert the chemical energy of a fuel to thermal energy, then to mechanical energy, and, finally, to electricity. In the last decade, fuel cells have emerged as one of the most promising technologies to meet the nation's energy needs for the 21st century. They produce electricity at efficiencies of 40 to 60% with negligible harmful emissions, and operate so quietly that they can be used in residential neighborhoods. Fuel cells are particularly well suited to the distributed power generation market because of these characteristics as well as their scalability, high efficiency, and modularity.

In the 1960s, fuel cells were developed for space applications that required strict environmental and efficiency performance. The successful demonstration of efficient and environmentally sensitive fuel cells in space led to their serious consideration for terrestrial applications in the 1970s. Due to the emergence of several new fuel cell types (e.g., solid oxide and molten carbonate), the last 10 to 15 years have seen a tremendous expansion and diversification of developers and manufacturers which has dramatically expanded the list of potential products and applications of fuel cells.

There are many challenges and technical hurdles, however, that the fuel cell community must face in order for fuel cells to be widely used in the distributed generation market. The first challenge is establishment of the market. Fuel cells could contribute to the establishment of a distributed generation market if they become more economically competitive with current technologies. The key challenge is to produce an ideal hydrogen-fueled engine (a fuel cell) that can cost-effectively produce power in the hydrocarbon-based economy of today. This is the most significant technical challenge with regard to integrating fuel cell systems with available infrastructure, reducing their capital cost through volume manufacturing, and achieving widespread use in various sectors.

6.1 Principles of Operation

Fuel cells are similar to batteries containing electrodes and electrolytic materials to accomplish the electrochemical production of electricity. Batteries store chemical energy in an electrolyte and convert it to electricity on demand until the chemical energy has been depleted. Applying an external power source can recharge depleted secondary batteries, but primary batteries must be replaced. Fuel cells do not store chemical energy but, rather, convert the chemical energy of a fuel to electricity. Thus, fuel cells do not need recharging and can continuously produce electricity as long as fuel and oxidant are supplied (Brown and Jones, 1999).

Figure 6.1 presents the basic components of a fuel cell, which include a positive electrode (anode), a negative electrode (cathode), and an electrolyte. Fuel is supplied to the anode, while oxidant is supplied to the cathode. Fuel is electrochemically oxidized on the anode surface, and oxidant is electrochemically reduced on the cathode surface. Ions created by the electrochemical reactions flow between the anode and cathode through the electrolyte. Electrons produced at the anode flow through an external load to the cathode, completing an electric circuit.

A typical fuel cell requires both gaseous fuel and oxidants. Hydrogen is the preferred fuel because of its high reactivity, which minimizes the need for expensive catalysts. Hydrocarbon fuels can be supplied, but typically require conversion to hydrogen prior to entering the fuel cell (for lower temperature fuel cells) or within the fuel cell (for higher temperature fuel cells). Oxygen is the preferred oxidant because of its availability in the atmosphere. As indicated in Figure 6.1, the electrolyte serves as an ion conductor. The direction of ion transport depends upon the fuel cell type, which determines the type of ion that is produced and transported across the electrolyte between the electrodes. The various fuel cell types are described in a later section.

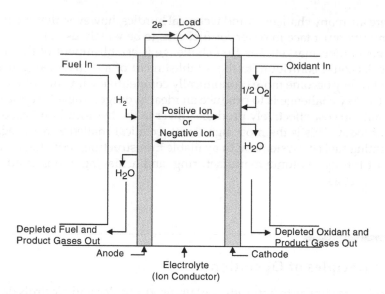

FIGURE 6.1
Fuel cell diagram.

6.1.1 Fuel Cell Stack

A single fuel cell is only capable of producing about 1 volt, so typical fuel cell designs link together many individual cells to form a stack that produces a more useful voltage. A fuel cell stack can be configured with many groups of cells in series and parallel connections to further tailor the voltage, current, and power produced. The number of individual cells contained within one stack is typically greater than 50 and varies significantly with stack design.

Figure 6.2 presents the basic components that comprise a fuel cell stack. These components include the electrodes and electrolyte of Figure 6.1 with additional components required for electrical connections and/or insulation and the flow of fuel and oxidant through the stack. These key components include current collectors and separator plates. The current collectors conduct electrons from the anode to the separator plate. The separator plates provide the electrical series connections between cells and physically separate the oxidant flow of one cell from the fuel flow of an adjacent cell. The channels in the current collectors serve as the distribution pathways for the fuel and oxidant. Often, the two current collectors and the separator plate are combined into a single unit called a bipolar plate.

6.1.2 Fuel Cell System

The preferred fuel for most fuel cell types is hydrogen. Hydrogen is not readily available, however, but the infrastructure for the reliable extraction, transport or distribution, refining, and/or purification of hydrocarbon fuels

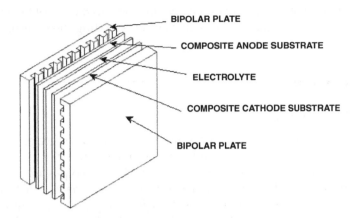

FIGURE 6.2
Isometric view of the basic components of a fuel cell stack.

is well established in our society. Thus, fuel cell systems that have been developed for practical applications to date have been designed to operate on hydrocarbon fuels. This typically requires the use of a fuel processing system, or "reformer," as shown in Figure 6.3. The fuel processor typically accomplishes the conversion of hydrocarbon fuels to a mixture of hydrogen-rich gases and, depending upon the requirements of the fuel cell, subsequent removal of contaminants or other species to provide pure hydrogen to the fuel cell.

In addition to the fuel cell system requirement of a fuel processor for operation on hydrocarbon fuels, a power conditioning or inverter system is needed. This is required for the use of current end-use technologies, which are designed for consuming alternating current (AC) electricity, and for grid connectivity in distributed power applications. Since the fuel cell produces direct current (DC) electricity, the power conditioning section is a requirement for fuel cell systems that are designed for AC-based distributed generation. In the

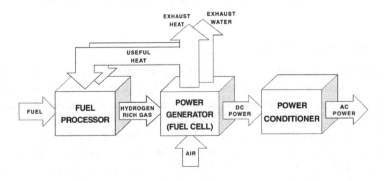

FIGURE 6.3
Schematic representation of a fuel cell system.

future, systems and technologies may be amenable to the use of DC electricity, which would allow significant cost savings by avoiding the inverter.

6.2 Fuel Cell Types

Five principle types of fuel cells are currently in various stages of commercial availability or undergoing research, development, and demonstration for distributed generation applications. These five fuel cell types are significantly different from each other in many respects; however, the key distinguishing feature is the electrolyte material, which is generally used to identify each of the five fuel cell types: (1) alkaline fuel cell (AFC), (2) molten carbonate fuel cell (MCFC), (3) phosphoric acid fuel cell (PAFC), (4) proton exchange membrane fuel cell (PEMFC), and (5) solid oxide fuel cell (SOFC).

6.2.1 Alkaline

Alkaline fuel cells (AFCs) were the first type of fuel cell to be widely used for space applications. AFCs contain a potassium hydroxide (KOH) solution as the electrolyte and operate at temperatures between 100 and 250°C (211 to 482°F). Higher temperature AFCs use a concentrated (85 wt%) KOH solution, while lower temperature AFCs use a more dilute KOH solution (35 to 50 wt%). The electrolyte is contained in and/or supported by a matrix (usually asbestos) that wicks the electrolyte over the entire surface of the electrodes. A wide range of electrocatalysts can be used in the electrodes (e.g., Ni, Ag, spinels, metal oxides, and noble metals). The fuel supplied to an AFC must be pure hydrogen. Carbon monoxide (CO) poisons an AFC, and carbon dioxide (CO_2) reacts with the electrolyte to form potassium carbonate (K_2CO_3). Even the small amount of CO_2 in the air (about 370 PPM) must be considered for operation of an AFC (Hirschenhofer et al., 1998).

6.2.2 Molten Carbonate

The electrolyte in a molten carbonate fuel cell (MCFC) is an alkali carbonate (sodium, potassium, or lithium salts, Na_2CO_3, K_2CO_2, or Li_2CO_3) or a combination of alkali carbonates that is retained in a ceramic matrix of lithium aluminum oxide ($LiAlO_2$). An MCFC operates at 600 to 700°C where the alkali carbonates form a highly conductive molten salt with carbonate ions ($CO_3^=$) providing ionic conduction through the electrolyte matrix. Relatively inexpensive nickel (Ni) and nickel oxide (NiO) are adequate to promote reaction on the anode and cathode, respectively, at the high operating temperatures of an MCFC (Baker, 1997).

MCFCs offer higher fuel-to-electricity efficiencies than lower temperature fuel cells, approaching 60%, and greater fuel flexibility. The higher operating temperatures of MCFCs make them candidates for combined-cycle applications in which the exhaust heat is used to generate additional electricity. When the waste heat is used for cogeneration, total thermal efficiencies can approach 85%.

The two leading U.S. MCFC developers are Fuel Cell Energy (Connecticut) and M-C Power Corporation (Illinois). In addition to the U.S. developers, MCFC technology is being developed in both Europe and Japan. Demonstration plants of up to 2 MW have been designed, constructed, and tested.

6.2.3 Phosphoric Acid

Phosphoric acid fuel cells (PAFCs) are the most mature fuel cell technology. PAFCs use a concentrated 100% phosphoric acid (H_3PO_4) electrolyte retained on a silicon carbide matrix and operate at temperatures between 150 and 220°C. Concentrated H_3PO_4 is a relatively stable acid that allows operation at these temperatures. At lower temperatures, problems with CO poisoning of the anode electrocatalyst (usually platinum) and poor ionic conduction in the electrolyte become problems (Hirschenhofer et al., 1998). The electrodes typically consist of Teflon™-bonded platinum and carbon (PTFE-bonded Pt/C).

PAFC fuel cells produced by the ONSI Corporation are the world's first commercially available fuel cell product (King and Ishikawa, 1996). Turnkey 200 kW plants are now available and have been installed at more than 180 sites in the United States, Europe, and Japan. Operating at 200°C, the PAFC plant also produces heat for domestic hot water and space heating with an electrical efficiency of 36 to 40%.

6.2.4 Proton Exchange Membrane

The proton exchange membrane fuel cell (PEMFC) is also known as the solid polymer or polymer electrolyte fuel cell. A PEMFC contains an electrolyte that is a layer of solid polymer (usually a sulfonic acid polymer whose commercial name is Nafion™) that allows protons to be transmitted from one face to the other (Gottesfeld and Zawadinski, 1998). PEMFCs require hydrogen and oxygen as inputs, though the oxidant may also be ambient air, and these gases must be humidified. PEMFCs operate at a temperature much lower than other fuel cells because of the limitations imposed by the thermal properties of the membrane itself (Appleby and Yeager, 1986). The operating temperatures are around 90°C. The PEMFC can be contaminated by carbon monoxide, reducing the performance and damaging catalytic materials within the cell. A PEMFC requires cooling and management of the exhaust water in order to function properly (Gottesfeld and Zawadinski, 1998).

The development of PEMFC technology is primarily sponsored by the transportation sector, which includes most automobile manufacturers, and several companies specializing in the advancement and manufacture of PEMFC technology (e.g., Ballard, Allied Signal, Siemens, IFC, H-Power, Plug Power, Avista Labs, Energy Partners, etc.).

6.2.5 Solid Oxide

Solid oxide fuel cells (SOFCs) are currently being demonstrated in various sizes from 1 kW up to 250 kW plants. SOFCs utilize a non-porous metal oxide (usually yttria-stabilized zirconia, Y_2O_3-stabilized ZrO_2) electrolyte material. SOFCs operate between 650 and 1000°C, where ionic conduction is accomplished by oxygen ions (O^-). Typically, the anode of an SOFC is cobalt or nickel zirconia ($Co-ZrO_2$ or $Ni-ZrO_2$), and the cathode is strontium-doped lanthanum manganite (Sr-doped $LaMnO_3$) (Singhal, 1997; Minh, 1993).

SOFCs offer the stability and reliability of all-solid-state ceramic construction. High-temperature operation, up to 1000°C, allows more flexibility in the choice of fuels and can produce very good performance in combined-cycle applications. SOFCs approach 60% electrical efficiency in the simple-cycle system and 85% total thermal efficiency in cogeneration applications (Singhal, 1997).

The flat plate and the monolithic designs are at a much earlier stage of development typified by subscale, single-cell, and short-stack development (kW scale). Companies pursuing these concepts in the U.S. are Allied-Signal Aerospace Company, Ceramatec, Inc., Technology Management, Inc., and Ztek, Inc. At least seven companies in Japan, eight in Europe, and one in Australia are developing SOFCs. Tubular SOFC designs are closer to commercialization and are being produced by Siemens Westinghouse Power Corporation (SWPC) and a few Japanese companies.

6.3 Comparison of Fuel Cell Types

Table 6.1 presents a summary comparison of the four primary fuel cell types under serious consideration for distributed power generation (e.g., MCFC, PAFC, PEMFC, and SOFC). Notice that the higher-temperature fuel cells do not require an external reformer. The PAFC and PEMFC units tend to use precious metal catalysts, while catalysts of the MCFC and SOFC units are typically nickel based. These differences lead to many variations in design and function which will be described in more detail in the next section.

6.3.1 Fuel Cell Design

A fuel cell power system embodies more than just the fuel cell stack, and the design of a fuel cell system involves more than the optimization of the fuel

TABLE 6.1

Comparison of Key Features of the Four Fuel Cell Types Under Serious Consideration for Distributed Power Generation (Hirschenhofer et al., 1998, with permission)

	MCFC	PAFC	PEMFC	SOFC
Electrolyte	Immobilized liquid molten carbonate	Immobilized liquid phosphoric acid	Ion exchange membrane	Ceramic
Operating temperature	650°C	205°C	80°C	800–1000°C now, 600–1000°C in 10–15 years
Charge carrier	CO_3^-	H^+	H^+	O^-
External reformer for CH_4 (below)	No	Yes	Yes	No
Prime cell components	Stainless steel	Graphite-based	Carbon-based	Ceramic
Catalyst	Nickel	Platinum	Platinum	Perovskites
Product water management	Gaseous product	Evaporative	Evaporative	Gaseous product
Product heat management	Internal reforming + process gas	Process gas + independent cooling medium	Process gas + independent cooling medium	Internal reforming + process gas

cell section with respect to efficiency or economics. It involves the minimization of the cost of electricity (or cogenerated product) within the constraints of the desired application. For most applications, this requires that the fundamental processes be integrated into an efficient plant with low capital costs. Often, these objectives are conflicting, so compromises or design decisions must be made. In addition, project-specific objectives such as desired fuel, emission levels, potential uses of rejected heat (electricity, steam, or heat), desired output levels, volume, or weight criteria, all influence the design of the fuel cell power system.

6.3.2 Fundamental Limitations

The ideal performance of a fuel cell depends upon the electrochemical reactions that occur within the fuel cell. Table 6.2 presents a summary of the electrochemical reactions that occur within the various fuel cell types. The lower-temperature fuel cells (AFC, PAFC, and PEMFC) all require noble metal (e.g., platinum) electro-catalysts to achieve practical reaction rates at the anode and cathode, and they typically require hydrogen fuel. The higher-temperature fuel cells (MCFC and SOFC) typically use nickel-based materials to accomplish the electrochemistry described in Table 6.2. In addition, as indicated in Table 6.2, higher-temperature fuel cells can electrochemically react with hydrogen as well as other fuels (e.g., CO and CH_4). Note that carbon monoxide poisons the noble metal catalysts of lower-temperature fuel cells, but serves as a source of fuel (H_2) for the higher-temperature fuel cells. Also note that the reactions of CO and CH_4 in Table 6.2 are presented as anodic electrochemical reactions. In reality, these reactions may not occur on the anode surface but, rather, through water–gas shift, and steam reformation chemical reactions likely produce hydrogen in the gas phase (Hirschenhofer et al., 1998).

TABLE 6.2

Fuel Cell Anode and Cathode Reactions

Fuel Cell Type	Anode Reactions	Cathode Reactions
Alkaline (AFC)	$H_2 + 2(OH)^- \rightarrow 2\,H_2O + 2e^-$	$1/2O_2 + H_2O + 2e^- \rightarrow 2(OH)^-$
Molten Carbonate (MCFC)	$H_2 + CO_3^{2-} \rightarrow H_2O + CO_2 + 2e^-$	
	$CO + CO_3^{2-} \rightarrow 2CO_2 + 2e^-$	$1/2O_2 + CO_2 + 2e^- \rightarrow CO_3^{2-}$
Phosphoric Acid (PAFC)	$H_2 \rightarrow 2H^+ + 2e^-$	$1/2O_2 + 2H^+ + 2e^- \rightarrow H_2O$
Proton Exchange Membrane (PEMFC)	$H_2 \rightarrow 2H^+ + 2e^-$	$1/2O_2 + 2H^+ + 2e^- \rightarrow H_2O$
Solid Oxide (SOFC)	$H_2 + O^{2-} \rightarrow H_2O + 2e^-$	
	$CO + O^{2-} \rightarrow CO2 + 2e^-$	
	$CH_4 + 4O^{2-} \rightarrow 2H_2O + CO_2 + 8e^-$	$1/2O_2 + 2e^- \rightarrow O^{2-}$

The ideal performance of a fuel cell is determined by the potential voltage level that it can theoretically produce. This potential voltage is called the Nernst potential and is defined by the Nernst equation. For the general reaction

$$aA + bB \rightarrow cC + dD$$

the Nernst equation can be expressed as

$$E = E_o + \left(\frac{RT}{2T}\right)\ln\left(\frac{P_A^a\,P_B^b}{P_C^c\,P_D^d}\right)$$

where E_o is the reversible standard potential for a cell reaction, E is the ideal equilibrium potential, T is temperature, T is Faraday's constant, and P is pressure. Therefore, for each of the fuel cell types, there is a theoretical voltage level that can be achieved which is determined by the Nernst equation for each of the electrochemical reactions that occur within the cell. Note that according to the Nernst equation, the ideal cell voltage can be increased by operation at higher pressures for a given temperature.

Table 6.3 presents fuel cell electrochemical reactions and their corresponding Nernst equations. The reaction of hydrogen and oxygen produces water, but when a carbon-containing fuel is used at the anode, carbon dioxide is also produced. For MCFCs, CO_2 is required at the cathode to maintain a constant carbonate concentration in the electrolyte. Because CO_2 is produced at the anode and consumed at the cathode in MCFCs, the partial pressure of CO_2 is included in both the anode and cathode Nernst equations of Table 6.3.

TABLE 6.3

Fuel Cell Reactions and Corresponding Nernst Equations

Fuel Cell Reaction	Nernst Equation
$H_2 + 1/2O_2 \rightarrow H_2O$	$E = E_o + (RT/2F) \ln [P_{H_2}/P_{H_2O}] + (RT/2F) \ln [P_{O_2}^{1/2}]$
$H_2 + 1/2O_2 + CO_2(c) \rightarrow H_2O + CO_2(a)$	$E = E_o + (RT/2F) \ln [PH_2/P_{H_2O}(P_{CO_2})a] + (RT/2F) \ln [P_{O_2}^{1/2}(P_{CO_2})]c$
$CO + 1/2O_2 \rightarrow CO_2$	$E = E_o + (RT/2F) \ln [P_{CO}/P_{CO_2}] + (RT/2F) \ln [P_{O_2}^{1/2}]$
$CH_4 + 2O_2 \rightarrow 2H_2O + CO_2$	$E = E_o + (RT/8F) \ln [P_{CH_4}/P^2_{H2O}P_{CO_2}] + (RT/8F) \ln [P_{O2}^{1/2}]$

The ideal standard potential of a hydrogen oxygen fuel cell is 1.229 volts. Figure 6.4 presents the ideal potential for each of the cell reactions versus temperature. Note that the ideal potential for some of the primary fuel cell reactions increases with decreasing temperature. This is very different from all of the typical generation technologies based upon heat engine designs, which exhibit decreased performance with reductions in temperature.

6.3.3 Practical Limitations

The reversible standard potentials presented in the previous section determine the fundamental limitations on the performance of fuel cell technologies. These fundamental limitations would suggest that voltage levels greater

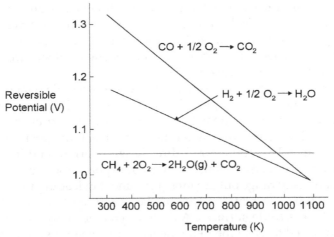

FIGURE 6.4
Reversible ideal potential for FC electrochemical reactions versus temperature.

than one volt could be achieved with fuel cells. This would correspond to very high efficiencies for the conversion of fuel chemical energy to electricity. However, there are practical limitations to the performance of fuel cells that are due to several physical processes (e.g., transport and chemical reaction) which do not occur without losses.

The physical processes associated with these losses include: (1) transport of reactants to the gas/electrolyte interface, (2) dissolution of reactant in electrolyte, (3) transport of reactant through electrolyte to the electrode surface, (4) pre-electrochemical homogeneous or heterogeneous chemical reaction, (5) adsorption of electro-active species onto the electrode, (6) surface migration of adsorbed species, (7) electrochemical reaction involving electrically charged species, (8) post-electrochemical surface migration, (9) desorption of products, (10) post-electrochemical reaction, (11) transport of products away from the electrode surface, (12) evolution of products from electrolyte, and (13) transport of gaseous products from electrolyte/gas.

The losses associated with these chemical and physical processes are generally manifested in three major fuel cell losses. In electrochemical terms, losses are often referred to as overpotentials (i.e., the potential over that observed to reach theoretical potential), polarizations, or overvoltages. The three major losses are:

1. Activation overpotential (polarization)
2. Ohmic overpotential (polarization)
3. Concentration overpotential (polarization)

These losses are irreversible and result in a practical cell voltage that is less than the ideal voltage. Activation polarization is generally a result of losses associated with slow chemical reactions (e.g., overcoming the activation energy of chemical reactions). Ohmic polarization is loss due to the flow of electricity through the fuel cell which resists the flow of electricity, and concentration polarization is caused by transport phenomena which lead to lower concentrations of reactants at the electrochemical surface than in the bulk flow (Appleby and Foulkes, 1989).

Figure 6.5 presents a comparison of the ideal and actual voltage versus current characteristics of a typical fuel cell. Note that activation polarization is dominant at lower current densities where electronic barriers must be overcome before significant current can flow and reactant species can be consumed. Ohmic polarization varies directly with current and increases over the whole range of current densities. Gas transport losses occur throughout the current density range but are most prominent in leading to concentration polarization when current densities are large and reactants are rapidly consumed at the electrode surfaces. A thermodynamic analysis of these losses can be performed to yield the dependence of the major losses on cell operating parameters, which is presented in the following sections.

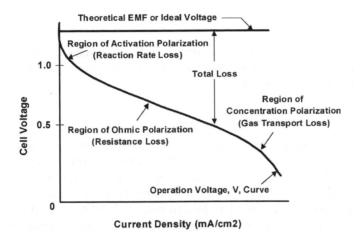

FIGURE 6.5
Ideal and actual fuel cell current and voltage characteristics.

6.3.3.1 Activation Polarization

Thermodynamic analyses indicate that activation polarization occurs when the rate of an electrochemical reaction on an electrode surface is controlled by slow electrode kinetics. The rate of electrochemical reactions, similar to chemical reactions, involves overcoming an activation barrier before chemical reaction can occur. In the case of electrochemical reactions, this activation energy ranges from 50 to 100 mV and is governed by the Tafel equation (Atkins, 1986) as follows:

$$\eta_{act} = \frac{RT}{\alpha n \mathcal{T}} \ln \left(\frac{i}{i_o} \right)$$

where α is the electron transfer coefficient, n is the number of electrons participating in the reaction, R is the universal gas constant, T is the temperature, $\mathcal{T}$ is Faraday's constant, i is the current, and i_o is the exchange current density. The exchange current density is a measure of the maximum current that can be drawn with negligible polarization (i.e., $\eta_{act} = 0$).

6.3.3.2 Ohmic Polarization

Ohmic polarization is caused by resistance to the flow of ions in the electrolyte and resistance to the flow of electrons through the electrodes. The dominant losses are those associated with flow of ions through the electrolyte. These losses can be reduced by decreasing the distance between the electrodes (shortening the ionic flow distance) and/or enhancing the ionic conductivity of the electrolyte material. Typically, the losses due to electrolyte and electrode resistance are lumped together. Ohm's law governs these losses as follows:

$$\eta_{ohm} = iR_{cell}$$

where i is the current and R_{cell} is the overall cell resistance which includes ionic, electrode, and interconnect resistances. The overall cell resistance can be obtained experimentally.

6.3.3.3   Concentration Polarization

Concentration polarization is caused by the formation of a reactant concentration gradient at the electrode surface. This concentration gradient is formed due to the rapid consumption of reactants at the electrode surface when the bulk fluid does not have sufficient time to replenish the reactants to their original concentration. This loss of potential can be caused by a number of physical mechanisms including (1) diffusion in the gas phase within the electrode pores, (2) solution of reactants into the electrolyte, (3) dissolution of the products out of the electrolyte, and (4) diffusion of reactants and products between electrolyte and electrochemical reaction sites. The overall effect of concentration overpotential is determined as follows:

$$\eta_{conc} = \frac{RT}{nT} \ln\left(1 - \frac{i}{i_L}\right)$$

where i_L is the limiting current, a measure of the maximum rate at which a reactant can be supplied to an electrode.

6.4   Efficiency

The efficiency of a fuel cell is closely related to the voltage level that can be practically produced. The total cell voltage includes contributions from the anode and cathode as well as the ohmic polarization. Each of the electrodes can be affected by activation and concentration overpotentials as follows:

$$V_{cathode} = E_{cathode} - \eta_{act,cathode} - \eta_{conc,cathode}$$

$$V_{anode} = E_{anode} - \eta_{act,anode} - \eta_{conc,anode}$$

The total cell voltage is then:

$$V_{cell} = V_{cathode} - V_{anode} - iR_{cell}$$

Thus, overall cell performance is reduced by five primary overpotentials as presented in the equations above. Typically, cathodic losses far exceed losses at the anode, primarily due to the stability of the oxygen molecule compared to the primary reactants participating in anodic electrochemistry (e.g., hydrogen). In addition, activation overpotentials are typically greater than ohmic losses, which are greater than concentration overpotentials for typical fuel cells at typical operating conditions. In general, the losses lead to practical cell voltages in the range of 0.6 to 1.0 volts.

Figure 6.4 presents the voltage levels achieved by each of the fuel cell types versus temperature as well as the reversible cell potential. Note that some of the highest voltage levels are achieved for the higher temperature fuel cell types even though the reversible potential decreases with increasing temperature. This is primarily due to reductions in activation overpotentials at higher temperatures.

The voltage levels presented in Figure 6.4 roughly correlate with system efficiency such that practical fuel cell efficiencies are greatest for alkaline fuel cells, followed by the high-temperature molten carbonate and solid oxide fuel cells. The lower-temperature proton exchange membrane fuel cells can achieve practical efficiencies that exceed those observed for phosphoric acid fuel cells but are less than either alkaline or higher-temperature fuel cell systems. When operated on natural gas, system fuel-to-electricity conversion efficiencies have been observed as presented in Table 6.4.

TABLE 6.4

Typical Efficiency Ranges for Fuel Cell Systems Operating on Natural Gas

Fuel Cell Type	Range of Fuel-to-Electricity Efficiency (Natural Gas Operation, LHV Basis)
Molten Carbonate (MCFC)	50–60%
Phosphoric Acid (PAFC)	38–45%
Proton Exchange Membrane (PEMFC)	33–45%
Solid Oxide (SOFC)	40–55%

6.5 Operating Parameters

Fuel cell performance is affected by operating parameters such as temperature, pressure, reactant gas concentrations, reactant utilizations, and current density. These parameters determine the ideal cell potential as well as the magnitude of the losses described above. There is a wide variety of operating points that can be selected for a specific design, and the modification of operating parameters can have beneficial effects on system performance in one case and detrimental effects in the other.

To illustrate just one of the design considerations, Figure 6.7 presents the typical relationship between voltage and power levels for a fuel cell. One would typically desire to operate a fuel cell at its maximum current density to reduce the required size and potentially reduce the cost of the fuel cell. However, as Figure 6.5 indicates, this leads to operation at lower voltage levels which lowers the available voltages and cell efficiency as indicated above. Typically, fuel cells are designed to operate at a point that yields a good compromise of low operating cost (higher cell efficiency at higher voltages and lower power densities) and low capital cost (less cell area due to higher power densities and lower voltages). This is further complicated by the fuel cell system requirements, which may dictate overall volume limitations or require certain levels of heat generation within the stack to provide the fuel reformer with sufficient heat to overcome the endothermic reforming reactions.

In the following paragraphs, further considerations are made for the effects of operating conditions such as temperature and pressure on fuel cell performance.

6.5.1 Temperature

The operating temperature of a fuel cell affects the change in entropy associated with the electrochemical reactions that occur within that cell. As operating temperature increases, the entropy change increases, leading to a reduction in the reversible potential of the fuel cell (see Figure 6.4). Therefore, with all other parameters unchanged, increases in operating temperature *reduce* fuel cell performance and efficiency. However, as shown in Figure 6.4, practical cell voltages tend to increase with increasing temperature due to reductions in some of the overpotentials. The increased performance of fuel cells with increasing operating temperature may be offset by practical limitations on materials, increased corrosion, electrode and catalyst degradation, and electrolyte loss.

6.5.2 Pressure

The direct effect of the operating pressure of a fuel cell is to change the concentration of reactants at the electrode surfaces. Thus, according to the Nernst equation, as operating pressure increases, partial pressures of reactants increase and the reversible cell potential increases. In addition, as operating pressure increases, gas solubility and mass transfer rates increase, reducing some overpotential associated with these processes. Electrolyte loss by vaporization is typically reduced and overall system efficiencies are increased with increasing operating pressure. However, parasitic losses, in particular those associated with the compression of the air (and fuel) stream, increase with increasing operating pressure. In addition, increased materials requirements (larger pipes, containment vessels) and increased controls hardware must be taken into consideration in the system design to ensure increased performance with increased operating pressure.

6.5.3 Gas Concentration

Reactant gas composition has a strong effect on the performance of fuel cells. The Nernst equation, presented earlier, indicates a logarithmic dependence of electric potential on the partial pressures of the reactants. These reactant partial pressures are directly proportional to reactant gas concentration (and pressure) through the ideal gas law. Therefore, cell voltages and cell efficiency increase with increasing reactant gas concentrations.

6.5.4 Utilization

Reactant gas utilization also strongly influences the performance of fuel cells. Unlike combustion systems, fuel cells are not typically designed to utilize 100% of the reactants but, rather, a certain fraction of the reactants to allow the presence of reactants along the entire reactive surface area. Without this consideration, reactants would be consumed at the end of the flow channels through the cell, with the final portion of the cell unable to produce a voltage and reducing the overall performance.

The overall voltage of any fuel cell is determined by the portion of the cell with the lowest reactant gas concentration, which changes within each stream as the reactants are utilized. A fuel cell adjusts to the minimum local Nernst potential because the electrodes are typically good electronic conductors leading to isopotential surfaces. Therefore, less than 100% utilization is desired, with some type of reactor (usually catalytic) used to consume the remainder of the fuel to recover thermal energy for use within the system. This understanding notwithstanding, cell efficiency is directly increased with increases in utilization.

6.5.5 Current Density

Finally, the operating current density of a fuel cell has an impact on cell performance. As indicated in previous sections, the polarizations (activation, concentration, and ohmic) are all affected by the operating current levels. Typically, activation losses are high when operating at low current levels, and concentration losses are high at very high current densities. Ohmic losses are directly proportional to operating current levels throughout the range of current densities.

6.6   Fuels and Fuel Processing

6.6.1   Primary Fuels

The primary fuels used in fuel cell systems today include natural gas, hydrogen, and methanol. The most common fuel used in fuel cell systems developed for distributed generation is natural gas. Natural gas is widely available in many countries at reasonable prices and is, therefore, the primary fuel of choice. Typically, a natural gas fuel processor is integrated into the system design for a fuel cell power plant. This integration requires the supply of heat to the fuel processor to overcome the endothermicity associated with reformation chemistry. This heat can be supplied by the fuel cell (e.g., exhaust flow into the fuel processor) or by a combustor (auxiliary or anode gas reactor). The most common strategy uses steam reformation over a catalyst, but many other reformation technologies are available, including partial oxidation and autothermal reformation. When steam reformation is used, steam must be supplied to the fuel processor; the steam can be provided through the fuel cell exhaust stream or a separate steam generator. Other fuels that can be used in fuel cell systems today include hydrogen, which can be used directly in all fuel cell types, and methanol, which can be used directly only in a direct methanol fuel cell but can easily be reformed for use in other fuel cell systems.

6.6.2   Secondary Fuels

Several other fuels have been demonstrated as viable candidates for use in fuel cells. These are listed in this chapter as secondary fuels, since they are not considered primary candidates for widespread application in fuel cells for distributed power generation. However, each of the following fuels has been demonstrated, to some extent, as a feasible fuel to utilize in fuel cells:

- Landfill gas
- Digester gas
- Gasoline
- Diesel
- JP-8 (military fuels)
- Dimethyl ether
- Ethanol
- New petroleum distillates
- Coal gasification products
- Naphtha

Depending upon the fuel selection above, many considerations must be made. Some fuels are more difficult to reform than others. Some require extensive processing before they can be used. Other fuels are only available in limited markets and can be considered opportunity fuels. In all cases, contaminants such as sulfur must be removed from the gases that enter the fuel cell.

6.6.3 Fuel Cell Stack Fuels

The primary fuels that can be directly utilized within fuel cell stacks to date are hydrogen, carbon monoxide, methanol, and dilute light hydrocarbons like methane, depending upon the fuel cell type. Table 6.5 presents various fuel cell types and the primary fuels that they are amenable to using. Note that the presence of sulfur is not tolerated by fuel cells in general and that the SOFC is the most inherently fuel-flexible of the fuel cell types. MCFC units are also quite fuel-flexible.

TABLE 6.5

Effects of Various Gaseous Reactants on Various Fuel Cell Types (Hirschenhofer et al., 1998, with permission)

Gas Species	PAFC	MCFC	SOFC	PEFC
H_2	Fuel	Fuel	Fuel	Fuel
CO	Poison (> 0.5%)	Fuel[a]	Fuel	Poison (> 10 ppm)
CH_4	Diluent	Diluent[b]	Fuel[a]	Diluent
CO_2 and H_2O	Diluent	Diluent	Diluent	Diluent
S as (H_2S and COS)	Poison (>50 PPM)	Poison (>0.5 PPM)	Poison (>1.0 PPM)	No studies to date

[a] In reality, CO with H_2O shifts to H_2 and CO_2, and CH_4 with H_2O reforms to H_2 and CO faster than reacting as a fuel at the electrode.
[b] A fuel in the internal reforming MCFC.

All fuel cells prefer hydrogen as the primary fuel. Methanol can be used directly in a certain type of PEMFC called a direct methanol fuel cell. Carbon monoxide is a poison for lower temperature fuel cells but is used as a fuel in the high temperature fuel cells (e.g., SOFC and MCFC). CO may not actually react electrochemically within these cells. It is commonly understood that CO is consumed in the gas phase through the water–gas shift reaction as follows:

$$CO + H_2O \rightarrow CO_2 + H_2$$

The hydrogen formed in this reaction is subsequently consumed electrochemically. Methane can be oxidized directly using a solid oxide fuel cell; however, high concentrations of CH_4 lead to severe coking problems. Therefore, only fuels containing dilute concentrations of CH4 can be oxidized directly in current SOFCs. In addition, the oxidation of CH4, like that of CO, may not actually occur at active electrochemical sites within an SOFC. Rather, CH4 is probably reformed within the cell through steam reformation chemistry as follows:

$$CH_4 + 2H_2O \xrightarrow{catalyst} CO_2 + 4H_2$$

6.6.4 Fuel Processing

Fuel processing depends on both the raw fuel and the fuel cell technology. The fuel cell technology determines what constituents are desirable and acceptable in the processed fuel. For example, fuel sent to a PAFC needs to be H_2-rich and have less than 5% CO, while both the MCFC and SOFC fuel cells are capable of utilizing CO. PEMFCs require a pure hydrogen stream with less than 10 ppm CO. In addition, SOFCs and internal reforming MCFCs are also capable of utilizing methane (CH_4) within the cell, whereas PAFCs are not. Contamination limits are also fuel cell technology specific and therefore help to determine the specific cleanup processes required.

Since the components and design of a fuel processing subsection depend on the raw fuel type, the following discussion is organized by the raw fuel being processed.

6.6.4.1 *Hydrogen Processing*

When hydrogen is supplied directly to the fuel cell, the fuel processing section becomes a simple fuel delivery system.

6.6.4.2 *Natural Gas Processing*

Natural gas is usually converted to H_2 and CO in a steam reforming reactor. Steam reforming reactors yield the highest percentage of hydrogen. In addition to natural gas, steam reformers can be used on light hydrocarbons such as butane and propane. In fact, with a special catalyst, steam reformers can also reform naphtha. Steam reforming reactions are highly endothermic and need a significant heat source. Often, the residual fuel exiting the fuel cell is burned to supply this requirement. Fuels are typically reformed at temperatures of 760 to 980°C. Partial oxidation reformers can also be used for converting gaseous fuels, but do not produce as much hydrogen as steam reformers.

Natural gas has sulfur containing odorants (mercaptans, disulfides, or commercial odorants) for leak detection. Since neither fuel cells nor reformer catalysts are sulfur tolerant, the sulfur must be removed. This is usually accomplished with a fixed or packed bed of zinc oxide or the possible use of a hydrodesulfurizer if required.

6.6.4.3 Liquid Fuel Processing

Liquid fuels such as distillate, naphtha, diesel, and heavy fuel oil can be reformed in partial oxidation, autothermal, and preferential oxidation reformers. All commercial partial oxidation reactors employ non-catalytic partial oxidation of the feed stream by oxygen in the presence of steam with flame temperatures of approximately 1300 to 1500°C.

Partial oxidation, autothermal reformation, and preferential oxidation fuel processing techniques use some of the energy contained in the fuel to convert these hydrocarbons to H_2 and CO. For example, the overall partial oxidation reaction for pentane is exothermic and is largely independent of pressure. The process is usually performed at elevated pressure in order to yield smaller equipment.

6.6.4.4 Coal Processing

Numerous coal gasification systems are available today. The most common systems are moving-bed or fixed-bed reactors, fluidized-bed reactors, and entrained-bed reactors, all of which use steam and air or oxygen to partially oxidize coal into a gaseous product. Heat required for gasification is essentially supplied by the partial oxidation of the coal. Overall, the gasification reactions are exothermic, so waste heat boilers are often utilized at the gasifier effluent. The temperature, and therefore composition, of the product gas is dependent upon the amount of oxidant and steam and the design of the reactor that each gasification process utilizes.

6.6.4.5 Gas Cleanup

Gasifiers typically produce contaminants, which need to be removed before entering the fuel cell anode. These contaminants include H_2S, COS, NH_3, HCN, particulates, tars, oils, and phenols. The contaminant levels are dependent on both the fuel composition and the gasifier employed. Gas cleanup equipment that efficiently and reliably removes the contaminant species for coal gasification products to the specifications required by fuel cells is yet to be demonstrated.

6.6.4.6 Other Solid Fuel Processing

Solid fuels other than coal can also be utilized in fuel cell systems. For example, biomass and refuse-derived fuels (RDF) can be integrated into a fuel cell system as long as the gas product is processed to meet the requirements of the fuel cell. The resulting systems would be very similar to the coal gas system with appropriate gasifying and cleanup systems.

6.7   Cogeneration

Although fuel cells are not heat engines, significant quantities of heat are still produced in a fuel cell power system which may be used to produce steam or hot water, or may be converted to electricity via a gas turbine, steam bottoming cycle, or some combination thereof.

6.7.1   Low- and High-Grade Heat

When small quantities of heat and/or low temperatures typify a fuel cell exhaust, the heat is either rejected or used to produce hot water or low-pressure steam. For example, in a PAFC cycle where the fuel cell operates at approximately 205°C, the highest pressure steam that can be produced is something less than 200 psia. At the other end of the spectrum is the SOFC, which operates at 1000°C and often has a cell exhaust temperature of approximately 815°C after air preheating. Gas temperatures of this level are capable of producing steam temperatures in excess of 540°C, which makes the SOFC more than suited for a steam-bottoming cycle.

Whenever significant quantities of high temperature waste heat are available, high-pressure steam can be generated or a combined-cycle or hybrid heat engine fuel cell approach can be considered. This can be accomplished in many different configurations and can include a large number of technologies (e.g., steam turbine, gas turbine, and Stirling engine) in combination with a fuel cell. In those cycles, heat engines utilize the high grade heat to produce electricity for dramatic increases in overall fuel-to-electricity efficiency.

A bottoming cycle simply adds a heat engine to the fuel cell exhaust for utilizing the heat produced in the fuel cell to produce electricity. This is typically a consideration when the exhaust of the fuel cell is available at low pressure. When a fuel cell operates under pressurized conditions, the high-temperature, high-pressure exhaust could potentially power a gas turbine whose exhaust could be utilized subsequently in a heat recovery steam generator and/or steam turbine.

Many of the cycles described above, which utilize the heat available in a fuel cell system, are called hybrid cycles. These cycles are described in more detail in the following section.

6.7.2   Hybrid Fuel Cell — Heat Engine

Hybrid fuel cell systems have been designed to obtain the highest possible fuel-to-electricity efficiency by using the heat produced by the electrochemical oxidation of fuels within a fuel cell to produce electricity. A hybrid system recovers the thermal energy in the fuel cell exhaust and converts it into additional electrical energy through a heat engine. Several heat engines have been considered for this type of system including gas turbines, steam turbines, and reciprocating engines. The only conversion device which has been tested in this role to date is a microgas turbine (or micro-turbine generator, MTG).

A microgas turbine fulfills this role with some particular synergistic benefits:

- MTGs require relatively low turbine inlet temperature which can be supplied by the exhaust of a high temperature fuel cell.
- MTGs operate at relatively low pressure ratios amenable to hybrid fuel cell applications.
- The fuel cell can be operated under pressurized conditions, improving its output and efficiency.
- Sufficient thermal energy is contained in the fuel cell exhaust to power the compressor (for fuel cell pressurization) and an electric generator (to produce additional electricity).
- The power density of the system can be increased.
- Overall system cost is lower on a $/kW basis.

The hybrid fuel cell–gas turbine concept integrates a high-temperature MCFC or SOFC with a gas turbine, air compressor, combustor, heat exchangers, and several balance-of-plant items to produce a hybrid system. Synergistic effects of the combined fuel cell–gas turbine lead to electrical conversion efficiencies of 72 to 74% lower heating value (LHV) for systems under 10 MW. Larger hybrid systems are being considered which may be able to achieve fuel-to-electricity efficiencies greater than 75%.

Figure 6.6 presents a schematic system diagram for a generic hybrid fuel cell–gas turbine system to illustrate the concept of hybrid fuel cell systems. Compressed air and fuel pass through a gas-to-gas heat exchanger (recuperator) which is typically used to recover heat from the combustion product gases leaving the gas turbine. The heated fuel and air streams pass into the anode and cathode fuel cell compartments of the fuel cell, respectively, where the electrochemical reactions take place. Fuel cell exhaust gases that already contain thermal energy from the electrochemical reactions are subsequently mixed and burned, raising the turbine inlet temperature. The thermal energy contained in this stream replaces that typically delivered by the conventional combustion section of the gas turbine engine. Expansion of the fuel cell exhaust gases through the gas turbine provides an inexpensive means for recovery of the fuel cell waste heat.

High-pressure hybrid systems or topping arrangements are the most likely of the hybrids to be commercialized in the near future. The simplest of these systems is a topping-cycle SOFC integrated with a recuperated gas turbine, as shown in Figure 6.6. This particular arrangement operates with an SOFC system that can capture and recirculate steam-laden anode exhaust gases to an internally integrated fuel reformer in order to produce hydrogen and

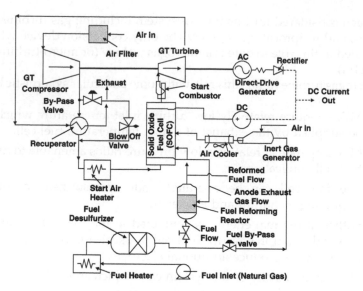

FIGURE 6.6
Generic hybrid solid oxide fuel cell gas turbine cycle schematic.

carbon monoxide, as shown in Figure 6.6. The potential benefits are obvious. High electrical conversion efficiencies are possible. The combination of fuel cells and heat engines provides a cost-effective new system with greater flexibility to meet the needs of the distributed power generation market.

Hybrid cycles are myriad. Typical fuel cell gas turbine configurations include topping cycles (where the fuel cell replaces a combustor and generator, and the gas turbine is the balance-of-plant) and bottoming cycles (where the fuel cell uses the gas turbine exhaust as an air supply and the gas turbine is balance-of-plant). In general, topping cycles lead to the highest efficiency systems with high oxygen concentration at the cathode, fewer cells required in the fuel cell stack as compared to low pressure systems, and higher power density. Bottoming cycles perform well depending on fuel cell type and are simple to integrate, easy to start, and simple to control. To achieve high efficiencies, most of the electricity of a hybrid system is produced in the fuel cell (typically between 70 and 80%).

6.8   Interconnection and Control

6.8.1   Power Conditioning

Power conditioning for a fuel cell power plant includes power consolidation, current control, direct current (DC) to alternating current (AC) inversion

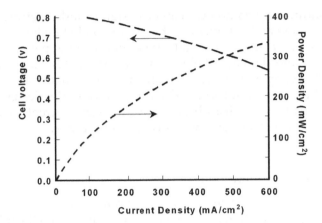

FIGURE 6.7
Typical relationship between voltage and power for fuel cells.

(unless the application is DC), and stepping the voltage up through a transformer. In addition, power quality aspects such as maintaining consistent voltage and frequency with low harmonic distortion, as well as the transient response of the power conditioning equipment, should be considered. When connected to the utility grid, additional considerations may include interconnection equipment and switch-gear, synchronization, real power ramp rate, and reactive power (VAR) support capabilities.

An important aspect of power conditioning equipment is the efficiency of the power conversion and conditioning. These efficiencies vary widely with system design but are typically on the order of 94 to 98% (Hirschenhofer et al., 1998). Fuel cells can be used to supply DC to power systems such as DC-driven motors, batteries, UPS systems, solenoids, controls, electronic equipment, or other DC equipment. In fact, most of the consumer electronics equipment (e.g., televisions, stereos, telephones, video cameras, etc.) and all computers and computer-based equipment and controls use DC electricity. In most cases, this DC voltage is provided by a connection to an AC supply followed by conversion of the AC voltage to DC voltage in an AC-to-DC converting power supply. Even with the direct use of fuel cell DC current, quality power conditioning equipment is required to maintain voltage levels and accommodate fluctuations in desired current flow. This could be accomplished at much lower cost and higher efficiencies. Fuel cells offer the opportunity of direct use of DC electricity and the elimination of many of these power supplies and concomitant energy losses in the future. However, since current technologies are already designed to accommodate the AC electricity that is standard, most fuel cell systems are designed with inverters to convert DC current to AC electricity. The power conditioning system that is required includes the capability of delivering real power and reactive power to the user. The power conditioning system also usually includes the provision of power to the fuel cell system auxiliaries and controls.

Power conditioning equipment can be designed and constructed in many different manners and can include the use of solid-state inverters, voltage transformers, and controls. Typically, fuel cell operating parameters such as fuel flow rate and demanded electrical output must be controlled by the power conditioning equipment to maintain power quality. In addition, the power conditioning equipment and wiring are typically designed to withstand utility grid or user disturbances in voltage and current including voltage spikes, voltage shorts, and overcurrent disturbances.

6.8.2   Utility Interconnection

A fuel cell can be designed and installed in a number of different modes. The various modes of operation include parallel operation with the utility grid to supply a user, direct connection to the utility grid, backup power to a normally grid-connected load, and connection to a dedicated load. The most commonplace installations to date are those of the first type. In all of the modes of operation listed above, except the last case, some interaction with the utility grid is required. Interconnection with the utility grid has many advantages including reliability improvement, increase in load factor, and reductions in electricity demand. Connection with the utility grid, however, requires that the power conditioning equipment provide:

- Synchronization with the grid,
- Voltage regulation to within ± 2%
- Frequency regulation to within ± 0.5%
- Reactive power supply adjustable between 0.8 lagging and 1.0 power factor without impact to power output
- System fault protection
- Suppression of ripple voltage feedback
- Suppression of harmonics to within IEEE 519 limits
- High efficiency
- High reliability
- Stable operation

Current inverter and power conditioning equipment is available to meet all of the above requirements with some need for cost reduction, which could be accomplished through volume manufacturing (Penner, 1995). However, a few technical challenges still exist with regard to the inverter system's capability to supply current transients associated with motor starts, the operation of overcurrent devices to clear equipment or cable faults, or other inrush currents (Hirschenhofer et al., 1998).

Whether a fuel cell is connected to a dedicated load or to the utility grid, the response of the fuel cell to system disturbances or load swings must be considered. The fleet of ONSI PC-25™ 200 kW power plants, the only commercial fuel cell fleet today, has demonstrated power conditioning equipment responses that should be characteristic of most current systems. These system responses are as follows (King and Ishikawa, 1996; Hirschenhofer et al., 1998):

1. No transient overload capacity beyond the power rating of the fuel cell
2. Load ramp rate of 0 to 100% in one cycle when operated independent of the grid
3. Load ramp rate of 10 kW/second when grid connected
4. Load ramp rate of 80 kW/second when operated independent of the grid and following the initial ramp up to full power

6.9 Dispatchability

The dispatch and control of fuel cell systems (as well as other distributed generation technologies) in the future may be accomplished through a virtual power exchange. The virtual power exchange could be comprised of a collection of computer programs running in real time to monitor and control (to some degree) the output of many distributed systems so that they appear to the independent system operator (ISO) and power exchange (PX) as a single dispatchable entity. This may become necessary because the present capabilities of the ISO and PX do not accommodate dispatch of many thousands of small generators.

The virtual power exchange concept includes the capability to (1) predict individual and aggregated building loads and coincident distributed generation asset availability; (2) make financial decisions in real time regarding which distributed generation units should be operated and when; (3) communicate with individual distributed generation units over the Internet, wireless fiber, and radio frequency links; and (4) diagnose individual distributed generation unit performance for reliability assessments in real time.

A key feature of aggregating many distributed generation devices is that overall efficiency can be optimized by allowing the distributed generation devices to run at optimal power levels for the particular unit. The integrated software package that could accomplish these tasks does not exist today, but each component has been used in other applications and could be integrated into a stand-alone package.

6.9.1   Control Techniques

Control of a fuel cell system must include an evaluation of safety, economics, and overall system (grid or local user) reliability and/or needs. The price of electricity, the impact to operations, and the cost of fuel and maintenance must all be taken into account. The goal of the control scheme is to determine whether or not the fuel cell should be operating during a particular period. Generally, a simple hour-ahead control method is sufficient if the start-up transients of the generator are not too inefficient. Chapter 7 presents some of the techniques that can be used to control on-site generation. Each of these could be applied to a fuel cell system. The control strategies include:

- Threshold control — always run if load demand is greater than a predetermined threshold
- Buy-back priority — used when power needs to be sold back to the utility
- Simple buy-back — power is sold back to the utility at a predetermined rate, measured by a separate meter
- Net metering — user pays utility for net power consumed
- Cooling/heating priority control — cogeneration
- Optimal control — minimizes cost over lifetime of fuel cell
- Complete optimization — optimizes fuel cell and complete system operation

6.9.2   Current Status

The ONSI PC-25™ PAFC units available for commercial sale are currently capable of being monitored and controlled remotely through a telephone line connection. This product is completely Internet-ready and allows safe and reliable operation without manned attention for several months at a time. Regular inspection of the units is recommended and can be accomplished on a periodic basis, with one service professional attending to the inspection of several units at a time. The typical fleet performance with this scenario has been remarkable, and, in most cases, the remote, unmanned control and operation of the ONSI units has been readily accepted by the regulatory groups responsible for siting and permitting of such installations. The ONSI Corporation has made great strides in the development and implementation of the technology throughout the world to set the stage for more widespread use of unmanned, remote operation and control of distributed power generation. It is expected that the type of Internet-ready technology, controls, siting, and operation contained within the ONSI systems and installations to date will be representative of all fuel cell systems that emerge in the marketplace of the future.

6.10 Fuel Cell Systems Costs

The cost of fuel cell systems has been dramatically reduced since their first use in space applications. Estimates of the installed cost of fuel cell systems applied to the Apollo and Gemini missions range in the millions of dollars per kilowatt of capacity (\$/kW). Since the first demonstration of fuel cells for distributed power generation, the installed cost of fuel cell systems has been dramatically reduced from the order of \$1 million/kW to the current price of about \$4000/kW.

The high cost of fuel cell systems is due to several factors. The first is the utilization of high-cost materials in the construction of fuel cell systems (e.g., noble metal catalyst materials). The amount of high-cost materials required for effective operation has been dramatically reduced in recent years. This is particularly true for proton exchange membrane fuel cell performance, which has been increased with a reduction in platinum catalyst loading. The second factor is complicated designs with increased instrumentation and controls not optimized for ease of manufacturing, and the third is labor intensive manufacturing processes. The fourth and most important factor for continued high cost of fuel cell systems to date is the lack of volume production, which impacts not only fuel cell stack material cost, design, and manufacture, but also the cost of all system and balance-of-plant items.

One of the best sources of information on fuel cell costs is available through the U.S. Department of Defense (DoD) Fuel Cell Program, which has installed more than 30 ONSI PC-25™ 200 kW fuel cells and is providing publicly available feedback and information regarding their installation and performance. This information is readily available at *www.dodfuelcell.com* (DoD Fuel Cell Program Web site, 1999). This fleet of PAFC units is managed by the Construction Engineering Research Laboratory (CERL) for the Defense Utilities Coordinating Council on behalf of all branches of the military. Information provided at this site includes not only cost information, but also information on reliability, availability, efficiency, installation requirements, and more.

Data from the DoD Fuel Cell Program indicate that the complete installed cost of commercially available fuel cells today is approximately \$1.1 million for an ONSI PC-25™ PAFC unit. At a rated output of 200 kW, this corresponds to \$5412/kW installed (DoD Fuel Cell Program Web site, 1999). All of these installations qualified for the DoD Fuel Cell Buy Down Program, managed by the Federal Energy Technology Center (FETC Web site, 1999), which reduced the cost of the installed fuel cells to approximately \$4400/kW. Note that these prices reflect purchases over the last seven years.

All fuel cell manufacturers require some savings due to volume manufacturing to reach the target prices for distributed power generation systems which are in the range of \$800 to \$1000/kW. At this price for installed cost, the increased efficiency of fuel cell systems compared to other distributed generation technologies can make them strictly cost competitive. With the additional reductions in pollutant emissions that fuel cell systems can provide, fuels cells portend a tremendously competitive distributed generation technology. Reaching the target price identified above with reliable fuel cell systems is the primary challenge facing the fuel cell community.

6.11   Technology Development and Barriers

Fuel cells have been shown to provide electricity from both hydrogen and fossil fuels at efficiencies greater than any other electric generating device. Emissions from fuel cells have been shown to be near zero for most pollutants of concern (e.g., NO_x, SO_x, CO, and hydrocarbons) depending upon the specific fuel cell technology and application. Fuel cells are scalable down to very small sizes while maintaining extraordinary fuel efficiency and environmental sensitivity. So why are fuel cells not playing a major role in today's market, and why do some experts not expect them to become widely utilized until at least ten years from now? There are several reasons, each of which is a critical and current area of research, development, and demonstration for fuel cell technologies.

6.11.1   Cost Reduction

The high capital cost for fuel cells is by far the largest factor contributing to the small market penetration of fuel cell technology. The high capital cost (on a $/kW basis) has led to an electric power sector wary of installing fuel cells or using them in other systems (e.g., transportation applications). Reducing the installed cost of fuel cell technology is perhaps the most important driver of fuel cell research, development, and demonstration (RD&D) today. This RD&D encompasses many aspects of the fuel cell industry and research community. Specific areas in which cost reductions are being investigated include (1) materials, (2) complexity of integrated systems, (3) temperature constraints, (4) manufacturing processes, (5) power density (footprint reduction), and (6) benefit from economies of scale (volume) through increased market penetration.

6.11.2   Fuel Flexibility

The ability of fuel cells to operate on widely available fossil fuels as well as handle variations in fuel composition reliably and without detrimental impact to the environment or the fuel cell is necessary. In addition, the capability of operating fuel cells on renewable and waste fuels is essential to capturing market opportunities for fuel cells.

The primary fuel used in a fuel cell is hydrogen, which can be obtained from natural gas, coal gas, methanol, landfill gas, and other fuels containing hydrocarbons. Increasing the fuel flexibility of fuel cells implies power generation that can be ensured even when a primary fuel source is unavailable. This will increase the initial market opportunities for fuel cells and enhance market penetration. Specific RD&D topics being addressed to increase the fuel flexibility of fuel cells include (1) non-traditional fuel storage (H_2), (2) transportation fuel reforming, (3) renewable fuels processing (reforming, gasifying, clean-up), (4) biogas operation, and (5) tolerance to gas supply variation.

6.11.3   System Integration

The development and demonstration of integrated fuel cell systems in grid-connected and transportation applications, as well as the development and demonstration of hybrid systems for achieving very high efficiencies, are important to the success of fuel cell technology. In order to minimize the cost of electricity, integrated fuel cell systems must be developed and demonstrated. For most applications, this requires that the fundamental processes be integrated into an efficient plant with capital costs kept as low as possible. Specific systems and system integration RD&D occurring today include (1) power inverters, (2) power conditioners, (3) hybrid system designs, (4) hybrid system integration and testing, (5) operation and maintenance issues, and (6) robust controls for integrated systems.

6.11.4   Endurance and Reliability

Fuel cells could be great sources of premium power if they could be demonstrated to have superior reliability and power quality and could be shown to provide power for long continuous periods of time. The power quality of fuel cells alone could provide the most important marketing factor in some applications and, coupled with longevity and reliability, would greatly advance fuel cell technology.

Although fuel cells have been shown to be able to provide electricity at high efficiencies with exceptional environmental sensitivity, the long-term performance and reliability of certain fuel cell systems have not been significantly demonstrated to the market. Research, development, and demonstration of fuel cell systems that will enhance the endurance and reliability of fuel cells are currently underway. The specific RD&D issues in this category include (1) endurance and longevity, (2) thermal cycling capability, (3) durability in installed environment (seismic, transportation effects, etc.), and (4) grid-connection performance.

6.12 Summary

Governmental regulations are a significant driver for the consideration of environmentally sensitive technologies. The consideration of fuel cells has benefited from governmental regulations because of the special characteristics of fuel cells (highly efficient and low polluting). If governmental regulations or credits provide additional incentives for the consideration of highly efficient systems — through carbon dioxide reduction credits to address global climate change — then fuel cells will benefit because of that high efficiency. However, the most significant drivers will likely be those provided by the global free market which is increasingly aware of fuel cell benefits, increasingly considerate of high efficiency technologies (e.g., the Kyoto protocol and Buenos Aires accords), and increasingly participatory in the development of cost-effective systems to solve the energy environmental challenge. This market will eventually produce lower-cost fuel cell systems that will outperform current technologies in every respect (life cycle cost, environmental sensitivity, etc.), which will lead to their widespread use in distributed power generation.

References

Appleby, A.J. and Foulkes, F.R., *Fuel Cell Handbook*, Van Nostrand Reinhold, New York, 1989.

Appleby, A.J. and Yeager, E.B., Solid Polymer Electrolyte Fuel Cells (SPEFCS), in Assessment of Research Needs for Advanced Fuel Cells, S.S. Penner, Ed., *Energy*, 11, 137, 1986.

Atkins, P.W., *Physical Chemistry*, 3rd ed., W.H. Freeman, New York, 1986.

Baker, B.S., Carbonate fuel cells — a decade of progress, 191st meeting of the Electrochemical Society, May 1997.

Brown, D.R. and Jones, R., An overview of stationary fuel cell technology, U.S. Department of Energy Contract No. DE-AC06-76RLO, document number PNNL-12147, February 1999.

Department of Defense Fuel Cell Web site, Construction Engineering Research Laboratory, U.S. Department of Defense, *www.dodfuelcell.com*, 1999.

FETC Web site, Federal Energy Technology Center, U.S. Department of Energy, www.fetc.doe.gov, 1999.

Gottesfeld, S. and Zawadinski, T., PEFC Chapter, *Advances in Electrochemical Science and Engineering*, Vol. 5, Alkire, R., Gerischer, H., Kolb, D., and Tobias, C., Eds., Weinheim, New York, 1998.

Hirschenhofer, J.H., Stauffer, D.B., Engleman, R.R., and Klett, M.G., *Fuel Cell Handbook*, 4th ed., U.S. Department of Energy, Reading, PA, 1998.

King, J.M. and Ishikawa, N., Phosphoric acid fuel cell power plant improvements and commercial fleet experience, Fuel Cell Seminar, November 1996.

Minh, N.Q., Ceramic fuel cells, *J. Am. Ceramic Soc.*, 76(3), 563, 1993.
Penner, S.S., Ed., Report of the DOE advanced fuel cell commercialization working
    group, DOE/ER/0643, DOE Contract No. DEFG03-93ER30213, March 1995.
Singhal, S.C., Recent progress in tubular solid oxide fuel technology, *Proceedings of
    the Fifth International Symposium on Solid Oxide Fuel Cells (SOFC-V)*, The Electro-
    chemical Society, Inc., Pennington, NJ, 1997.

7

Principles of Control of Distributed Generation Systems

Peter S. Curtiss

CONTENTS

This chapter summarizes some of the methods used to control distributed generation (DG) at the local building or campus level. This problem is important because the control mode can make the difference between a profitable

DG installation and one that is not profitable, all else being equal. Distributed power offers many different services. For example, combustion-based DG can provide significant amounts of heat to a building's space and water heating loads if cogeneration is used. When coupled with absorption cooling, the available heat can also be used to supplement the building's conventional cooling system (Kreider and Curtiss, 2000). The trade-off, of course, is that the cost of gas consumption increases while the cost of grid electricity decreases. The optimum control of such systems is the subject of this chapter. Optimal control maximizes the financial return on DG system investment.

This chapter does not discuss internal controls provided by manufacturers of DG equipment. Safety items, combustion control, alarms, and many other features of control systems are the province of hardware providers. This chapter addresses the best way to control DG systems that have properly engineered local control systems that ensure nominal operation of DG hardware. Given that, the principles here can be used to maximize revenue to the DG system owner. Figure 7.1 is a schematic diagram of a combustion-based generator. It is assumed that this generator is capable of providing heat recovery to a building.

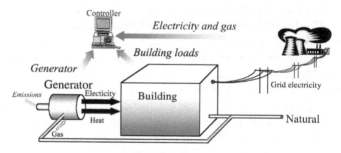

FIGURE 7.1
Components of a typical distributed generation system.

7.1   Control Techniques

The control techniques chosen for DG will depend on the type of equipment installed. In the case of wind or solar power generation, the main goal is to produce as much energy from the system as possible to recover the installation cost. For combustion-based processes, however, the costs of fuel and maintenance must also be taken into account. The fundamental goal of the control scheme is to determine whether or not the on-site generation should be operating during a particular hour.* Generally, a simple hour-ahead

---

* Here the time internal is taken to be an hour, but any period can be used. For example, many demand rates use 15-minute periods.

control method is sufficient if the start-up transients of the generator are short. This section, adapted from Curtiss (2000), describes some of the techniques that can be used to control on-site generation.

7.1.1 Threshold Control

In threshold control, the generators run whenever a building's electrical load is greater than a predetermined threshold. The number of generators initially installed is equal to the difference between the annual peak and the threshold divided by the nominal power output of each installed unit:

$$Number\ installed\ =\ \frac{kW_{PEAK} - kW_{THRESHOLD}}{kW_{PER\ UNIT}}$$

If the electrical load of the building is greater than the threshold, then the number of generators operating is equal to the number required to reduce the grid load to the threshold limit:

$$Number\ operating\ =\ \frac{kW_{BUILDING} - kW_{THRESHOLD}}{kW_{PER\ UNIT}}$$

A problem with this control method is deciding where to assign the threshold limit during system design. A high limit means that the generator is used only for peak shaving and the number of operating hours may be small. A low limit forces the generators to run more often and is akin to base loading. A threshold of zero indicates that the generators will try to operate whenever possible. This specific case is referred to as *always-on* control.

Thresholds are established during design by finding the best financial return given energy requirements as they are known during design. The trade-off is between capital investment in DG hardware versus savings in utility bills, both taken over the economic lifetime of the DG equipment. Of course, process or building usage often changes drastically during system lifetime, and the installed DG controller must be intelligent enough to operate the DG system optimally no matter how building usage has evolved from the initial design upon which the initial threshold was based.

7.1.2 Buyback Priority

Buyback priority is used in cases where the operator wishes to produce electricity and sell any or all of the produced power back to the utility. There are two versions of buyback control; one takes advantage of a simple buyback rate and the other responds to *net metering,* whereby the value of produced power is used to offset the traditional electrical bill.

7.1.2.1   Simple Buyback

In the case of simple buyback, the generators use the threshold control scheme as previously described. If the buyback cost is greater than the equivalent cost of gas, then all the generators run and the excess is sold to the utility or power exchange. The number of generators installed depends on the projected income the process operator expects to earn from selling electricity. This control method finds the incremental sum of all fuel used to get the total cost for the hour:

$$Total\ Cost = \Delta\$kWh_{GRID} + \Delta\$Btu_{GRID} - \Delta\$kWh_{BUYBACK}$$

The $\Delta\$$ term implies that the gas and electric costs are evaluated on an incremental monthly (i.e., billing period) basis except for real time pricing rates. For example, the change of the grid electricity bill is

$$\Delta\$kWh_{GRID} = M\$(kWh_1, kWh_2,..., kWh_{N-1}, kWh_N)$$
$$- M\$(kWh_1, kWh_2,..., kWh_{N-1})$$

where $M\$$ is the monthly bill amount (including consumption and demand fees, surcharges, and taxes) based on N hourly electricity use values for that billing period. This allows the bills to be calculated, including any time-of-use and block components. Unfortunately, these latter components also affect the linearity of the cost function — the cost function is not necessarily linear under these conditions. That is, the electricity used and the utility bill are not related in a simple linear fashion.

The algorithm for determining whether or not to use buyback, therefore, should (1) determine the loads on the building for a given hour, (2) calculate the total cost function for all integral numbers of generators operating, from zero to the number installed, and (3) determine which number of operating generators minimizes the total cost function. That is the number of generators that will operate that hour to maximize financial benefit.

7.1.2.2   Net Metering Control

In the net metering scenario, the electrical meter runs backwards if excess electricity is produced on site. If the meter reaches zero, buyback rates apply. As with the buyback priority control, the incremental sum of all fuel uses is calculated to get the total cost for the hour:

$$Total\ Cost = \Delta\$kWh_{GRID} + \Delta\$Btu_{GRID} - \Delta\$kWh_{BUYBACK}$$

The $\Delta\$$ terms are the incremental costs as discussed with buyback priority control. Consequently, the control algorithm is the same as in buyback priority with the exception that the $\Delta\$kWh_{GRID}$ term here refers to the adjusted (i.e., rolled back) meter usage, and the $\Delta \$kWh_{BUYBACK}$ amount is decreased by the kWh that go into reducing $\Delta \$kWh_{GRID}$.

If the monthly sum is positive (i.e., more electricity has been used from the grid than produced on site), then the monthly bill is based on simple aggregation of hourly consumption plus demand and fees. Otherwise, the customer is refunded the value of excess electricity produced as dictated by the buyback rate.

7.1.3 Cooling/Heating Priority Control

In some cases, the DG units will be deployed as cogenerators to satisfy a cooling load (either through auxiliary absorption cooling or direct mechanical or electrical connection to conventional cooling equipment) or a heating load (through heat recovery). In this mode of control, the generators operate primarily to satisfy these loads, and the satisfaction of the electrical load is a secondary benefit. The number of generators installed is sufficient to meet the annual peak thermal load, and the control algorithm has the generators operating as required to meet the thermal load of the building. No consideration is given to the value of electricity to determine control actions.

7.1.4 Optimal Control

Ideally, on-site generation is operated using an algorithm that reduces the operating cost such that the cost to the building operator is minimized every hour. If the building is subject to a real-time pricing rate schedule, then the optimization can be trivial; the costs of grid electricity and locally produced electricity are compared at each hour; and, when the former is more expensive, the on-site generators are operated. However, more conventional rate structures such as block rates and time-of-use rates, with accumulation over a billing period, can make the calculation of instantaneous "next kWh" costs much more difficult. In that case, the electricity bill C_{ELEC} at any given hour is:

$$
C_{ELEC} = \Phi_{KWH}
\begin{bmatrix}
kWh_{BLDG}(1) - kWh_{GEN}(1) \\
kWh_{BLDG}(2) - kWh_{GEN}(2) \\
\cdots \\
kWh_{BLDG}(k) - kWh_{GEN}(k)
\end{bmatrix}
+ \Phi_{KW}
\begin{bmatrix}
kW_{BLDG}(1) - kW_{GEN}(1) \\
kW_{BLDG}(2) - kW_{GEN}(2) \\
\cdots \\
kW_{BLDG}(k) - kW_{GEN}(k)
\end{bmatrix}
$$

where Φ_{KWH} is the utility function used to calculate the bill based on consumption, Φ_{KW} is the function used for demand, $kWh_{BLDG}(1)$ is the total electric load at hour 1, $kWh_{GEN}(1)$ is the kWh offset from the on-site generation equipment at hour 1, and so forth. The calculation must be performed for each hour of the billing period to account for variations in the hourly load, any time-of-use components of the utility rate, and any ambient temperature or solar dependencies of the generation equipment. If the generators use natural gas to produce electricity (an internal combustion engine, combustion turbine, microturbine, or fuel cell), then a similar calculation is performed for the gas consumption. Assuming no demand component for gas, the total gas bill up to hour k of the billing period is given as

$$C_{GAS} = \Phi_{GAS} \begin{bmatrix} Btu_{BLDG}(1) + Btu_{GEN}(1) \\ Btu_{BLDG}(2) + Btu_{GEN}(2) \\ \cdots \\ Btu_{BLDG}(k) + Btu_{GEN}(k) \end{bmatrix}$$

where Btu_{GEN} is the incremental gas consumption of the generation equipment at each hour. Note that kWh_{GEN}, kW_{GEN}, and Btu_{GEN} can have zero values at any hour depending on whether or not the generation equipment is operating for that hour. To determine if the generators should operate at hour $k+1$, the total cost $C_{ELEC} + C_{GAS}$ should be evaluated twice: once using values for the terms kWh_{GEN}, kW_{GEN}, and Btu_{GEN} based on the estimated generator performance, and then again with these values set to zero. If the former is greater than the latter, the generators should not be run for that hour.

7.1.5 Complete Optimization

The procedure just described is sufficient for performing an optimization based on a single type of generation equipment without accounting for any other inputs. To be truly optimal, however, the algorithm should account for any different capacities of generators installed, any utility incentives, and the variable operation and maintenance costs experienced during operation. The structure and calculation methods used for the electricity and gas utility rate schedules must be known. The optimization routine must also be able to keep track of all data acquired during a given billing period and provide cost estimates for the current hour. Any utility-sponsored incentives and rebates should be tallied along with the method of their application (e.g., by kWh produced, kW installed, etc.)

At each hour of the billing period, the optimization routine determines the number of generators that should run for that hour. This requires a prediction of the building load data for that hour, including whole building kWh use, whole building Btu use, kWh used for domestic water heating, Btu used for domestic water heating, kWh used for space heating, Btu used for space heating, and kWh used for space cooling.

The electrical and thermal output from each generation device in the building must then be determined. This may require monitoring of the ambient temperature, wind speed, and insolation.

One then must examine the benefit of operating each generator, accounting for any generators that may already be operating and for any part-load ratio (PLR) characteristics of generators that are not operating at full load. The cost function in the analysis includes all of the costs of providing on-site electrical and thermal energy. This cost is compared with that for grid consumption, and the lower of the two is chosen. To properly assess these costs, the grid electricity consumption kWh_{GRID} is adjusted by the decrease of grid electricity consumption due to on-site power generation:

$$kWh_{GRID} = kWh_{BLDG} - kWh_{GEN} - kWh_{COOL}$$

where kWh_{BLDG} is the building electrical load and kWh_{GEN} is the amount of electricity produced from on-site generators. The term kWh_{COOL} is non-zero if the generator provides direct cooling through absorption cooling and must be corrected for the nominal efficiency of the conventional cooling equipment:

$$kWh_{COOL} = \frac{\sum \dot{Q}_{COOL}}{COP_{COOL}}$$

where the summation is taken over all devices that provide supplemental cooling, and the COP is the average over all cooling equipment in use. If the DG equipment includes any gas-fired devices, the incremental cost of natural gas consumption must also be taken into account:

$$Btu_{GRID} = Btu_{BLDG} + Btu_{GEN} - Btu_{DG\text{-}EXHAUST}$$

where Btu_{BLDG} is the building load and Btu_{GEN} is the consumption of gas by the generators:

$$Btu_{GEN} = \sum W_{GEN}(PLR)$$

and where the summation is taken over all devices that convert gas to electricity. The work term must also be corrected by the part-load efficiency of any generators that are not at full load. The term $Btu_{DG\text{-}EXHAUST}$ represents any credit that can be applied due to exhaust heat recovery from the generators that precludes the use of conventional space or water heating sources. As with the cooling term, this credit is adjusted by the nominal efficiency of the conventional sources:

$$Btu_{HEAT} = \frac{\sum \dot{Q}_{DG-EXHAUST}}{\eta_{HEAT}}$$

The total operating cost can now be calculated from the incremental rates, incentives, and maintenance costs. The result is:

$$
C_{TOTAL} = \Phi_{KWH} \begin{bmatrix} kWh_{BLDG}(1) - kWh_{GEN}(1) \\ kWh_{BLDG}(2) - kWh_{GEN}(2) \\ \ldots \\ kWh_{BLDG}(2) - kWh_{GEN}(k) \end{bmatrix}
$$

$$
+ \Phi_{KW} \begin{bmatrix} kWh_{BLDG}(1) - kWh_{GEN}(1) \\ kWh_{BLDG}(2) - kWh_{GEN}(2) \\ \ldots \\ kWh_{BLDG}(k) - kWh_{GEN}(k) \end{bmatrix}
$$

$$
+ \Phi_{GAS} \begin{bmatrix} Btu_{BLDG}(1) + Btu_{GEN}(1) \\ Btu_{BLDG}(2) + Btu_{GEN}(2) \\ \ldots \\ Btu_{BLDG}(k) + Btu_{GEN}(k) \end{bmatrix}
$$

$$
- \Phi_{CRED} \left[\sum kWh_{GEN} \right]
$$

$$
+ \Phi_{O\&M} \left[kW_{INST} + \sum kWh_{GEN} \right]
$$

where Φ_{CRED} represents a positive cash flow based on any utility incentives provided including transmission loss credits, wheeling charge credits, voltage support credits, etc. The term $\Phi_{O\&M}$ is used to account for any operation and maintenance costs that arise from operating the generation equipment. The two operation and maintenance (O&M) terms are those associated with regular maintenance independent of power production and that which depends on power production (i.e., number of hours operated), respectively.

Finally, an hourly cost matrix is compiled that represents all reasonable combinations of generation available to a building. The combination of equipment with the lowest cost is chosen and operated for that hour.

7.2   System Modeling

As evident from the above description, correct optimal control is complex. Various data are necessary to properly assess the benefits of operating generators. The data needs for DG control are described below.

A *weather simulation* module should provide hourly site temperatures, solar radiation, wind speed, etc. This model can take advantage of recent measured conditions and forecasts. In most cases, weather needs to be predicted for only an hour or two into the future. A *process or building simulation* algorithm should generate hourly loads based on the historic load shape and the actual consumption and demand for a given billing period. Actual data are needed because few processes or buildings are used as originally designed. Energy uses based on the designed configuration will be erroneous. Neural networks offer one method of making accurate predictions based on measured data. Linear regressions and first principles models should also be considered.

The *equipment simulation* should use any site-specific values (e.g., atmospheric pressure), the weather data from the weather simulation, and the predicted building loads to determine the amount of building electrical and thermal loads able to be offset by on-site production. This algorithm should also allocate energy savings into the respective constituent categories such as space heating, plug loads, etc. For all but renewable-powered DG, the algorithm must also calculate the increased consumption of fuel.

Table 7.1 summarizes the information required to implement the various control techniques. Weather data are not explicitly included in this list but may be required to estimate those values marked with an asterisk. Generator capital costs need not be known because they are involved in the initial selection and installation. Once the plant is built, it is the task of the controller to control that plant. Capital costs are not a variable that can be adjusted.

TABLE 7.1

Data Required for Each Control Method

Input Value	Threshold Control	Buyback Priority	Heating/Cooling Priority	Optimal Control
Electricity rate schedule		X		X
Natural gas rate schedule		X		X
Utility incentives (including buyback)		X		X
Predicted building electrical load*	X	X		X
Predicted building heating load*			X	X
Predicted building/equipment cooling load*			X	X
Predicted generator electrical output*	X	X		X
Predicted generator heat output*				X
Generator operation and maintenance costs				X

Note: Values marked with an asterisk may require use of a weather model.

7.3 Examples of Control Operation

This section compares the performance of threshold control versus optimal control for three different buildings in different rate regions. In all cases, an hourly microturbine (MT) simulation was performed for an entire year. The MT systems had installed costs of $1000 per kW and an expected system lifetime of 10 years. The turbines were further assumed to be generic 50 kW generators operating with heat recovery where any available heat was used to supplement the space heating load if one existed each hour. A nominal heating efficency of 80% was assumed for the conventional space heating equipment in each building. Operation and maintenance costs were taken as $4.75 per installed kW per year plus an additional $0.007 per kWh produced. The discount, inflation, and tax rates were taken as 7, 2.4, and 33%, respectively. The cost of energy was assumed to be *decreasing* at the rate of 1.5% per year for electricity and 0.6% per year for natural gas. No buyback or other utility incentives were included in the analysis. The values assumed for all variables are conservative.

7.3.1 Sit-Down Restaurant in San Francisco

The first case considered is a sit-down restaurant in San Francisco. The total annual utility bill for this restaurant is $35,400 with about 80% of this cost coming from electricity usage. Figure 7.2 shows example weekday electrical load shapes used in the analysis of this restaurant. The peak load is 56 kW. The restaurant analysis used rates quoted by a large utility in the Bay Area.

Table 7.2 summarizes the annual performance of the different control algorithms for this establishment. Due to the relatively small size of the load and the relatively high cost of electricity, the results do not vary significantly between one control algorithm and the other. The simple threshold control provides an adequate return on investment regardless of where the threshold is set. The always-on control (i.e., a threshold of zero) suffers from the turbine running at low loads when the part-load performance is inefficient.

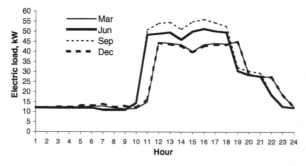

FIGURE 7.2
Selected weekday electrical load shapes for a restaurant in San Francisco.

TABLE 7.2

Summary of Microturbine Performance in San Francisco Restaurant

Control Method	First Year Savings ($ thousands)	Payback Period (years)	Internal Rate of Return	Change in Annual Energy Costs	
				Electricity ($ thousands)	Natural Gas ($ thousands)
Threshold 45 kW	$10.7	4.6	12%	–$16.2	$4.6
Threshold 30 kW	$11.3	4.4	13%	–$20.5	$7.9
Threshold 15 kW	$11.9	4.2	15%	–$23.5	$10.1
Always-on	$10.4	4.8	11%	–$27.6	$15.4
Optimal	$12.5	4.0	16%	–$22.9	$9.0

The optimal controller produces the highest rate of return and the smallest payback period of all options considered because it uses DG assets most intelligently.

7.3.2 Supermarket in Chicago

This example examines a 24-hour supermarket in Chicago with a peak load of 167 kW, an annual electricity bill of $67,500 (average of 5.7¢ per kWh), and an annual gas bill of $566 (average of $5.90 per MMBtu) calculated using actual rates from a large utility in Chicago; the rates are time-of-use rates. Figure 7.3 shows selected weekday electrical load shapes for this facility. Note that the load shapes are much flatter than those for the restaurant presented in the previous example. Two different scenarios were examined for the Chicago supermarket: peak shaving and base-load control. Table 7.3 shows the comparative results from the analysis.

The peak shaving scenario used a single 50 kW microturbine with heat recovery, while the base-load scenario had three turbines. This example illustrates how optimal control can recognize the periods when it is expensive to purchase grid electricity. Threshold control does not recognize these and, in fact, will tend to operate during the periods of high load and corresponding high electricity costs. Optimal control, on the other hand, takes advantage of the time-of-use component and shows a clear benefit over the threshold control method. The rate of return on this investment is not as attractive as with the previous example due to the relatively low cost of electricity. As in the first case, optimal control performs best with either of the two control objectives. The difference between optimal control and the best threshold control is significant.

7.3.3 Medium-Sized Office in New York City

The final example looks at a medium-sized office building in New York City. The peak electrical load experienced by this building is about 370 kW during the month of July. The building annual energy bill is $144,300 for electricity

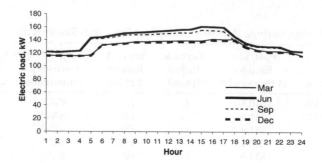

FIGURE 7.3
Selected weekday electrical load shapes for a supermarket in Chicago.

TABLE 7.3

Summary of 150 kW Microturbine Performance in a 24-Hour Supermarket in Chicago

Control Objective	Control Method	First Year Savings ($ thousands)	Payback Period (years)	IRR	Change in Annual Energy Costs	
					Electric ($ thousands)	Gas ($ thousands)
Peak shaving; one turbine installed	Threshold 140 kW	$3.6	> 10	–6%	–$9.9	$5.2
	Threshold 130 kW	$7.0	7.1	3%	–$19.4	$10.3
	Threshold 120 kW	$5.5	9.1	–2%	–$22.5	$14.1
	Threshold 110 kW	$4.1	> 10	–19%	–$24.5	$16.9
	Optimal	$9.3	5.4	9%	–$17.1	$6.4
Base-load; three turbines installed	Always-on	$8.0	> 10	–16%	–$64.1	$47.4
	Optimal	$25.2	6.0	7%	–$46.9	$17.8

and $11,100 for natural gas using rates from a large New York utility. With these rates, the average electricity cost is 11.8¢ per kWh and the annual average gas cost is $5.68 per MMBtu. Example load shapes for this building are shown in Figure 7.4.

Three different scenarios were examined. The first studies the effects of a single 50 kW microturbine installed on the building used for peak shaving. This yielded the results shown in Table 7.4. As would be expected, the performance of the various control techniques is not significantly different over the range of threshold levels. The exception is when a very high threshold is used and the turbine runs only part of the year. The second scenario (Table 7.5)

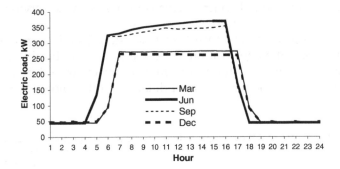

FIGURE 7.4
Selected weekday electrical load shapes for a medium-sized office in New York City.

TABLE 7.4

Summary of 50 kW MT Performance in a Medium-Sized Office in New York City

Control Method	First Year Savings ($ thousands)	Payback Period (years)	IRR	Change in Annual Energy Costs	
				Electricity ($ thousands)	Gas ($ thousands)
Threshold 300 kW	$2.9	> 10	–8%	–$6.8	$3.3
Threshold 250 kW	$11.5	4.3	14%	–$20.1	$7.3
Threshold 200 kW	$11.5	4.3	14%	–$20.1	$7.3
Threshold 150 kW	$11.4	4.4	14%	–$20.9	$8.2
Threshold 100 kW	$11.4	4.4	14%	–$21.7	$8.9
Threshold 50 kW	$11.8	4.2	14%	–$24.0	$10.4
Always-on	$10.9	4.6	11%	–$36.6	$22.6
Optimal	$11.9	4.2	14%	–$25.0	$11.2

used 200 kW of installed generation capacity, roughly equivalent to half of the annual peak electrical load of the building. In that case, the economics of the always-on control are the same as the optimal control, indicating that the turbines should be operated as much as possible. The third scenario (Table 7.6) used 400 kW of installed generation capacity — enough to more than cover the entire building load. As with the previous scenario, the economics improve the more the turbines operate. This seems reasonable when comparing the relative prices of electricity and gas: the cost of gas is equivalent to 1.94¢ per kWh, which is more than six times cheaper than the electricity. This allows for even relatively inefficient gas-to-electricity conversion equipment to be operated at a profit.

Note that optimal control operates the turbines differently from always-on control (refer to the change of annual energy costs in Tables 7.5 and 7.6). The difference between annual gas costs indicates that the turbines operate for fewer hours with optimal control. This is important when trying to reduce maintenance costs and increase system lifetime.

In certain cases, optimal control does not significantly improve the economics of the installed generation system, and it is doubtful that the incremental cost of implementing an optimal controller would be money well spent. However, in other cases, the optimal controller allowed the system to

TABLE 7.5

Summary of 200 kW MT Performance in a Medium-Sized Office in New York City

Control Method	First Year Savings ($ thousands)	Payback Period (years)	IRR	Change in Annual Energy Costs	
				Electricity ($ thousands)	Gas ($ thousands)
Threshold 300 kW	$4.2	> 10	−17%	$11.4	$5.7
Threshold 250 kW	$16.4	> 10	−4%	$32.0	$13.0
Threshold 200 kW	$27.9	7.2	4%	$51.2	$19.7
Threshold 150 kW	$33.8	5.9	8%	$61.8	$23.9
Threshold 100 kW	$40.3	5.0	11%	$74.1	$28.8
Threshold 50 kW	$46.5	4.3	14%	$86.3	$34.0
Always-on	$46.9	4.2	14%	$101.9	$47.5
Optimal	$47.2	4.2	14%	$94.5	$40.7

TABLE 7.6

Summary of 400 kW MT Performance in Medium-Sized Office in New York City

Control Method	First Year Savings ($ thousands)	Payback Period (years)	IRR	Change in Annual Energy Costs	
				Electricity ($ thousands)	Gas ($ thousands)
Threshold 300 kW	$3.2	> 10	−23%	$11.4	$5.7
Threshold 250 kW	$15.5	> 10	−12%	$32.0	$13.0
Threshold 200 kW	$27.4	> 10	−6%	$52.4	$20.3
Threshold 150 kW	$36.3	> 10	−2%	$68.4	$26.6
Threshold 100 kW	$47.1	8.5	1%	$88.4	$34.6
Threshold 50 kW	$59.4	6.7	5%	$112.6	$45.0
Always-on	$68.1	5.9	7%	$141.0	$62.6
Optimal	$68.3	5.8	8%	$140.9	$62.2

be profitable when other control methods did not. This is expected to be true in areas with time-of-use electric rates and block demand rates. In such a location, the use of optimal control is necessary to guarantee savings with distributed generation.

7.4   Virtual Generation Plant (VGP) Control

A special case arises when there are numerous buildings with DG installations or several buildings are operated under a conjunctive billing agreement and one or more of the buildings has on-site generation capabilities. This is termed a *virtual generation plant* (VGP) because the aggregation of power plants will need to be treated as one unit. Under these circumstances, the control of a particular set of generators will depend on the individual loads of the various buildings. A global optimization must be performed similar to the method discussed earlier. On a larger scale, an entire geographical area with thousands of DG units will also be treated as a single dispatchable VGP to a power exchange or independent system operator (ISO). Without the ability to optimally control an aggregation of DG systems, the future of the industry will be curtailed.

Figure 7.5 shows an example of applying a single generator to a group of buildings. In Figure 7.5, a single primary school is assumed to have a conjunctive billing agreement with five surrounding apartment buildings. The hourly electric load shapes are shown for the school (solid line) and for the sum of the apartment buildings (dotted line). The corresponding bars show the aggregated load. In this scenario, a single 50 kW generator at the school operates whenever the aggregated load is greater than 200 kW. The contribution of the generator is represented by the top, darker portion of each bar.

The combined load shape for these buildings benefits from the different schedules of these two building types. The load factor has been increased, and the total combined peak grid load is not significantly higher than that of the school operating independently. By sharing the costs of operating the generator in addition to saving on any peak costs, the contribution of each building to the total cost is decreased. One concept for the VGP is shown in Figure 7.6. The splits between computational equipment located at the central site and at each DG unit are noteworthy.

7.4.1   Basic Goals for VGP Control

An optimization minimizes a cost (or maximizes a benefit) function. In the case of the VGP, the cost function is the financial cost of satisfying the electrical and thermal loads of a building or group of buildings from the viewpoint of the VGP owner/operator. The required goals listed in the following sections are those needed by a VGP that is capable of aggregating DG units.

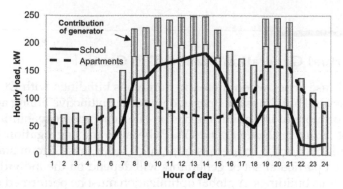

FIGURE 7.5
Example of VGP benefits.

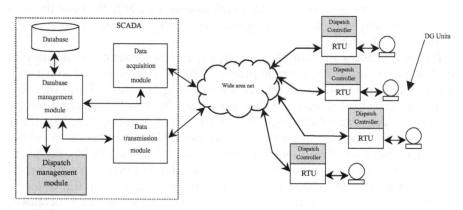

FIGURE 7.6
VPP schematic diagram.

7.4.1.1   *Minimization of Required Data Transfer between DG Units and Central Site*

The system must provide optimization and control in real time. Consequently, intensive data processing takes place locally at the generation site or at the campus where DG equipment is installed. Another reason to minimize the amount of transmitted information is to reduce the risk of data corruption and unauthorized interception.

7.4.1.2   *Reliability*

Each DG installation must provide reasonable dispatch control regardless of Internet, radio frequency (RF), fiber, cable, or telephonic connection status. This means that each site will continue to operate independently in the event of a loss of communication with the central station and that the DG equipment will not default to an all-on or all-off mode. This implies that some degree of optimization capability must reside within the local controller.

Thus, the local controller will need to be able to estimate the DG electrical and thermal output, the building electric and thermal loads, and the utility bills.

7.4.1.3   Conjunctive Billing

Small- and medium-sized buildings may be on different rate schedules than larger buildings. While small buildings may not be subject to demand charges, for example, consumption charges can be significantly higher than those for larger buildings. One option that may become more available in a competitive utility market is aggregating the loads of a number of smaller buildings into one bill, thereby allowing these buildings to take advantage of beneficial rates.

7.4.1.4   Selection of DG Priorities

Certain applications (supermarkets, hotels, hospitals, etc.) may value the thermal energy from a DG system more highly than the electrical power. Therefore, the control system must account for an assignment of priorities for energy from DG systems.

7.4.1.5   Buyback and Retail Wheeling

The VPP must be able to dispatch power reliably to an ISO for use in a utility service area. Therefore, the VPP software must be able to make buy and sell decisions in real time. A description of these costs is given in Curtiss et al. (1999).

7.4.1.6   Cost Function Minimization

The cost function for the DG control is the sum of all operating costs. Components of the cost equation that increase the operating costs include:

- Grid electricity consumption at meter (in kWh)
- Grid electricity demand (in kW)
- Grid gas consumption at meter (in either energy content or volume used)
- Grid gas consumption by DG equipment (in either energy content or volume used)
- Operation and maintenance costs (variable as a function of run time; see below for fixed O&M costs)

Components of the cost equation that decrease the operating costs include:

- Grid electricity consumption reduction due to local DG
- Grid gas consumption reduction from DG cogeneration (if used)

- Grid electricity consumption reduction from supplementary cooling supplied by DG cogeneration (if used)
- Revenue from sale of ancillary services (per VAR, energy, or other unit)
- Utility incentives (per kWh produced locally)
- Utility electricity buyback (per kWh produced locally)
- Utility electricity buyback (per kW produced on demand)
- Retail wheeling income (per kW produced)
- Retail wheeling income (per kW produced on demand)

The cost function is the sum of the listed cost components.

7.4.1.7   Arbitrary Load Shapes

Thermal loads and electrical loads that depend on weather, occupancy, and time of day must be superimposable on the VPP.

References

Kreider, J. F. and Curtiss, P., Distributed electrical generation technologies and methods for their economic assessment, *ASHRAE Trans.*, 106(1), 2000.

Curtiss, P., Control of distributed electrical generation systems, *ASHRAE Trans.*, 106(1), 2000.

Curtiss, P., Cohen, D., and Kreider, J. F., A methodology for technical and financial assessment of distributed generation in the U.S., Proceedings of the ASME ISEC Annual Conference, Maui, April 1999.

8

Economic and Financial Aspects of Distributed Generation

Ari Rabl and Peter Fusaro

CONTENTS

0-8493-0074-6/01/$0.00+$1.50
© 2001 by CRC Press LLC

We do not live in paradise, and our resources are limited. Therefore, it behooves us to try to reduce the costs of energy to a minimum — subject, of course, to the constraint of providing energy demands for an industrial process or maintaining the desired indoor environment and services to a building. In the final analysis, the minimum costs of energy systems drive both design and DG system operation. Although succinctly stated, finding the optimum is subject to uncertainties such as future energy prices, future rental values, future equipment performance, and future and different uses of a building.

This chapter covers three topics: (1) economics; that is, life cycle costs of equipment and energy flows, (2) optimization of initial DG system design and ongoing operation, and (3) financial considerations having to do with buying and selling electricity. Considered first are the basics of engineering economics, sometimes called microeconomics. Costs and benefits of a given DG system design are found using this approach. A handful of equations handle most microeconomic analyses. For those who prefer to omit the algebraic details, charts and graphs of all key parameters are also provided.

8.1   Comparing Present and Future Costs

8.1.1   The Effect of Time on the Value of Money

Before one can compare first costs (i.e., capital costs) and operating costs, one must apply a correction, because a dollar (or any other currency unit) to be paid in the future does not have the same value as a dollar available today. This time dependence of money is due to two quite different causes. The first is inflation, the well known and ever present erosion of the value of our currency. The second reflects the fact that a dollar today can buy goods to be enjoyed immediately or can be invested to increase its value by profit or interest. Thus, a dollar that becomes available in the future is less desirable than a dollar today; its value must therefore be discounted. This is true even without inflation. Both inflation and discounting are characterized in terms of annual rates.

The discussion begins with inflation. To avoid confusion, subscripts have been added to the currency signs, indicating the year in which the currency is specified. For example, during the mid-1980s and 1990s, the inflation rate r_{inf} in Western industrial countries was around $r_{inf} = 4\%$. Thus, a dollar bill in 1999 is worth only $1/(1 + 0.04)$ as much as the same dollar bill one year before.

$$1.00\ \$_{1999} = \frac{1}{1 + r_{inf}}\ \$_{1998} = \frac{1}{1 + 0.04}\ \$_{1998} = 0.96\ \$_{1998}$$

Actually, the definition and measure of the inflation rate are not without ambiguities, since different prices escalate at different rates and an average inflation rate depends on the mix of goods assumed. Probably the most common measure is the consumer price index (CPI), which was arbitrarily set at 100 in 1983. Its evolution is shown in Figure 8.1 along with two specific indices of interest to the DG designers and analysts who measure inflation in the cost of equipment and the cost of construction.

In terms of the CPI, the average inflation rate* from year *ref* to year *ref + n* is given by

$$(1 + r_{inf})^n = \frac{CPI_{ref+n}}{CPI_{ref}} \tag{8.1}$$

Suppose $1.00\ \$_{1995}$ has been invested at an interest rate $r_{int} = 10\%$, the nominal or market rate, as usually quoted by financial institutions. Then, after one year this dollar has grown to $1.10\ \$_{1999}$, but it is worth only $1.10\ \$_{1998}/1.04 = 1.06\ \$_{1998}$. To show the increase in the real value, it is convenient to define the real interest rate r_{int0} by the relation

$$1 + r_{int0} = \frac{1 + r_{int}}{1 + r_{inf}} \tag{8.2}$$

or

$$r_{int0} = \frac{r_{int} - r_{inf}}{1 + r_{inf}}$$

---

* For simplicity, the equations are written as if all growth rates were constant. Otherwise, the factor $(1 + r)^n$ would have to be replaced by the product of factors for each year $(1 + r_1) \dots (1 + r_n)$. Such a generalization is straightforward but tedious and of dubious value in practice as it is chancy enough to predict average trends without trying to guess a detailed scenario.

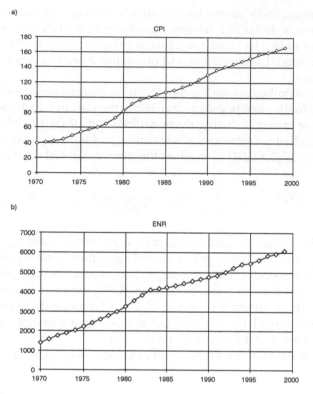

FIGURE 8.1
Various cost indices.

The simplest way of dealing with inflation is to eliminate it from the analysis right at the start by using so-called *constant currency* and expressing all growth rates (interest, energy price escalation, etc.) as real rates net of inflation, relative to constant currency. After all, one is concerned about the real value of cash flows, not about their nominal values in a currency eroded by inflation. Constant currency is obtained by expressing the *current* or *inflating* currency of each year (i.e., the nominal value of the currency) in terms of equivalent currency of an arbitrarily chosen reference year *ref*. Thus the current dollar of year *ref* + *n* has a constant dollar value of

$$\$_{ref} = \frac{\$_{ref+n}}{(1 + r_{inf})^n} \tag{8.3}$$

A *real growth rate* r_0 is related to the *nominal growth rate* r in a way analogous to Eq. 8.2:

$$r_0 = \frac{r - r_{inf}}{1 + r_{inf}} \tag{8.4}$$

For low inflation rates one can use the approximation

$$r_0 \approx r - r_{inf} \tag{8.5}$$

if r_{inf} is small. The constant dollar approach offers several advantages. Having one variable less, it is simpler and clearer. More importantly, the long term trends of real growth rates are fairly well known, even if the inflation rate turns out to be erratic. For example, from 1955 to 1980, the real interest rate on high quality corporate bonds has consistently hovered around 2.2% despite large fluctuations of inflation (Jones, 1982), while the high real interest rates of the 1980s were a short term anomaly. Riskier investments such as the stock market may promise higher returns, but they, too, tend to be more constant in constant currency.

Likewise, prices tend to be more constant when stated in terms of real currency. For example, the market price (price in inflating currency) of crude oil reached a peak of $36 in 1981, ten times higher than the market price during the 1960s, while in terms of constant currency, the price increase over the same period was only a factor of four. Crude oil during an oil crisis is, of course, an example of extreme price fluctuation. For other goods, the price in constant currency is far more stable (it would be exactly constant in the absence of relative price shifts among different goods). Therefore, it is instructive to think in terms of real rates and real currency.

8.1.2 Discounting of Future Cash Flows

As mentioned above, even if there were no inflation, a future cash amount F is not equal to its present value P; it must be discounted. The relationship between P and its future value F_n n years from now is given by the discount rate r_d, defined such that

$$P = \frac{F_n}{(1 + r_d)^n} \tag{8.6}$$

The greater the discount rate, the smaller the present value of future transactions. To determine the appropriate value of the discount rate, one has to ask at what value of r_d one is indifferent between an amount P today and an amount $F_n = P/(1 + r_d)^n$ a year from now. That depends on the circumstances and on individual preferences. Consider a consumer who would put her money in a savings account with 5% interest. Her discount rate is 5%, because by putting $1000 into this account, she, in fact, accepts the alternative of $(1 + 5\%) \times \$1000$ a year from now. If, instead, she would use that money to pay off a car loan at 10%, then her discount rate would be 10%; paying off the loan is like putting the money into a savings account which pays at the loan interest rate. If the money would allow her to avoid an emergency loan at 20%, then her discount rate would be 20%.

The situation becomes more complex when there are several different investment possibilities offering different returns at different risks, such as savings accounts, stocks, real estate, or a new business venture. By and large, if one wants the prospect of a higher rate of return, one has to accept a higher risk. Thus, a more general rule would state that the appropriate discount rate for the analysis of an investment is the rate of return on alternative investments of comparable risk. In practice, that is sometimes quite difficult to determine, and it may be desirable to have an evaluation criterion that bypasses the need to choose a discount rate. Such a criterion is obtained by calculating the profitability of an investment in terms of an unspecified discount rate and then solving for the value of the rate at which the profitability goes to zero. That method, called *internal rate of return method*, will be explained later.

Just as with other growth rates, one can specify the discount rate with or without inflation. If F_n is given in terms of constant currency, designated as F_{n0}, then it must be discounted with the real discount rate r_{d0}. The latter is, of course, related to the market discount rate r_d by

$$r_{d0} = \frac{r_d - r_{inf}}{1 + r_{inf}} \tag{8.7}$$

according to Eq. 8.4. Present values can be calculated with real rates and real currency or with market rates and inflating currency; the result is readily seen to be the same because multiplying the numerator and denominator of Eq. 8.6 by $(1 + r_{inf})_n$ yields

$$P = \frac{F_n}{(1 + r_d)^n} = \frac{F_n(1 + r_{inf})^n}{(1 + r_{inf})^n(1 + r_d)^n}$$

which is equal to

$$P = \frac{F_{n0}}{(1 + r_{d0})^n}$$

since

$$F_{n0} = \frac{F_n}{(1 + r_{inf})^n} \tag{8.8}$$

according to Eq. 8.3.

The ratio P/F_n of present and future value is called the present worth factor, which is designated here with the mnemonic notation

$$(P/F,r,n) = P/F_n = (1 + r)^{-n} \tag{8.9}$$

It is plotted in Figure 8.2. Its inverse

$$(F/P, r, n) = \frac{1}{(P/F, r, n)} \tag{8.10}$$

is called the compound amount factor. These factors are the basic tool for comparing cash flows at different times. Note that the so-called end-of-year convention has been chosen here by designating F_n as the value at the end of the nth year. Also, annual intervals have been assumed, generally an adequate time step for engineering economic analysis; accountants, by contrast, tend to work with monthly intervals, corresponding to the way most regular bills are paid. The basic formulas are the same, but the numerical results differ slightly because of differences in the compounding of interest; this point will be explained more fully later when we pass to the continuous limit by letting the time step approach zero.

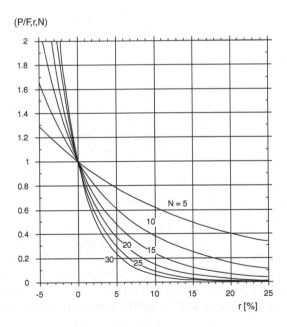

FIGURE 8.2
The present worth factor (P/F,r,N) as function of rate r and number of years N.

Example

What might be an appropriate discount rate for analyzing the energy savings from a proposed new DG-based cogeneration plant for a university campus? Consider the fact that from 1970 to 1988, the endowment of the university has grown by a factor of 8 (current dollars) due to profits from investments.

GIVEN
growth factor in current dollars = 8.0 and increase in CPI = 118.3/38.9 = 3.04 from Figure 8.1 over $N = 18$ years

FIND
real discount rate r_{d0}

Solution

There are two equivalent ways of solving for r_{d0}. The first is to take the real growth factor, 8.0/3.04, and set it equal to $(1 + r_{d0})^N$.

The result is $r_{d0} = 5.52\%$. The second method is to calculate the market rate r_d by setting the market growth in current dollars equal to $(1 + r_d)^N$ and calculating the inflation by setting the CPI increase equal to $(1 + r_{inf})^N$. One finds that $r_d = 12.246\%$ and $r_{inf} = 6.371\%$.

Then Eq. 8.7 can be solved for r_{d0}, with the result

$$r_{d0} = \frac{0.12246 - 0.06371}{1 + 0.06371} = 5.52\%$$

the same as before.

COMMENTS
Choosing a discount rate is not without pitfalls. For the present example, the comparison with the real growth of other long term investments seems appropriate; of course, there is no guarantee that the endowment will continue growing at the same real rate in the future.

8.1.3   Equivalent Cash Flows and Levelizing

It is convenient to express a series of payments that are irregular or variable as equivalent equal payments in regular intervals; in other words, one replaces nonuniform series by equivalent uniform or level series. This technique is referred to as levelizing. It is useful because regularity facilitates understanding and planning. To develop the formulas, one must calculate the present value P of a series of N equal annual payments A. If the first payment occurs at the end of the first year, its present value is $A/(1 + r_d)$. For the second year it is $A/(1 + r_d)^2$, etc. Adding all the present values from year 1 to N gives the total present value

$$P = \frac{A}{1+r_d} + \frac{A}{(1+r_d)^2} + \ldots + \frac{A}{(1+r_d)^N} \qquad (8.11)$$

This is a simple geometric series, and the result is readily summed to

$$P = A\frac{1-(1+r_d)^{-N}}{r_d} \quad for \ \ r_d \neq 0 \qquad (8.12)$$

For zero discount rate this equation is indeterminate, but its limit $r_d \to 0$ is A/N, reflecting the fact that the N present values all become equal to A in that case. Analogous to the notation for the present worth factor, the ratio of A and P is designated by

$$A/(P, r_d, N) = \begin{cases} \dfrac{r_d}{1-(1+r_d)^{-N}} & for \ \ r_d \neq 0 \\[2mm] \dfrac{1}{N} & for \ \ r_d = 0 \end{cases} \qquad (8.13)$$

This is called the capital recovery factor and is plotted in Figure 8.3. For the limit of long life, it is worth noting that $(A/P,r_d,N) \to r_d$ if $r_d > 0$. The inverse is known as the series present worth factor since P is the present value of a series of equal payments A.

With the help of the present worth factor and capital recovery factor, any single expense C_n that occurs in year n, for instance, a major repair, can be expressed as an equivalent annual expense A that is constant during each of the N years of the life of the system. The present value of C_n is $P = (P/F,r_d,n)$ Cn and the corresponding annual cost is

$$A = (A/P, r_d, N)(P/F, r_d, n)C_n$$
$$= \frac{r_d}{1-(1+r_d)^{-N}}(1+r_d)^{-n}C_n \qquad (8.14)$$

A very important application of the capital recovery factor is the calculation of loan payments. In principle, a loan can be repaid according to any arbitrary schedule, but in practice, the most common arrangement is based on constant payments in regular intervals. The portion of A due to interest varies, but to find the relationship between A and the loan amount L, one need not worry about that. First consider a loan of amount L_n that is to be repaid with a single payment F_n at the end of n years. With n years of interest at loan interest rate r_l, the payment must be

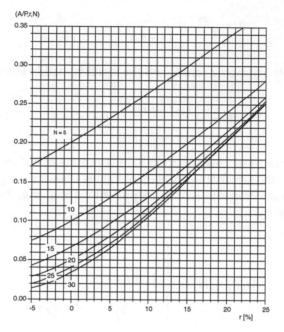

FIGURE 8.3
The capital recovery factor $(A/P,r,N)$ as function of rate r and number of years N.

$$F_n = L_n (1 + r_l)^n$$

Comparison with the present worth factor shows that the loan amount is the present value of the future payment F_n, discounted at the loan interest rate.

A loan that is to be repaid in N equal installments can be considered as the sum of N loans, the nth loan to be repaid in a single installment A at the end of the nth year. Discounting each of these payments at the loan interest rate and adding them gives the total present value, which is equal to the total loan amount

$$L = P = \frac{A}{1 + r_l} + \frac{A}{(1 + r_l)^2} + \dots + \frac{A}{(1 + r_l)^N} \qquad (8.15)$$

This is just the series of the capital recovery factor. Hence, the relationship between annual loan payment A and loan amount L is

$$A = (A/P,r_l,N)\, L \qquad (8.16)$$

Now the reason for the name capital recovery factor becomes clear: it is the rate at which a bank recovers its investment in a loan.

Some payments increase or decrease at a constant annual rate. It is convenient to replace a growing or diminishing cost with an equivalent constant or levelized cost. Suppose the price of energy is p_e at the start of the first year, escalating at an annual rate r_e while the discount rate is r_d. If the annual energy consumption Q is constant, then the present value of all the energy bills during the N years of system life is

$$P_e = Qp_e\left\{\left(\frac{1+r_e}{1+r_d}\right)^1 + \left(\frac{1+r_e}{1+r_d}\right)^2 + \ldots + \left(\frac{1+r_e}{1+r_d}\right)^N\right\} \qquad (8.17)$$

assuming the end-of-year convention described above. As in Eq. 8.3, a new variable $r_{d,e}$ is introduced and defined by

$$1 + r_{d,e} = \frac{1+r_d}{1+r_e} \qquad (8.18)$$

or

$$r_{d,e} = \frac{r_d - r_e}{1+r_e} \quad (\approx r_d - r_e \text{ if } r_e \ll 1), \qquad (8.19)$$

which allows writing P_e as

$$P_e = (P/A, r_{d,e}, N)\, Q\, p_e \qquad (8.20)$$

Since $(A/P, r_d, N)$ is the inverse of $(A/P, r_d, N)$, this can be written as

$$P_e = (P/A, r_d, N) Q \left[\frac{(A/P, r_d, N)}{(A/P, r_{d,e}, N)} p_e\right]$$

If the quantity in brackets were the price, this would be just the formula without escalation. Let us call this quantity the *levelized* energy price $\bar{p}_e$:

$$\bar{p}_e = \frac{(A/P, r_d, N)}{(A/P, r_{d,e}, N)} p_e \qquad (8.21)$$

This allows calculation of the costs as if there were no escalation. Levelized quantities can fill a gap in our intuition which is ill prepared to gauge the effects of exponential growth over an extended period. The levelizing factor

$$levelizing\ factor = \frac{(A/P, r_d, N)}{(A/P, r_{d,e}, N)} \qquad (8.22)$$

tells us, in effect, the average of a quantity that changes exponentially at a rate r_e while being discounted at a rate r_d, over a lifetime of N years. It is plotted in Figure 8.4 for a wide range of the parameters.

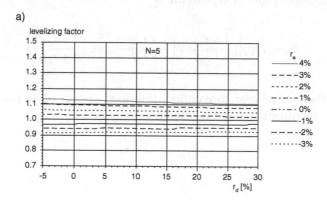

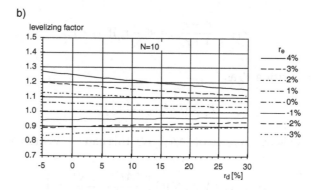

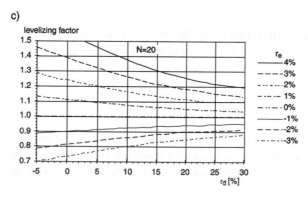

FIGURE 8.4
Levelizing factor $(A/P, r_d, N)/(A/P, r_{d,e}, N)$ as function of r_d and $r_{d,e}$. (a) $N = 5$ yr, (b) $N = 10$ yr, (c) $N = 20$ yr.

Several features may be noted in Figure 8.4. First, the levelizing factor increases with cost escalation r_e, being unity if $r_e = 0$. Second, for a given escalation rate, the levelizing factor decreases as the discount rate increases, reflecting the fact that a high discount rate de-emphasizes the influence of high costs in the future.

8.1.4 Discrete and Continuous Cash Flows

The above formulas suppose that all costs and revenues occur in discrete intervals. That is common engineering practice, in accord with the fact that bills are paid in discrete installments. Thus, growth rates are quoted as annual changes even if growth is continuous. It is instructive to consider the continuous case.

Let us establish the connection between continuous and discrete growth by way of an apocryphal story about the discovery of e, the basis of natural logarithms. Before the days of compound interest, a mathematician who was an inveterate penny pincher thought about the possibilities of increasing the interest he earned on his money. He realized that if the bank gives interest at a rate of r per year, he could get even more by taking the money out after half a year and reinvesting it to earn interest on the interest as well. With m such compounding intervals per year, the money would grow by a factor

$$(1 + r/m)^m$$

and the larger the m, the larger this factor. Of course, he looked at the limit $m \rightarrow \infty$ and found the result

$$\lim_{m \rightarrow \infty} \left(1 + \frac{r}{m}\right)^m = e^r \qquad (8.23)$$

with $e = 2.71828$.

At the end of one year, the growth factor is $(1 + r_{ann})$ with annual compounding at a rate r_{ann}, while with continuous compounding at a rate r_{cont}, the growth factor is $exp(r_{cont})$. If the two growth factors are to be the same, the growth rates must be related by

$$1 + r_{ann} = exp(r_{cont}) \qquad (8.24)$$

Compounding with m compounding intervals at rate r_m is equivalent to annual compounding if one takes

$$1 + r_{ann} = (1 + r_m/m)^m \qquad (8.25)$$

8.1.5   The Rule of Seventy for Doubling Times

Most of us do not have a good intuition for exponential growth. As a helpful tool, the rule of seventy for doubling times is, therefore, presented here. The doubling time N2 is related to the continuous growth rate r_{cont} by

$$2 = exp(N_2\, r_{cont}) \qquad (8.26)$$

This shows that the product of doubling time and growth rate in units of percent is very close to 70 years. Alternatively,

$$N_2 = \frac{ln(2)}{r_{cont}} = \frac{0.693\ldots}{r_{cont}} \qquad (8.27)$$

8.2   The Life Cycle Cost

Having developed the previous tools for discounted cash flow (DCF) analysis, we are ready to use them to find the life cycle costs (LCCs) of DG system investments. There are a number of terms involved in the final expression for the LCC. These will each be discussed in the following pages.

8.2.1   Cost Components

A rational decision is based on the true total cost. That is the sum of the present values of all cost components, and it is called life cycle cost. The cost components relevant to DG analyses are capital cost (total initial investment) net of tax credits; energy costs, for example, gas fuel for microturbines; costs for maintenance, including major repairs; resale value; insurance; and taxes.

There is some arbitrariness in this assignment of categories. One could make a separate category for repairs, or one could include energy among operation and maintenance (O&M) cost as is done in some industries. There is, however, a good reason for keeping energy apart. In DG analyses, energy costs dominate O&M costs and can grow at a different rate. Furthermore, electric rates usually contain charges for peak demand in addition to charges for energy. As a general rule, if an item is important, it merits separate treatment.

Quite generally, when comparing two or more options, there is no need to include terms that would be the same for each. For instance, when choosing between two microturbine manufacturers, one can restrict one's attention to the costs associated directly with the turbines (capital cost, energy, maintenance) without worrying about the electrical distribution system if that is not affected.

Finally, in some cases, it becomes necessary to account for the effects of taxes, due to tax deductions for interest payments and depreciation; hence, these items are discussed first, before the equation for the complete system cost is presented.

8.2.2 Principal and Interest

In the U.S., interest payments are deductible from income tax, while principal payments for the reimbursement of a loan are not. A tax paying investor, therefore, needs to know what fraction of a loan payment is due to interest. The present value P_{int} of the total interest payments is found by discounting each year's interest payment. It can be shown that P_{int} is given by

$$P_{int} = \sum_{n=1}^{N_l} \frac{1 - (1 + r_l)^{n-1-N_l}}{(1 + r_d)^n} \, (A/P, r_l, N_l) \, L \tag{8.28}$$

in which L is the loan amount, r_l is the loan interest rate and N_l is the loan period in years.

Using the formula for geometric series, this can be transformed to

$$P_{int} = \left\{ \frac{(A/P, r_l, N_l)}{(A/P, r_d, N_l)} - \frac{(A/P, r_l, N) - r_l}{(1 + r_l)(A/P, r_d, N_l)} \right\} L \tag{8.29}$$

with

$$r_{dl} = (r_d - r_l)/(1 + r_l)$$

If the incremental tax rate is τ, the total tax payments are reduced by τP_{int} (assuming a constant tax rate; otherwise, the tax rate would have to be included in the summation).

8.2.3 Depreciation and Tax Credit

U.S. tax law allows business property to be depreciated. This means that for tax purposes, the value of the property is assumed to decrease by a certain amount each year, and this decrease is treated as a tax deductible loss. For economic analysis, one needs to express the depreciation as an equivalent present value. The details of the depreciation schedule have been changing with the tax reforms of the 1980s and 1990s. Instead of trying to present the full details, which can be found in the publications of the Internal Revenue Service, we merely note the general features. In any year n, a certain fraction

$f_{dep,n}$ of the capital cost (minus salvage value) can be depreciated. For example, in the simple case of straight line depreciation over N_{dep} years,

$$f_{dep,n} = 1/N_{dep} \qquad (8.30)$$

for straight line depreciation. To obtain the total present value, one multiplies by the present worth factor and sums over all years from 1 to N:

$$f_{dep} = \sum_{n=1}^{N_{dep}} f_{dep,n}(P/F, r_d, n) \qquad (8.31)$$

For straight line depreciation, the sum is

$$f_{dep} = \frac{(P/A, r_d, N_{dep})}{N_{dep}} \qquad (8.32)$$

for straight line depreciation. A further feature of some tax laws is the tax credit. For instance, in the U.S. for several years around 1980, federal tax credits were granted for certain renewable energy systems. Today, many states have tax credits for DG and renewable energy systems. If the tax credit rate is τ_{cred} for an investment C_{cap}, the tax liability is reduced by $\tau_{cred} C_{cap}$.

Example
A fuel cell system costs $100,000 and is depreciated with straight line depreciation over 5 years, the salvage value after 5 years being $10,000. Find the present value of the tax deduction for depreciation if the incremental tax rate is $\tau = 40\%$, and the discount rate r_d is 15%.

GIVEN

$$\begin{aligned}
C_{cap} &= 100 \text{ k\$} \\
C_{salv} &= 10 \text{ k\$} \\
N &= 5 \text{ yr} \\
\tau &= 0.4 \\
r_d &= 0.15
\end{aligned}$$

FIND

$$\tau \times f_{dep} \times (C_{cap} - C_{salv})$$

LOOKUP VALUES

$$f_{dep} = \frac{(P/A, 0.15, 5 \text{ yr})}{5 \text{ yr}} = \frac{(1/0.2983)}{5} = 3.3523/5 = 0.6705,$$

from Eq. 8.32.

Solution

For tax purposes, the net amount to be depreciated is the difference

$$C_{cap} - C_{salv} = (10 - 1) \text{ k\$} = 9 \text{ k\$}$$

and with straight line depreciation, $1/N_{dep} = 1/5$ of this can be deducted from the tax each year. Thus, the annual tax is reduced by

$$\tau \times (1/5) \times 90 \text{ k\$} = 0.40 \times 18 \text{ k\$} = 7.2 \text{ k\$}$$

for each of the five years.
The present value of this tax reduction is

$$\tau \times f_{dep} \times (C_{cap} - C_{salv}) = 0.40 \times 0.6705 \times 90 \text{ k\$} = 24.1 \text{ k\$}$$

COMMENT

The present value of the reduction would be equal to 5×7.2 k\$ = 36 k\$ if the r_d were zero. The discount rate of 15% reduces the present value by almost a third to $0.6705 = f_{dep}$.

8.2.4 Demand Charges

The cost of producing electricity has two major components: fuel and capital (for power plant and distribution system). As a consequence, the cost of electricity varies with the total load on the grid. To the extent that it is practical, deregulated electric utilities will try to base the rate schedule on their production cost. Even without full deregulation, rates for large customers contain two items: one part of the bill is proportional to the energy and another part is proportional to the peak demand.* If the monthly demand charge is p_{dem} and the energy charge is p_e, a customer with monthly energy consumption Q_m and peak demand P_{max} will receive a total bill of

$$monthly \ bill = Q_m \ p_e + P_{max} \ p_{dem} \tag{8.33}$$

There are many small variations from one utility company to another. In most cases p_e and p_{dem} depend on time of day and time of year, being higher during the system peak than off-peak. In regions with extensive air conditioning, the system peak occurs in the afternoon of the hottest days. In regions with much electric heating, the peak is correlated with outdoor temperature. Some companies use what is called a ratcheted demand charge; it has the effect of basing the demand charge on the annual, rather than monthly, peak.

---

* Of course, utility bills can contain up to several dozen different charges, but two of the largest are for energy and demand if real-time pricing is not used. For true RTP rates, the energy and demand charges are combined into one energy-based charge.

With the advent of deregulated utilities, the pricing of electricity will be simpler. Bids at a power exchange will include both energy and demand charges. A separate auction will be used for ancillary services such as voltage support, VAR support, black start, and spinning reserve. The operation of these markets is the subject of the last section of this chapter dealing with financial dimensions of DG power.

8.2.5 The Complete Cost Equation

The total cost of producing electricity with a DG system is developed in this section. The accounting can be done before or after taxes; the former counts the cash payments, the latter their net (after-tax) values. The two modes differ by a factor $(1 - \tau)$ where τ is the income tax rate. For example, if the market price of fuel is 5\$/GJ and the tax rate (federal plus state) $\tau = 40\%$, then the before-tax cost of fuel is 5\$/GJ and the after-tax cost $(1 - \tau) \times 5\$/GJ = 3\$/GJ$. Stated in terms of after-tax cost, the complete equation for the life cycle cost of a DG investment can be written in the form:

$$C_{life} =$$

$$C_{cap}\{(1 - f_l) \qquad\qquad \text{down payment}$$

$$+ f_l \frac{(A/P, r_l, N_l)}{(A/P, r_d, N_l)} \qquad\qquad \text{cost of loan}$$

$$- r f_l \left[\frac{(A/P, r_l, N_l)}{(A/P, r_d, N_l)} - \frac{(A/P, r_l, N_l - r_l)}{(1 + r_l)(A/P, r_{d,l}, N_l)}\right] \qquad \text{tax deduction for interest}$$

$$- \tau_{cred} \qquad\qquad \text{tax credit}$$

$$- \tau f_{dep}\} \qquad\qquad \text{depreciation}$$

$$- C_{salv}\left(\frac{1 + r_{inf}}{1 + r_d}\right)^N (1 - t) \qquad\qquad \text{salvage}$$

$$+ Q p_e \frac{1 - t}{(A/P, r_{d,e}, N)} \qquad\qquad \text{cost of energy}$$

$$+ P_{max} p_{dem} \frac{1 - t}{(A/P, r_{d,dem}, N)}\Bigg\} \qquad\qquad \text{cost of demand}$$

$$+ A_M \frac{1 - t}{(A/P, r_{d,M}, N)}\Bigg\} \qquad\qquad \text{cost of maintenance} \quad (8.34)$$

where

A_M = annual cost for maintenance (in first year $)
C_{cap} = capital cost (in first year $)
C_{salv} = salvage value (in first year $)
f_{dep} = present value of depreciation, as fraction of C_{cap}
f_l = fraction of investment paid by loan
N = system life (yr)
N_l = loan period (yr)
p_e = energy price (in first year $/GJ)
Q = annual energy consumption (GJ)
r_d = market discount rate
r_e = market energy price escalation rate
$r_{d,e}$ = $(r_d - r_e)/(1 + r_e)$
r_{dem} = market demand charge escalation rate
$r_{d,dem}$ = $(r_d - r_{dem})/(1 + r_{dem})$
r_{inf} = general inflation rate
r_l = market loan interest rate
$r_{d,l}$ = $(r_d - r_l)/(1 + r_l)$
r_M = market escalation rate for maintenance costs
$r_{d,M}$ = $(r_d - r_M)/(1 + r_M)$
τ = incremental tax rate
τ_{cred} = tax credit

If there are several forms of energy involved, e.g., gas and electricity, the term $Q\,p_e$ is to be replaced by a sum over the individual energy terms. Many other variations and complications are possible: for instance, the salvage tax rate could be different from τ.

Example

Find the life cycle cost of operating a 100 ton chiller (COP = 3) under the following conditions. This calculation would be the first step in analyzing if an alternative absorption chiller powered by the exhaust heat of an SO fuel cell would be feasible.

GIVEN
    system life N = 20 yr
    loan life N_l = 10 yr
    depreciation period N_{dep} = 10 yr, straightline depreciation
    discount rate r_d = 0.15
    loan interest rate r_l = 0.15
    energy escalation rate r_e = 0.01
    demand charge escalation rate r_{dem} = 0.01
    maintenance cost escalation rate r_M = 0.01

inflation $r_{inf} = 0.04$
loan fraction $f_l = 0.7$
tax rate $\tau = 0.5$
tax credit rate $\tau_{cred} = 0$
capital cost (at \$400/ton) $C_{cap} = 40$ k\$
salvage value $C_{salv} = 0$
annual cost of maintenance $A_M = 0.8$ k\$/yr (= 2% of C_{cap})
capacity *100 ton* = 351.6 kW$_t$
peak electric demand *351.6 kW$_t$/COP* = 117.2 kW$_e$
annual energy consumption Q = *100 kton.h* = 351.6 MWh$_t$
electric energy price p_e = 10 cents/kWh$_e$ = 100 \$/MWh$_e$
demand charge p_{dem} = 10 \$/kWe.month, effective during 6 months of
the year
The rates are market rates.

DISCOUNTED CASH FLOW PARAMETER VALUES

$r_{d,l} = 0.0000$ $(A/P,r_l,N_l) = 0.1993$
$r_{d,e} = 0.1386$ $(A/P,r_d,N_l) = 0.1993$
$r_{d,dem} = 0.1386$ $(A/P,r_{d,l},N_l) = 0.1000$
$r_{d,M} = 0.1386$ $(A/P,r_d,N) = 0.1598$
                         $(A/P,r_{d,e},N) = 0.1498$
$(1 + r_{inf})/(1 + r_d) = 0.9043$ $(A/P,r_{d,dem},N) = 0.1498$
$f_{dep} = 0.502$ from Eq. 8.32 $(A/P,r_{d,M},N) = 0.1498$

Solution
Components of C_{life} [all in k\$] from Eq. 8.34:

down payment	12.0
cost of loan	28.0
tax deduction for interest	–8.0
tax credit	0.0
depreciation	–10.0
salvage value	0.0
cost of energy	39.1
cost of demand charge	23.5
cost of maintenance	2.7
TOTAL = C_{life}	87.3

From this example, one can observe that (1) when examining various options for DG, a spreadsheet is recommended, and (2) the cost of energy and demand is higher than the capital cost. This is why one might consider DG-produced cogeneration as an alternative to mechanical cooling.

8.2.6 Cost per Unit of Delivered Electrical Energy

The marketing of DG-produced power (see the final section of this chapter) requires that the cost of power be known. After using Eq. 8.34 to find the discounted cash flow, it is an easy matter to find the power cost. This cost is the basis for bids in power exchange sales of DG-produced power. If a power exchange is not involved but control decisions for use of DG are made locally, one still must know the cost of local power for comparison with grid power.

The cost of on-site power is calculated as the ratio of levelized annual cost and the annual delivered good, namely, electricity. The levelized annual cost is obtained by multiplying the life cycle cost from Eq. 8.34 by the capital recovery factor for discount rate and system life. There appear two possibilities: the real discount rate r_{d0} and the market discount rate r_d. The quantity $(A/P,r_{d0},N)\ C_{life}$ is the annual cost in constant dollars (of the initial year), whereas $(A/P,r_d,N)\ C_{life}$ is the annual cost in inflating dollars. The latter is difficult to interpret because it is an average over dollars of different real value. Therefore, we levelize with the real discount rate because it expresses everything in first year dollars, consistent with the currency of C_{life}. Thus, we write the annual cost in initial dollars as

$$A_{life} = (A/P,r_{d0},N)\ C_{life} \tag{8.35}$$

The effective total cost per delivered energy is therefore

$$\textit{effective cost per energy} = A_{life}/Q \tag{8.36}$$

where Q equals the annual delivered electricity in units of kWh/yr (assumed constant, for simplicity).

The reader may wonder why C_{life} is not simply divided by the total electricity (NQ) delivered by the system over its lifetime. That would not be consistent because C_{life} is the present value, while (NQ) contains goods (and, thus, monetary values) that are associated with future times. One must allocate goods and costs within the same time frame. That is accomplished by dividing the levelized annual cost by the levelized annual good — the latter is equal to Q because we have assumed that the consumption is constant from year to year.

Example

What is the cost per kWh for a 30% efficient, 50 kW microturbine that consumes natural gas costing $3.00 per million Btu (0.010 $/kWh)? The turbine operates 8000 hr/yr.

GIVEN
Using Eq. 8.34, C_{life} is found to be 5.5 k$/yr and $Q = 400{,}000$ kWh/yr.

Solution

The power cost = A_{life}/Q = 0.013 \$/kWh.

The cost of fuel = (1kWh gas/0.3kWh electricity)(\$0.01/kWh gas)

= \$0.033/kWh

Finally, the total of capital-based and fuel costs is

$$C_{energy} = 0.013 + 0.333 = \$0.046/kWh$$

This cost is that at the DG busbar. Transmission costs and other adders will increase the price delivered to an off-site load.

8.3   Economic Evaluation Criteria

8.3.1   Life Cycle Savings

Having determined the life cycle cost of each relevant DG design alternative, one can select the best, i.e., the one that offers all desirable features at the lowest life cycle cost. Frequently, one takes one design as reference, for example the situation without DG, and considers the difference between it and each alternative DG design. The difference is called life cycle savings relative to the reference case

$$S = - DC_{life} \text{ with } DC_{life} = C_{life} - C_{life,ref} \tag{8.37}$$

Often, the comparison can be quite simple because the only terms that need to be considered are those that are different between the designs under consideration. For simplicity, the equations of this section are written only for an equity investment without tax. Then the loan fraction f_l in Eq. 8.27 is zero and most of the complications of that equation drop out. Of course, the concepts of life cycle savings, internal rate of return, and payback time are perfectly general, and tax and loan can readily be included.

A particularly important case is the comparison of two designs that differ only in capital cost and operating cost: often, the one that saves electricity costs has higher capital + fuel cost (otherwise, the choice would be obvious, without any need for an economic analysis). Setting $f_l = 0$ and $\tau = 0$ in Eq. 8.34 and taking the difference between the two designs, one obtains the life cycle savings as

$$S = \frac{-DQp_e}{(A/P, r_{d,e}, N)} - DC_{cap} \tag{8.38}$$

where

$DQ = Q - Q_{ref}$ = difference in annual energy consumption
$DC_{cap} = C_{cap} - C_{cap,ref}$ = difference in capital cost
$r_{d,e} = (r_d - r_e)/(1 + r_e)$

If the reference design has higher consumption and lower capital cost, DQ is negative and DC_{cap} is positive with this choice of signs.

8.3.2 Internal Rate of Return

Life cycle savings are the true savings if all the input is known correctly and without doubt. But future energy prices and system performance are uncertain, and the choice of the discount rate is not clear-cut. An investment in DG equipment is uncertain, and it must be compared with competing investments that have their own uncertainties. The limitation of the life cycle savings approach can be circumvented if one evaluates the profitability of an investment by itself, expressed as a dimensionless number. Then, one can rank different investments in terms of this index of profitability and in terms of risk. General business experience can serve as a guide for expected profitability as function of risk level. Among investments of comparable risk, the choice can then be based on profitability.

More precisely, the profitability is measured as so-called internal rate of return r_r, defined as that value of the discount rate r_d at which the life cycle savings S are zero:

$$S(r_d) = 0 \text{ at } r_d = r_r \tag{8.39}$$

For an illustration, take the case of Eq. 8.38 with energy escalation rate $r_e = 0$ (so that $r_{d,e} = r_d$), and suppose an extra investment DC_{cap} is made to provide annual energy savings $(-DQ)$. The initial investment DC_{cap} provides an annual income from energy savings:

$$annual\ income = (-DQ)\ p_e \tag{8.40}$$

If DC_{cap} were placed in a savings account instead, bearing interest at a rate r_r, the annual income would be

$$annual\ income = (A/P,r_r,N)\ DC_{cap} \tag{8.41}$$

The investment behaves like a savings account whose interest rate r_r is determined by the equation

$$(A/P,r_r,N)\ C_{cap} = (-DQ)\ p_e \tag{8.42}$$

Dividing by $(A/P,r_r,N)$, we see that the right and left sides correspond to the two terms in Eq. 8.38 for the life cycle savings

$$S = \frac{-DQp_e}{(A/P, r_d, N)} - DC_{cap} \qquad (8.43)$$

and that r_r is, indeed, the discount rate r_d for which the life cycle savings are zero; it is the internal rate of return. Now the reason for the name is clear; the internal rate of return is the profitability of the project by itself, without reference to an externally imposed discount rate. When the explicit form of the capital recovery factor is inserted, one obtains an equation of the Nth degree, generally not solvable in closed form. Instead, one must resort to iterative or graphical solution. (There could be up to N different real solutions, and multiple solutions can indeed occur if there are more than two sign changes in the stream of annual cash flows, but the solution is unique for the case of interest here — an initial investment that brings a stream of annual savings.)

Example

What is the rate of return for an energy conservation system costing an extra $30,000 but which saves $7481? The economic period of analysis is 20 years and the energy escalation rate is 0%.

GIVEN

$$S = \frac{-DQp_e}{(A/P, r_{d,e}, N)} - DC_{cap}$$

with $r_{d,e} = r_d$ (because $r_e = 0$), $N = 20$ yr
$(-DQ_{cap})\, p_e = \$7481$ and $DC = \$30,000$

Solution

$S = 0$ for

$$(A/P, r_r, N) = \frac{-DQp_e}{DC_{cap}} = \frac{7481}{30000} = 0.2494$$

with $N = 20$.

By iteration or using the solver feature of a calculator, one can readily find

$r_r = 0.246$ (24.6%)

8.3.3 Payback Time

The payback time N_p is defined as the ratio of extra capital cost DC_{cap} over first year savings:

$$N_p = \frac{DC_{cap}}{first\ year\ savings} \tag{8.44}$$

The inverse of N_p is sometimes called *return on investment*. If one neglects discounting, one can say that after N_p years, the investment has paid for itself and any revenue thereafter is pure gain. The shorter the N_p, the higher the profitability. As selection criterion, the payback time is simple, intuitive, and obviously wrong because it neglects some of the relevant variables. There has been no lack of attempts to correct for that by constructing variants such as a discounted payback time (in contrast to which Eq. 8.44 is sometimes called simple payback time), but the resulting expressions become so complicated that one might as well work directly with life cycle savings or internal rate of return.

The simplicity of the simple payback time is, however, irresistible. When investments are comparable to each other in terms of duration and function, the payback time can give an approximate ranking that is sometimes clear enough to discard certain alternatives right from the start, thus avoiding the effort of detailed evaluation.

To justify the use of the payback time, let us recall Eq. 8.42 for the internal rate of return and note that it can be written in the form

$$(A/P,r_r,N) = 1/N_p \text{ , or } (P/A,r_r,N) = N_p \tag{8.45}$$

The rate of return is uniquely determined by the payback time N_p and the system life N. This equation implies a simple graphical solution for finding the rate of return if one plots $(P/A,r_r,N)$ on the x-axis vs. r_r on the y-axis as in Figure 8.5. Given N and N_p, one simply looks for the intersection of the line $x = N_p$ (i.e., the vertical line through $x = N_p$) with the curve labeled N; the ordinate (y-axis) of the intersection is the rate of return r_r. This graphical method can be generalized to the case where the annual savings change at a constant rate r_e.

Example

Find the payback time for the previous example and check the rate of return graphically.

GIVEN

> first year savings $(-DQ)$ p_e = $7481
>
> N = 20 yr
>
> extra investment DC_{cap} = $30,000

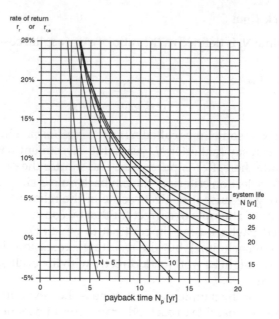

FIGURE 8.5

Relationship between rate of return r_r, system life N, and payback time N_p. If r_e = escalation rate of annual savings, the vertical axis is the variable $r_{r,e}$ from which r_r is obtained as $r_r = r_{r,e} (1 + r_e) + r_e$.

Solution

$N_p = 30000/7481 = 4.01$ yr

It is independent of r_e. Then, $r_r = 0.246$ for $r_e = 0$, from Figure 8.5, and $r_r = 0.271$ for $r_e = 2\%$.

Generally, a real (i.e., corrected for inflation) rate of return above 10% can be considered excellent if there is low risk — a look at savings accounts, bonds, and stocks shows that it is difficult to find better. From the graph, we see immediately that $r_{r,e}$ is above 10% if the payback time is shorter than 8.5 yr (6 yr), for a system life of 20 yr (10 yr). And $r_{r,e}$ is close to the real rate of return if the annual savings growth is close to the general inflation rate.

8.4   Optimization

Optimization in the context of this book means selecting the DG system configuration that maximizes financial benefit to its owner. In principle, the process of optimizing the design of a distributed generation system for a building or campus of buildings is simple: evaluate all possible design variations and select the one with the largest life cycle benefit to the system owner. Who would not want to choose the optimum? In practice, it would be a daunting task to find the true optimum among all conceivable designs. The difficulties, some of which have already been discussed, are

- The enormous number of possible design variations (DG system types, building configurations, electrical use systems, types and models of HVAC equipment, and control modes)
- Uncertainties (costs, future energy prices, reliability, occupant behavior, and future uses of buildings)

Fortunately, there is a certain tolerance for moderate errors, as shown below. That greatly facilitates the job, because one can reduce the number of steps in the search for the optimum. Also, within narrow ranges, some variables can be suboptimized without worrying about their effect on others.

Some quantities are easier to optimize than others. The optimal controller example in Chapter 7 looked at various combinations of microturbines for several buildings. The energy consumption system was assumed to be fixed and only the optimal operation of the DG system was examined. Presumably, the building designer and DG system designer had already selected the optimal configuration for the example buildings. It was the job of the optimal controller to maximize savings given the hand that was dealt, i.e., the already existing design.

8.4.1 A Simple Example

It is instructive to illustrate the optimization process with a very simple example: the thickness of insulation in a building wall. The annual heat flow Q across the insulation is

$$Q = A \, k \, D/t \tag{8.48}$$

where

A = area (m^2)
k = thermal conductivity (W/m.K)
D = annual degree-seconds (K·s)
t = thickness of insulation (m)

The capital cost of the insulation is

$$C_{cap} = A \, t \, p_{ins} \tag{8.49}$$

with p_{ins} = price of insulation ($/m^3). The life cycle cost is

$$C_{life} = C_{cap} + Q \frac{p_e}{(A/P, r_{d,e}, N)} \qquad (8.50)$$

where p_e = first year energy price, and $r_{d,e}$ is related to discount rate and energy escalation rate. We want to vary the thickness t to minimize the life cycle cost, keeping all the other quantities constant. (This model is a simplification that neglects fixed cost of insulation as well as possible feedback of t on D). Eliminating t in favor of C_{cap}, one can rewrite Q as

$$Q = K/C_{cap} \qquad (8.51)$$

with a constant

$$K = A^2 k D p_{ins} \qquad (8.52)$$

Then, the life cycle cost can be written in the form

$$C_{life} = C_{cap} + P K/C_{cap} \qquad (8.53)$$

where the variable

$$P = \frac{p_e}{(A/P, r_{d,e}, N)} \qquad (8.54)$$

contains all the information about energy price and discount rate. K is fixed, and the insulation investment C_{cap} is varied to find the optimum. C_{life} and its components are plotted in Figure 8.6. As t is increased, capital cost increases and energy cost decreases; C_{life} has a minimum at some intermediate value. Setting the derivative of C_{life} with respect to C_{cap} equal to zero yields the optimal value C_{capo}:

$$C_{capo} = \sqrt{KP} \qquad (8.55)$$

It is rare, indeed, when a simple equation is the result of an optimization study. In both DG design and DG system operation, the optimal situation can only be identified reliably using computer simulation tools. For design one needs to jointly simulate loads and DG performance including weather effects with specified economics. The result will be the system design that appears to maximize performance for the system selected for final design. During operation, a similar simulation tool would be used for operation of the system rather than for design.

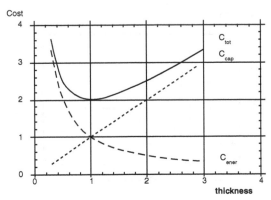

FIGURE 8.6
Optimization of insulation thickness. Insulation cost = C_{cap}, energy cost = C_{ener}.

8.4.2 The Cost of Misoptimization

What is the penalty for not optimizing correctly? In general, the following causes could prevent correct optimization:

- Insufficient accuracy of the algorithm or program for calculating the performance
- Incorrect information on economic data (e.g., the factor P in Eq. 8.55)
- Incorrect information on technical data (e.g., the factor K in Eq. 8.55)
- Unanticipated changes in the use of the building and the resulting changes in electrical load

Misoptimization would produce a different design at a value C_{cap} different from the true optimum C_{capo}. For the example of insulation thickness, the effect on the life cycle cost can be seen directly with the solid curve in Figure 8.6. For example, a ±10% in C_{capo} would increase C_{life} by only +1%. Thus, the penalty is not excessive for small errors for this simple example.

This relatively large insensitivity to misoptimization is a feature much more general than the insulation example. As shown by Rabl (1985), the greatest sensitivity likely to be encountered in practice corresponds to the curve

$$\frac{C_{life,true}(C_{capo,guess})}{C_{life,true}(C_{capo,true})} = \frac{x}{1 + \log(x)} \quad \text{("upper bound")} \quad (8.56)$$

also shown in Figure 8.7 with the label "upper bound." Even here, the minimum is broad; if the true energy price differs by ±10% from the guessed price, the life cycle cost increases only 0.4% to 0.6% over the minimum. Even when the difference in prices is 30%, the life cycle cost penalty is less than 8%.

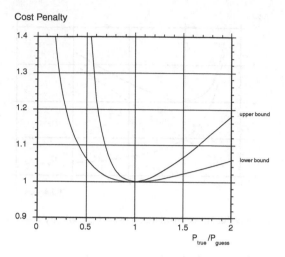

FIGURE 8.7
Life cycle cost penalty versus energy price ratio.

Errors in the factor K (due to wrong information about price or conductivity of the insulation material) can be treated the same way, because K and P play an entirely symmetric role in the above equations. Therefore, curves in Figure 8.7 also apply to uncertainties in other input variables.

The basic phenomenon is universal: any smooth function is flat at an extremum. The only question is, how flat? For energy investments, that question has been answered with the curves of Figure 8.7. We can conclude that misoptimization penalties are definitely less then 1% (10%) when the uncertainties of the input variables are less than 10% (30%).

8.5   Basics of Electrical Energy Financial Transactions

At the end of the day, the decision to use DG or not boils down to one consideration: costs vs. benefits. If benefits are greater, the energy user will consider and adopt a DG system if the DG benefits exceed those of competing power sources. This section describes how bulk electricity produced by significant numbers of installations of DG systems can be marketed in the U.S. DG systems in buildings, for example, will be able to produce significant amounts of power not needed in the host building or building campus. The marketing of such power and ancillary power services into the U.S. power markets will be a key feature of economic feasibility of DG systems. A number of concepts of energy markets are described here in basic terms. These need to be understood by the DG community if a significant market for DG hardware is to exist in the U.S. For more details, the reader is referred to Fusaro (1998).

If DG-produced power is ever to make a difference in the U.S. electricity sector, it must be present in sufficient amounts and must be dispatchable as if it were produced by hundreds of generators rather than hundreds of thousands that will actually produce the power physically. Aggregation of many small generators into a number of large ones involves the virtual power plant concept (VPP) described in Chapter 7. Observe that retail markets in electricity do not exist in the U.S. as of this book's publication. It is unclear if a uniform approach will be adopted nationwide, or if the states will have their own solutions to what is, of course, a multi-state market.

8.5.1 Forward Markets

A forward market for energy develops because there is a time interval between the day a deal is made and the day it matures (i.e., a client buys now for delivery in the future). Oil in the spot market is priced on a transaction-by-transaction basis. For example, oil in the paper market is contracted for physical delivery at a specific date in the future; during this time interval, the contract can be bought and sold over and over again. In this way, forward markets are used to hedge forward physical supply. Forward markets in gas and electricity operate on the same conceptual basis. The success of a forward contract depends on its liquidity and the performance of the market players.

Liquidity is the ease with which the commodity can be bought and sold in the market. *Performance* refers to the ability of the market players to comply with the terms of the contract. The contract must be satisfied through physical delivery or cash settlement at the time of delivery. A wide range of players actively participate in the forward oil markets, including oil traders, major producers, refiners, investment banks, and major oil companies. These parties all provide liquidity to the market by guaranteeing performance.

In the electricity markets, forward and over-the-counter (OTC) swap markets exist, without a futures contract. Trading between forwards and futures in the same or similar commodities is often used to offset positions. However, there are major differences between the two markets. While both use standardized contracts which are traded regularly, futures contracts are traded on organized and regulated exchanges, while forward contracts trade on the off-exchange, or unregulated, markets. Another important difference is that forward markets depend on market makers who, on an informal basis, are expected to perform, but future delivery is not guaranteed. The futures market guarantees performance through a clearinghouse and formal delivery procedure. Also, forward markets often deal in larger trading lots than futures markets. In fact, the ability of the forward markets to absorb larger lots without moving the market represents a significant liquidity advantage over the futures market.

Forward markets are generally used by a relatively small group of well-financed players, in contrast to the wider range of participants in futures. Forwards are similar to other off-exchange instruments in this way and have evolved into price benchmarks that are used in two- to six-month price swaps or OTC options transactions. In natural gas markets, some analysts feel that the major breakthrough that occurred with gas futures trading for the gas industry was the acceptance of the simple concept of selling natural gas forward. In electricity, it is not the concept of selling electricity forward, which has been going on for many years, but the concept that electricity is now a fungible commodity that is changing that industry's frame of reference.

8.5.2 Futures Markets

The introduction of financial futures — futures on currencies and interest rates — during the 1970s transformed the futures markets, which had been trading agricultural commodities for more than 100 years. Financial futures brought new participants and new strategies to the futures markets, and many more types of risk could now be hedged. The commodity concept has broadened to include energy, beginning with heating oil futures in 1978; many of the strategies that were devised for financial futures have been adapted and applied to the energy markets. Energy futures contracts are used by producers, refiners, and consumers to hedge against price fluctuations in these volatile markets. Another function of energy futures contracts has been to protect the inventory value of crude oil, refined products, and natural gas.

The oil futures market developed to allow oil traders to offset some of their risk by taking a position on the futures market opposite their physical position. A producer, who has the physical commodity to sell, hedges by selling futures. The producer's position is then long cash and short futures. A consumer, who needs to buy the commodity, hedges by buying futures. The consumer's position is then short cash and long futures.

A futures contract is an agreement between two parties, a buyer and seller, for delivery of a particular quality and quantity of a commodity at a specified time, place, and price. Futures can be used as a proxy for a transaction in the physical cash market before the actual transaction takes place. The commodity exchanges set margin requirements for hedgers and speculators. The buyer or seller of a futures contract is required to deposit with the clearinghouse a percentage of the value of the contract as a guarantee of contract fulfillment. The margin deposit made when the position is established is called the original margin. Hedge margins generally are lower than speculative margins. At the end of each trading day, each position is marked-to-market, i.e., the margin requirement for each account is adjusted based on the day's price changes. If the value of the position has decreased, the holder of the contract will have to make an additional deposit, called the variation margin. This fulfilling of margin requirements on a daily basis means that the extent of default in the futures market is limited to one day's price change.

Although futures contracts require physical delivery, most do not go to physical delivery because they are settled financially. Therefore, futures should generally be used as a price risk management tool, not as a source of physical supply. This basic concept is still a stumbling block to futures trading, as many potential futures players think that physical delivery is a key component of futures trading, when, in fact, the vast majority of futures deals are settled financially.

The three major energy futures exchanges are the NYMEX, IPE, and SIMEX (Singapore International Monetary Exchange). For electricity and natural gas, only the NYMEX and IPE will be significant. Besides launching two electricity futures contracts on March 29, 1996, the NYMEX launched electricity options one month later on April 26, 1996. An option gives the holder the right, but not the obligation, to buy or sell an underlying instrument, in this case a futures contract, at a specific price within a specified time period. Options can be used to establish floor and ceiling price protection for commodities and offer a risk management alternative to futures contracts.

There are two types of options: puts and calls. A put is an option to sell the futures at a fixed price. A call is an option to buy the futures at a fixed price. For hedgers, a put establishes a minimum selling price but does not eliminate the opportunity to receive higher market prices. A call establishes a maximum buying price but does not eliminate the opportunity to buy at lower market prices. Options contracts can be used in combination with futures to form hedging strategies for exchange-traded commodities. They will be widely used in electricity price risk management.

The NYMEX natural gas futures contract, launched on April 3, 1990, has rapidly established itself as an effective instrument for natural gas price discovery in North America. NYMEX launched natural gas options in October 1992, adding another tool to manage risk in that volatile market. Natural gas options have enhanced liquidity in, and can buffer against, price volatility in the rapidly changing North American gas market. A second natural gas futures contract was launched on August 1, 1995 by the Kansas City Board of Trade, which is oriented to the western U.S. market. Two more NYMEX natural gas contracts were launched during 1996: one for the Permian Basin on May 31, 1996, and one for Canada's Alberta market later in the year. These multiple natural gas contracts represent the regional nature of North American natural gas markets, which will also be the case for electricity futures.

8.5.3 Spread Trading

Spreads are another means to limit price risk in rapidly changing markets. A spread is the simultaneous purchase and sale of futures or options contracts in the same or related markets. Intramarket, intermarket, and interexchange arbitrages are commonly used. Purchasing a contract in one expiration of a futures contract while selling a contract in a different expiration would be an intramarket spread. This strategy would be useful where there are seasonal variations in demand for a commodity.

The objective of a spread trade is to profit from a change in the price differential between the contracts or between futures and options. In natural gas, the wide basis risk of the North American markets has created an active market in spreads trading for different locations, as various locations are priced as a differential to NYMEX. In electricity, it is anticipated that there will be an active market in seasonal spreads and in spreads between contracts covering different delivery points as well as intercommodity spread trading, particularly between natural gas and electricity, called the spark spread.

8.5.4 Exchange for Physical

Because energy futures are primarily financial management tools, deliveries are rare. Most futures contracts are liquidated before expiration. However, some contracts do result in delivery. Because futures contracts are standardized as to location and quality, the holders of contracts for delivery usually need to deviate from the terms of the futures contract. To accommodate this, the exchange-for-physical (EFP) has become the preferred method of delivery because it offers more flexibility. It has become an extension of futures and spot market trading.

In an exchange-for-physical, a long futures position held by the buyer of the commodity is transferred to the seller. The firm chooses its trading partner, the product to be delivered, and the timing of the delivery. An EEP allows buyers and sellers to negotiate a cash market price and to adjust the basis, as well as the exact time the physical exchange takes place. Exchanges-for-physical are an important element in the success of the natural gas futures contract because they add greatly to liquidity.

In electricity, a standardized contract has been developed to facilitate electricity trade at locations other than the futures contracts standard delivery locations of the California–Oregon border and the Palo Verde switchyard.

8.5.5 Price Swap

A price swap is an exchange of cash flows, one at a fixed rate and the other at a floating rate. A buyer and a seller agree to exchange the value of the commodity at a given price, quantity, and time period with no physical commodity actually being exchanged. The swap can be short term (1 to 3 months) or long term (6 months to 30 years) in duration. The contracting parties agree to pay each other the difference between an agreed-upon fixed price and a price index that fluctuates. The price index is an average market price that will reflect the volatility of the market during the term of the agreement. Payments are made at predetermined times, such as monthly, quarterly, or semiannually. The swap provider or market maker, either an oil or gas trading company or financial institution, can match both sides (fixed and floating) of the transaction or assume the price risk itself. Swaps may not require any up front payment. The cash settlement aspect of price swaps allows energy producers and consumers to make the physical transaction with a party of their own choosing yet still receive price protection. Frequently, parties negotiate a premium payable by one party to the other to make the swap palatable.

There are many components of a swap that must be agreed upon before a binding contract is written. Generally, the following elements are needed to structure a commodity swap: the commodity; the quantity to be swapped each month, quarter, or year; the duration of the swap; the fixed price against which the index (market) price will be evaluated to derive the difference; the index basis such as McGraw-Hill's Power Markets Week, Dow Jones Markets, Gas Daily, NYMEX, or IPE; and settlement procedures regarding payment including creditworthiness standards.

Two parties enter into a swap contract because they have different price expectations. One party agrees to pay the other a fixed price and the other party agrees to pay the other a floating price. If in a settlement period the index price is above the fixed price, the floating price payer pays the other party the difference. If the index price is below the fixed price, the fixed price payer pays the other party the difference. No transfer of the physical commodity takes place. By separating pricing from supply, two markets are created, one for the physical commodity and the other for its price.

Electricity rate price swaps are hedging vehicles that allow a participant to fix the price of electricity for a specified time period, ensuring the participant's financial position against adverse price movements during that time. Although the International Swap Dealers Association (ISDA) has established a standardized energy risk-management swap, each individual agreement is customized to meet the needs of the participants. Unlike futures contracts, there is no concern over liquidity or expiration date. Another way of looking at swaps is to view them as a series of forward contracts that are wrapped into one contractual agreement.

Price swap agreements can also include provisions to manage credit risk within the transaction. This is an area where banking expertise can be effectively applied. Money center banks can offer structured derivative products (swaps) as a transaction linked to the company's financing in such a way that the market value of the company's assets and the market value of the financing can be positively correlated, reducing asset/liability mismatch. This means that the cost of financing can be lowered by bundling its borrowing with a swap transaction, creating a commodity-linked security that is not dependent on commodity price swings.

Firms writing swaps hedge their risk by taking a long position in a short-term delivery when their contract is appreciating as it nears expiration, or by taking a short position if the contract is losing value. The underwriters can run matched books by transferring excess risk to the futures market to balance their portfolios, or they can run unbalanced books when it is opportune for them; this strategy is used primarily by companies active in derivatives because they possess the physical product and believe they know the market. Unbalanced books can also be run by banks that have the liquidity to do so.

Swap agreements can be customized to meet the internal cash flow requirements of energy producers and consumers. They are a form of price insurance, because the intermediaries assume most of the risk and are responsible for managing their own books. These financial arrangements are extremely discreet transactions because of the competitive nature of the energy markets and the necessity to protect the client's positions in the physical and futures markets. The swaps market is by nature a very private business with few deals made public, partially because of low liquidity in the secondary markets, although the recent entrance of swaps brokers as intermediaries has brought greater price transparency by providing more competitive bid/ask price quotes. Also, real-time market electronic news services, such as Reuters, Dow Jones Markets, Bridge News, and Bloomberg, provide swaps quotes for energy commodities.

Price swaps offer both short- and long-term solutions for energy price risk. They can also be a very effective method of dealing with basis risk, the differential between the price of the physical commodity to be bought or sold and the futures or other index price. Basis can vary widely in the oil, natural gas, and electricity markets; therefore, basis risk requires active management by the swaps underwriter.

Price swaps can be viewed as complementary to futures. They allow hedging to go beyond the 6 to 9 months of true liquidity in the futures markets. Some brokers actually consider them longer-dated futures contracts. While futures cover only a small number of products, swaps and OTC options offer an almost infinite variety of customized arrangements for different products and time periods. For example, besides exchange-traded oil contracts, swaps can be made for naphtha, jet fuel, non-exchange traded crudes such as Dubai and Tapis, and vacuum gasoil (VGO), to name a few. And in electricity, price swaps and OTC options have developed initially without a futures contract. Swaps and futures contracts are similar in that the transactions have the same goals, but swaps can be used for longer time periods. Shorter-dated swaps, for 1 to 3 months, are actively written as are longer-dated swaps, from 6 months to as far out as 10 to 12 years forward for oil, 30 years for natural gas, and up to 20 years for electricity.

Because longer-term instruments are likely to involve wider price movements and can hedge against a wider variety of price risks than futures, their transaction costs can be higher, although replication of deals and greater competition have begun to bring costs down. Swaps, options, and other off-exchange financial instruments are considered hybrid financial products because they combine hedging strategies of commodities with long-term price risk management. Swaps fill the void left by futures contracts, where liquidity more than six months out becomes problematic.

8.5.5.1   Why Use Energy Swaps?

An energy producer that is constantly selling energy into the open market is exposing its revenue stream to the volatility of the market. The company's risk can be neutralized by converting the variable market price that it receives on its sales to a fixed price. A swap is then set up in which the company receives fixed payments from the buyer, based on the fixed price, and pays a variable amount to the buyer, based on the index. An energy consumer, who is concerned about rising prices, takes the other side of the swap in order to pay a fixed price and receive a variable price.

For the producer, the swap agreement provides income stability by eliminating the effects of market price fluctuations on its income stream. Risk is reduced through the swap. While it can be argued that the producer is giving up the opportunity to benefit from rising prices, the producer is protected from losses due to falling prices. Likewise, the energy consumer, by taking the opposite side of the swap transaction, insulates himself from rising prices while giving up the opportunity to benefit from declining prices.

Figure 8.8 illustrates a commodity swap. This example shows the relationship between the two sides of a swap transaction and the role of the intermediary, in this case a bank. Of course, the natural match between the producer and consumer's hedging needs could mean that the two parties could simply enter into a series of forward transactions without the services of an intermediary. However, the risk exposures of the producer and consumer usually do not match precisely, and intermediaries can add value by standing in between and assuming the risk.

8.5.5.2   *Different Types of Swaps Users*

Since most types of energy swap transactions were originally written for crude oil, a brief review of some of the variations is useful before discussing applications to electricity. Swaps and options, like futures, can be used to hedge crude oil production and product prices, carry inventory costs, and finance projects. Oil price swaps protect the value of oil in the ground and allow producers to hedge long-term price risk for their assets while selling production at market prices. Producers receive a fixed payment based on the contracted price with the third party. The transaction is made for an agreed-upon period of time. Banks also can arrange oil producer swaps to pay down debt and tie in interest rate and currency risk as well. Oil price swaps are particularly good instruments for exploration and production companies. These companies need to incorporate price floors into their development loans to mitigate their repayment risk by ensuring that cash flow requirements are met. This same concept can be applied to utility-generating capacity. The swap maximizes borrowing capacity by ensuring a future cash flow, protected from falling prices, that will cover loan payments.

8.5.6   Caps and Floors

Caps and floors are similar to swaps but offer price protection in a different manner. First, an upfront cash premium must be paid. Second, one side will

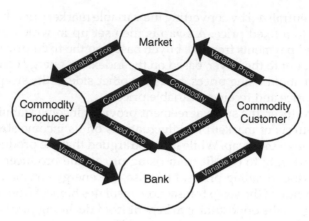

FIGURE 8.8
Energy swap pathways.

always benefit from a change in prices: consumers benefit from falling prices, and producers benefit from rising prices. Caps and floors are complementary trading strategies. A cap sets a maximum or ceiling price for energy consumers seeking full protection from rising prices with no loss of benefit if prices decline. Airlines are natural users of caps because of the price volatility of jet fuel and the need to limit fuel costs. A floor sets a minimum or bottom price and is used by energy producers seeking full protection from falling prices with no loss of benefit if prices rise. Oil, natural gas, and electricity producers are natural users of price floors. Caps and floors can be combined in a single structure called a collar, whereby players can get the advantage of managing both upside and downside price risks.

8.5.7 Price Protection Programs

Large industrial electricity consumers are interested in protecting themselves from unanticipated price spikes. Their goal is to fix electricity acquisition costs, which are variable, to lock in future profits for their products. A swap agreement puts production costs under control and locks in price margins. The financial intermediary takes positions on guaranteed differentials between natural gas and electricity to hedge the risk it has assumed. Since natural gas is by far the largest input to distributed power generation, it also makes it an effective hedge for electricity transactions at the present time because there is little liquidity for electricity futures.

8.5.7.1 *Natural Gas Swaps*

Because of the high volatility of North American natural gas prices (40% annualized price volatility) and the increasing use of natural gas futures to manage short-term price risk, the natural gas swap market took off during

1992 and continued to grow through the 1990s. Both natural gas producers and consumers, such as utilities and cogenerators, are becoming involved in swaps to protect themselves from adverse price movements. The emergence and growth of natural gas swap brokers during the latter half of 1992 only confirmed the high level of activity in this market. Electricity swap brokers emerged significantly during 1995 to further market price transparency.

8.5.7.2   New Types of Energy Swaps Developing Rapidly

Rapid development and adoption of interest rate and currency financial tools have, in turn, brought new swap instruments to the energy markets. These instruments include basis swaps, location swaps, weather swaps, and calendar swaps. Basis and location swaps allow positions in different energy markets to be hedged and are popular in the natural gas market. Calendar swaps allow prices to be locked in for seasonal or annual periods. Swaps have become more price transparent as real-time market news services, such as Dow Jones Markets, post bid and ask quotes for energy swap deals. Also, a new swap product has been developed that includes the right to extend or shorten the duration of the original agreement, creating volume flexibility.

8.5.8   Options

Two classes of options are increasingly used in the energy trade. Exchange-traded options and OTC options, traded outside the regulated exchanges, both exist. The buyer of an option has the right, but not the obligation, to buy or sell an underlying commodity at a fixed price, called the strike price. All options have expiration dates that define the time in which this right can be exercised. A put is an option to sell a commodity at the strike price. The buyer of a put option has the right to sell the underlying commodity. The seller of a put option must stand ready to buy the underlying commodity. A call is an option to buy a commodity at the strike price. The buyer of a call option has the right to buy the underlying commodity. The seller of a call option must stand ready to sell the underlying commodity.

OTC options offer a distinct advantage over forwards, futures, and swaps in that the need to arrange back-to-back transactions is eliminated. There is no need to match buyer or seller. This offers a particular advantage to producers and consumers because there is little chance of doing back-to-back transactions due to mismatched cash flow of the market makers. This greater flexibility makes OTC options easier to market than the other instruments.

While options have an upfront cost or premium, these costs can often be built into the transaction when it is used for longer-term project finance. Longer-term options can range in length from one to five years. Most option deals are written for three years or less, however. Options can be used to set floor prices for producers, making them ideally suited to long-term project finance since they lock in cash flow.

Options premiums have two components: intrinsic value and time value. Intrinsic value is the difference between the strike price and the price of the underlying commodity. It cannot be less than zero. A put has positive intrinsic value, or is "in the money," when the market price is less than the strike price — exercise will be profitable. It is "at the money" when the market price equals the strike price. It is "out of the money" when the market price is greater than the strike price. A call is in the money when the market price is greater than the strike price. It is at the money when the market price equals the strike price. It is out of the money when the market price is less than the strike price.

Time value is a function of time to expiration and the volatility of the price of the underlying commodity. The value of an option declines each day until expiration because as time to expiration diminishes, the probability of the option expiring in the money also diminishes. Greater volatility will increase the price of an option because it increases the probability that the option will expire in the money. Volatility is a measure of how much prices fluctuate. Past volatility can be measured and used as a gauge, but future volatility cannot be predicted with certainty. However, if the market is operating efficiently, the prices of active, at-the-money options can be considered fair. Because the strike price, the price of the underlying commodity, and the time to expiration are known, an implied volatility value can be derived from the price of the option. Market sentiment will affect how the actual price of the option compares to the theoretical price: the expectations of the buyers and sellers will make some options relatively more valuable than others.

As a rule, the less liquid the commodity, the higher the options premium because the seller is assuming more risk. Energy options prices are very volatile, reflecting the nature of the underlying energy markets. Options volatility comes in when there is more certainty in the energy markets. Because of this high price volatility, options would help to smooth some of the erratic price movements that occur with pricing due to seasonality and unexpected events in energy markets. In structuring longer-term risk management agreements, options are used to reduce risk and smooth volatility for energy producers and users.

Options portfolios require active management to adjust to changing market conditions. Options can also be used to structure indexed deals, which should gain popularity in coming years. Indexed deals are transactions that use an agreed-upon index such as in-trade publications, McGraw-Hill's Power Markets Week, NYMEX, Dow Jones' COB Index, or any other agreed-upon index.

The dynamism of options can be illustrated by a trading technique called stacking and rolling. In this strategy, a portfolio manager buys options to hedge a swap agreement as far forward as possible, usually 6 months to 3 years, and then rolls over the hedge each month to adjust the portfolio. Options can be bought on a daily basis for shorter-term deals to further fine-tune the portfolio. Sophisticated computer software for modeling options pricing is employed to optimize the series of transactions. This constant management factor, with its additional transaction costs, makes stacking and rolling with OTC options more expensive than hedging with futures, but costs often can be embedded in the deal. If not properly managed, this strategy can lead to disastrous results, as was evidenced by the Metallgesellschaft oil trading fiasco in December 1993.

8.5.8.1 Average Price Options

Asia options, or average price options, are increasingly being used by oil and gas underwriters. Average price options have a settlement period that more closely follows real-life transactions, as suppliers and consumers buy and sell crude oil or products in the spot month over the course of the option's term. Average price options are used to hedge oil production because production coming out of the ground and used by refiners is somewhat constant. Average price options are also used in jet fuel swaps to reduce basis risk. Basis risk is of great concern in the jet fuel market because of the widespread locational demands of the airline industry.

8.5.8.2 The Dealer's Book

The dealer's book is the record of all cash market positions and associated futures hedges. Dates of futures and cash positions never exactly match, but they are usually close enough so that liquidity is not a problem. Positions in distant months, which are less liquid, are later switched to more active positions in near months, and as distant months become near as time goes by, their positions in near months are switched as they expire. Because futures and cash are not perfectly matched, a gap exposure is always present. Often, dealers manage this risk by using other hedges. Rebalancing the book transfers excess risk from the intermediaries. The dealer's book may be thought of as a dynamic document that is actively managed and continually modified to meet changing market conditions.

Companies active in the physical and paper markets are often comfortable running unbalanced books because they are capable of supplying the underlying physical commodity. And since most electricity deals are still physical, some paper players are inactive at present. Banks are now more likely to run unbalanced books due to increased liquidity in the markets. The book gradually increases in size as more commodity-indexed business is added. Swap transactions add complexity and result in more active trading because such arrangements are for longer time periods, use more financial tools, and are customized to each client's needs. For these reasons, swaps can require higher transaction fees than futures trading, although replication of transactions and increased underwriting competition are driving down the costs to participants.

8.5.8.3 *Summary*

The futures exchanges provide financial integrity to the commodity markets and guarantee the performance of their contracts. This is not an unimportant point, for, in the U.S., no energy futures contract has ever defaulted. Performance is guaranteed by brokers, the clearinghouse, and the exchange. All transactions are matched and offset at settlement. Although there is a perception among the public that futures exchanges exist primarily for speculation, hedging is their true function. The creditworthiness of the exchanges and their standardized, fungible contracts is essential to off-exchange trading, where counter-party performance is a key component of swaps agreements. In off-exchange transactions, the market maker assumes the functions of an exchange. If a transaction unravels, the market maker must perform. Large banks, major oil companies, and large traders not only can make a market, but because of their creditworthiness and performance capability, their customers have confidence in their ability to stand behind their transactions. In the evolving electricity markets, electric utilities are starting to assume the market-making function.

Cash settlement, rather than physical delivery, is an obvious advantage of OTC markets over physical or forward markets for electricity in terms of performance. No one can squeeze a cash-settled contract, as sometimes occurs in the physical market. Cash settlement also offers a better hedging vehicle, since physical delivery alters supply and demand. Physical delivery at contract expiration can be a recurring problem. These problems do not exist in derivative products because they are cash settled, which enhances their performance function.

8.5.9 Banking Aspects

Multinational banks and energy companies have always had a strong business relationship due to the large capital needs and worldwide operations of the oil industry. Banks have provided services related to taxation, cash management, and foreign exchange and interest rate risk management. Many energy transactions are not new business, but are merely renewals of funding for existing facilities; these capital requirements are high and provide profitable margins for lending banks. Moreover, relationship banks can offer cross-border financial services, including currency swaps, as part of any commodity-linked transaction.

Because the oil and gas industries are cyclical and have high price volatility, price risk management is an element that will increasingly be incorporated into new energy lending. While many energy companies have not yet utilized these tools due to unfamiliarity, the benefits of customized risk management, including interest rate and currency risk reduction, are becoming apparent. Hedging will, at a minimum, allow the industry more flexibility through better cash flow management.

Banks are becoming more involved in oil and gas swaps as part of overall debt management strategies for their clients. Because of the high transaction costs of writing swaps and options, costs will be embedded in more deals as part of project prefinancings. This is a major area of new business development in the natural gas utilities market in North America. Commodity price swaps will be applied as a tool for other areas of project finance. Electric utilities have high financial and lending needs. As electricity becomes a commodity, more of these deals will include a commodity financing component.

8.6   Nomenclature

A	= annual payment
A_{life}	= levelized annual cost
A_M	= annual cost for maintenance (in first year $)
$(A/P,r,N)$	= capital recovery factor
C	= cost at size S
C_{cap}	= capital cost (in first year $)
C_{life}	= life cycle cost
CPI	= consumer price index
C_r	= cost at a reference size S_r
C_{salv}	= salvage value (in first year $)
f_{dep}	= present value of total depreciation, as fraction of C_{cap}
$f_{dep,n}$	= depreciation during year n, as fraction of C_{cap}
f_l	= fraction of investment paid by loan
I_n	= interest payment during nth year
L	= loan amount
m	= exponent of relation between cost and size of equipment
N	= system life (yr)
n	= year
N_2	= doubling time
N_{dep}	= depreciation period (yr)
N_p	= payback time (yr)
N_l	= loan period (yr)
p_e	= energy price
$\overline{p_e}$	= levelized energy price
$(P/F,r,n)$	= present worth factor

p_{dem}	= demand charge (\$/kW.month)
p_{ins}	= price of insulation (\$/m³)
P_{int}	= present value of interest payments
P_{max}	= peak demand (kW)
P_n	= principal during nth payment period n
r_0	$= (r - r_{inf})/(1 + r_{inf})$
r_d	= market discount rate
rd,e	$= (r_d - r_e)/(1 + r_e)$
$r_{d,l}$	$= (r_d - r_l)/(1 + r_l)$
$r_{d,M}$	$= (r_d - r_M)/(1 + r_M)$
r_e	= market energy price escalation rate
r_{inf}	= general inflation rate
r_l	= market loan interest rate
r_M	= market escalation rate for maintenance costs
r_r	= internal rate of return
S	= life cycle savings ($= - C_{life} + C_{life,ref}$)
S	= size of equipment
s	= annual savings
t	= thickness of insulation (m)

Greek

τ	= incremental tax rate
τ_{cred}	= tax credit

Subscripts

The subscript $_0$ designates real growth rates r_0, related to the corresponding nominal (or market) rates r.

The subscript $_{ann}$ designates annual growth rates, related to the corresponding continuous rates (with subscript $_{cont}$) by $1 + r_{ann} = exp(r_{cont})$.

References

Fusaro, P. C., *Energy Risk Management*, McGraw-Hill, New York, NY, 1998.

Jones, B. W., *Inflation in Engineering Economic Analysis*, Wiley Interscience, New York, NY, 1982.

Rabl, A., Optimizing investment levels for energy conservation: individual versus social perspective and the role of uncertainty, *Energy Econ.*, October 1985, 259.

Additional Reading

ASHRAE, Analysis of survey data on HVAC maintenance costs, ASHRAE Technical Committee 1.8 Research Project 382, 1985.

Boehm, R. F., *Design, Analysis of Thermal Systems*, John Wiley & Sons, New York, NY, 1987.

BOMA, BOMA Experience Exchange Report: Income/Expense Analysis for Office Buildings, Building Owners & Managers Association International, Washington, D.C., 1987.

M&S Equipment Cost Index, published periodically in *Chem. Eng.*, 2000.

Dickinson, W. C. and Brown, K. C., Economic analysis for solar industrial process heat, Report UCRL-52814, Lawrence Livermore Laboratory, Livermore, CA, 1979.

Economic Indicators, published periodically by United States Government Printing Office, Washington, D.C.

EIA, Energy Review, published in monthly and annual editions by the Energy Information Administration, U.S. Department of Energy, Washington, D.C, 2000.

EN-R Construction Cost Index, published periodically in *Eng. News Rec.*, 2000.

Konkel, J. H., *Rule of Thumb Cost Estimating for Building Mechanical Systems: Accurate Estimating and Budgeting Using Unit Assembly Costs*, McGraw-Hill, New York, NY, 1987.

Lippiat, B. C. and Ruegg, R. T., Energy prices and discount factors for life cycle cost analysis 1990, Report NISTIR 85-3273-4, Annual Supplement to NBS Handbook 135 and NBS Special Publication 709, National Institute of Standards and Technology, Applied Economics Group, Gaithersburg, MD, 1990.

Riggs, J. L., *Engineering Economics*, McGraw-Hill, New York, NY, 1982.

Ross, M. and Williams, R. H., *Our Energy: Regaining Control*, McGraw-Hill, New York, NY, 1980.

Statistical Abstract of the United States, published annually by the U.S. Department of Commerce, Washington, D.C., 2000.

9

The Regulatory Environment

Morey Wolfson

CONTENTS

The electric regulatory environment is as inevitable as the electricity market. Informed observers are certain that the electric power industry will continue to be regulated, albeit reregulated in fundamentally different ways than in the recent past. Electric regulators will set rules that provide the operational and financial framework for interaction between the local monopoly distribution electric utility company ("Disco") and customers in their franchised service territories. The incentives provided to Discos, and the rules by which they play, will be prime determinants of the pace and scale of deployment of distributed generation (DG). Electric regulators will provide the forum and make the decisions that could, at one end of the spectrum, lock grid-connected DG out of the market, or, at the other end, throw the doors open for rapid deployment of a wide array of DG.

0-8493-0074-6/01/$0.00+$1.50
© 2001 by CRC Press LLC

This chapter provides a high-level overview of the electric regulatory environment, aimed at informing the reader on background and issues to consider when working on DG issues. Electric utility industry restructuring (EUIR), which is creating a transition from traditional cost of service regulation to market-based service, will be discussed, along with potential implications that the transition may have for DG. The chapter concludes with projections on how DG regulatory issues might be addressed in the near future.

9.1   Electric Utility Industry Regulation: A Brief Background

Numerous volumes have been written describing the U.S. system of electric utility regulation. A shorthand description is offered here for the purpose of orientation for new entrants to the regulatory environment. Electric regulation is a complicated interplay of federal, state, local, economic and environmental, legislative and administrative jurisdictions. These jurisdictions govern owners of electric utility generation, transmission, and distribution services provided by federal, state, investor-owned, cooperatively-owned, municipal-owned, and independent entities. Electric industry regulations are continually evolving. New regulations typically follow proven technological, economic, and informed public opinion developments that present a mandate for the legislative and regulatory structure to change. Regulatory decisions apply pertinent local, state, and federal laws to persuasive argument reflecting ever-changing technical, economic, and market conditions and circumstances. The regulatory changes then have direct impacts on the marketplace.

For many, their first impression of the regulatory arena may be the hearing room at a state public utilities commission (PUC). If this is the impression, it may be because slightly over three-fourths of electric customers are served by investor-owned utilities, all of whom are regulated by PUCs. However, the regulatory arena also encompasses the city council chamber, the rural electric cooperative board room, the halls of Congress, the state legislative hearing room, a meeting of the Federal Energy Regulatory Commission (FERC), a conference at a federal energy laboratory, and literally dozens of other locations. For the limited purpose of this chapter, the focus will be primarily on PUCs. It is relevant to note that many other local jurisdictions (such as customer-owned cooperative and municipal utilities) often use PUC rules as model regulations for their jurisdictions.

The PUC is the forum for practitioners, including economists, financial analysts, engineers, social scientists, environmentalists, and their legal counselors. These practitioners intervene on behalf of residential, commercial, industrial, fuel supplier, electric utility, municipal, environmental, and other constituencies. Quasi-legislative or quasi-judicial administrative hearings in contested cases and rulemaking proceedings lead to PUC decisions, which establish new or refined policies. Depending on state laws, individuals may participate *pro se* (on their own), without legal counsel, if they do not represent a corporation.

PUCs are administrative agencies consisting of commissioners, often three to five (or sometimes seven) in number. In most states, the commissioners are appointed by the governor, subject to legislative confirmation, to a term of service (typically four years). In a minority of states, the commissioners are elected officials. Many states have laws that require a balance of political party representation on the PUC. With the exception of a few states where commissioners derive their authority from the state constitution, most PUCs derive their authority from the state legislatures. Since PUCs usually derive their authority and receive their funds from legislatures, this affects their relationship to the legislature, especially on matters of major public policy.

Commissions retain an expert staff to assist in analyzing applications by utilities or complaints from consumers. Staff members are often divided into a trial group and an advisory group. The trial group has "party" status in contested hearings, having the same rights (and often more, such as audit power) as other parties in the process. The commissioners' advisors and counselors assist the commissioners in the deliberative process, including assistance in writing decisions. Administrative law judges are an important part of most PUCs, hearing cases and rendering recommended decisions that become final decisions according to certain procedures.

PUC proceedings ensure participants due process rights, including the right to discovery and the opportunity for cross-examination of witnesses. PUC commissioners' decisions must be based upon the evidence submitted on the record. The decisions are subject to internal appeal procedures and appeal to the judicial system to ensure that they are just and reasonable, not arbitrary and capricious.

Since state legislatures typically charge PUCs with the responsibilities to regulate local electric, local gas, telecommunications, some transportation, and water distribution companies, they are typically quite busy places. However, now that state and federal competitive telecommunications policies rely on PUCs to establish and enforce new telecom rules, these agencies are busier than ever. The results of this on regulatory attention to DG will be addressed in greater detail in this chapter.

9.2   Historical Highlights

Over a fifty year period bounding the turn of the century, PUCs were created by almost every state legislature. They were born out of the social and political concern that natural monopolies, such as railroads, should be prohibited from exercising undue discrimination on captive customers. During this period, the electric industry was still very young, local in scope, small in scale, and lightly regulated. As managerial and technological changes brought about an increasing opportunity for utilities to exercise market power through their expanding scope and scale, the political and legislative structure strengthened its regulatory influence over utilities. As holding companies steadily abused their concentration of economic power, and as the Great Depression created a political climate to address market abuse, federal laws were passed in the 1930s (the Federal Power Act, the Public Utilities Holding Company Act, and others). The New Deal established the Rural Electrification Administration and created major federal electric developments (such as the Tennessee Valley Authority and the Bonneville Power Authority). These developments represented an alternative to the private and municipal owners of electric infrastructure.

Post-World War II technological improvements and expansion of the economy were accompanied by a stable and pervasive influence of local, state, and federal regulation of the electric power industry. The investor-owned electric industry was satisfied with this arrangement, as regulation provided predictability of capital recovery to finance and profit from a growing and reliable electric infrastructure. If price to consumers, particularly residential consumers, was the major criterion to judge, regulation was a success through the 1950s and 1960s. This period has been referred to as the Golden Age of Regulation (Hirsch, 1999).

Several major energy tremors occurred in the 1970s that led to changes evident in today's regulatory environment. Conventional coal-fired generating stations reached a point when larger size no longer resulted in a lower unit price of electricity. Domestic oil production had already peaked. Environmental awareness began to have an influence on policy making. Costly nuclear power plants were built, often with severe economic and environmental consequences. The economy suffered from double-digit inflation, producing substantial cost overruns. Already reeling from high oil embargo-induced gasoline prices, the public pleaded with utility regulators to take a more active role in protecting consumers from escalating electric rates. Consumer advocates achieved success across the country by undertaking full-time residential and small business consumer advocacy appearances before PUCs.

Utility planners in the early 1970s were unprepared for what would transpire in the mid- and late 1970s. Utility planners applied the same 7% per year electric growth rates that characterized the previous decades and forged ahead with construction of large nuclear and coal-fired generators. This led to excess capacity and excess cost. Precipitated by the Organization of Petroleum Exporting Countries' (OPEC) 1973 oil embargo, all fuel costs were on the rise. Despite the warning signals, utilities did not anticipate the potential for industrial conservation in response to higher electric rates. Utilities did not anticipate double-digit inflation, and they underestimated the depth of concern about environmental quality.

In response, electric regulators often agreed with the public outcry that utilities should be held accountable for their decisions. Regulators responded to excess capacity by applying "used and useful" criteria to disallow full recovery on certain investments. Utility managers learned lessons that are now common in utility culture: be cautious when considering major investments in generation, do not overestimate electric demand, and do not underestimate the influence that aroused consumers have on regulators.

State regulators were not alone in response to the utilities' problems. Social and political pressure focused federal attention on energy policy. The Carter Administration encouraged, and Congress passed, a series of electric power reforms in 1978. These changes included requirements that state regulators examine their rate structures to ensure that they did not result in energy waste. Of greatest importance to this chapter, in 1978 Congress passed the Public Utility Regulatory Policies Act (PURPA),* which contained a provision (Section 210,17{c}(1)) requiring investor-owned utilities to interconnect with smaller systems (typically combustion-turbine plants) owned by independent power producers (IPPs). The law created qualifying cogeneration facilities, which produce heat and electricity from fossil fuels. Such facilities would need to meet guidelines, to be determined by the FERC, regarding energy efficiency, types of fuel used, reliability, and other characteristics. These non-utility generators were paid a rate that matched the avoided cost that the utilities would have spent on their next (typically large nuclear or coal-fired) generating station. Although some tried, utilities and regulatory bodies could not fully dodge this Congressional mandate. Retail electric rates on the rise in the 1970s continued to move upward as projected utility generation costs pushed higher, resulting in high avoided costs paid to the IPPs.

Some states, such as California, established very high avoided-cost payments, which attracted the IPPs. Other states, because of either a larger customer-owned utility presence that was exempt from PURPA or lower avoided costs, were not attractive to the IPPs. The high avoided cost payments to IPPs resulted in important impacts. IPP development spurred the transformation of fossil-fuel plants from boilers to cleaner, often smaller, and always more efficient aeroderivative and combustion turbine plants. About 10% of all U.S. electric capacity is now composed of non-utility generation.** PURPA and production tax credits provided a financial platform for renewable energy developers, particularly wind developers, to gain manufacturing and operational experience, leading to significant declines in cost.***

High nuclear generation cost overruns and avoided cost payments to IPPs in the 1980s set the stage for competition in the late 1990s. States with high electric rates (often those states that had a high penetration nuclear generation and IPPs) deliberately sought out market mechanisms as a way to lower

---

* In his article on PURPA in the Aug./Sept. 1999 issue of the *Electricity Journal*, Richard Hirsch said that: "... though some observers expected only a few hundred megawatts of capacity to be offered in California, entrepreneurs signed up for more than 15,000 MW."
** Some states such as California, New England states, Texas, and Michigan have a much higher penetration of IPPs, particularly compared to some states in the South and Midwest.
*** See America Wind Energy Association.

customers' bills.* There is a correlation between the early tier of states that decided to restructure the electric industries and those states with high rates from high-cost nuclear generation and high-price IPPs. By the late 1990s, reaction to high PURPA payments was a major point of contention in Congress. Congress debated whether PURPA should be terminated, either through stand-alone legislation or as part of what has become elusive comprehensive restructuring legislation.**

Not long after the electric jolts of the late 1970s and early 1980s,*** regulators adopted a more rational way of planning for new capacity additions. Regulators developed (first in Wisconsin and California, then in dozens of states) least cost planning, or integrated resource planning (IRP). The new approach featured public participation in a planning process, ultimately leading to a more cost-effective, competitive acquisition of both supply-side and demand-side electric resources. But just as IRP was starting to be normative across the country, the regulatory structure started to de-integrate. The further that states moved across the EUIR continuum, the faster the underpinnings of IRP were being removed. IRP is now being removed by legislation or regulators, to the gratitude of the utilities but to the dismay of energy-efficiency advocates and beneficiaries of PUC-oversight of competitive bidding. In short, as EUIR is established in the states, a corresponding elimination or diminution of IRP is occurring.

During the heyday of IRP in the early 1990s, electric regulators were taking a more proactive position relative to electric planning, and their focus was primarily on generation and, to a lesser extent, transmission. Pertinent to this chapter, the focus in most states was not, and is still not, on distribution and certainly not on distributed generation. However, with the coming disintegration of the vertically-integrated electric utility structure, regulators and the market are expected to focus their attention more on distribution.† A review of electric utility asset utilization reveals that distribution assets are utilized a much smaller fraction of the time compared to electric generation asset utilization. This suggests that there is room for an increase in distribution efficiency. Distribution companies (Discos) in the evolving structure could be expected to spend more on distribution relative to generation and transmission. How they will spend that distribution money and whether their investment strategy includes DG are open questions.

---

* California's Public Utilities Commission, responding to a combination of too slow a recovery from the economic slump of the late 1980s, high electric rates, base closures, etc., set out on a course, beginning in 1993, to consider, then introduce, major structural changes to the California electric industry.
** See PURPA Reform Group, Washington, D.C.
*** Consider the debacles of the Washington Power Supply System, the Seabrook and Shoreham reactors, Three Mile Island, bankruptcies, and near-misses caused by synthetic fuel subsidies gone awry.
† The March 1, 1999, issue of *Public Utilities Fortnightly* is nearly completely dedicated to a discussion of this matter (*pur_info@pur.com*).

Before the advent of the EUIR debate, Congress yielded to 14 years (starting in 1978) of pressure for a comprehensive review of energy (including electricity) policy by passing the Energy Policy Act on October 25, 1992 (EPAct).* EPAct continued, and liberalized, the federal policy of allowing companies other than electric utilities to construct electric power plants and compete with utility-owned generation. EPAct required that interstate transmission line owners allow all electric generators access to their lines. The objective of these initiatives at the federal level was to further the efficiency of competitive wholesale electric markets. In effect, transmission lines became common carriers.

In 1996, after two or three years of piecemeal utility filing of tariffs required by EPAct, FERC opened what was called a mega-NOPR (notice of proposed rulemaking), that resulted in FERC Orders 888 and 889. The rules strengthened the open access transmission intent of EPAct. Tariffs were required to be filed that permitted any generator of electricity — utility-affiliated or independent — to have non-discriminatory access to transmission lines that have available capacity. As a follow up, the FERC was not satisfied with the progress made to create independent system operators (ISOs), which were first encouraged by EPAct. An NOPR was issued on the ISO, or Regional Transmission Organization, in the spring of 1999. The cost and benefit of creating multi-state ISOs to dispatch power by independent entities throughout the country is entangled in the broader EUIR debate and the state–federal relationship. The outcome of ISO developments could have an influence on distributed generation, as key questions of price transparency in wholesale markets will have a bearing on the transferability of these pricing rules on the distribution market.

EPAct moved federal electricity policy further along the competition continuum, but stopped short of mandating retail wheeling or direct access, which permits end-use customers to select their wholesale supplier of electricity. Since there is not a bright line that distinguishes between federal and state jurisdiction, and Congress did not want to define that distinction precisely in EPAct, the question of retail competition was left to the states, who have regulatory authority for approximately 90% of the electric infrastructure (the remaining being interstate transmission, which is regulated by the FERC).

---

* EPAct contained provisions including Energy Efficiency (Buildings, Utilities, Appliance and Equipment, Industrial, State and Local Assistance, Federal Energy Management), Natural Gas, Alternative Fuels, Electricity (Exempt Wholesale Generators, Federal Power Act, Interstate Commerce in Electricity), High-Level Radioactive Waste, Uranium, Renewable Energy, Coal, Strategic Petroleum Reserve, Global Climate Change, Reduction of Oil Vulnerability, Environment, Indian Energy Resources, and Nuclear Plant Licensing.

9.3   Electric Utility Industry Restructuring and DG

The fate of distributed generation is tied to how the provisions of EUIR leg-islation are written and implemented in restructured states and specifically whether the distribution utility rules in any state, restructured or not, are opened to allow market access by new technologies.

By way of background, the intense EUIR debate began soon after the pas-sage of EPAct. The early tier of EUIR states, such as California and New England, had high electric rates. These were followed by many other states* that saw opportunities to initiate EUIR through a grand deal. Simply stated, the grand deal typically will trade away traditional electric regulation in the long term for legislatively mandated lower rates (sometimes financed by divestiture of utility generating assets) in the short term. Accompanying state EUIR initiatives were high-profile Congressional proposals suggesting that a date-certain deadline be established to avoid a patchwork of separate state EUIR laws. Over half of the U.S. population now lives in states where cus-tomer choice is available.

Even in states that have opened supply to customer choice, EUIR change does not necessarily open markets for DG. Nonetheless, the move to customer choice offers the promise of new market choice and, in that way, joins at least four major groups in support of restructuring. Would-be DG vendors com-prise one of those groups, even though they do not receive open access to the distribution system under any of the restructuring regimes in place so far.

The first group that sees opportunity consists of new competitive electrical supply entrants who hope to compete for customers under new wholesale supply arrangements, but using the same distribution wires as the utility they replace. The second group is comprised of the DG developers who hope to enter what has been a relatively closed electric business. The Distributed Power Coalition of America** says, "electric restructuring will help the con-cept of distributed power because the buyers and sellers of electricity will have to be more responsive to market forces. Making sure that stranded costs do not impede distributed power is one of the Coalition's goals." As the argu-ment goes, DG will finally become more feasible in an open market where prices are unbundled and transparent. One requirement of free markets (an informed buyer) will place an economic value on electric reliability, quality, and delivery at time of day.

The third group that sees opportunity consists of investor-owned utilities that were initially ambivalent about EUIR but now have their long-sought prize in sight: pricing freedom. This new-found freedom will mean that their investment strategies will be governed more by the market, less by regulators.

---

* See the National Conference of State Legislatures for state-by-state accounting of EUIR (*www.ncsl.org*).
** The Distributed Power Coalition of America advocates for distributed generation, primarily at the national level (*www.dcpa.com*).

Whether this freedom causes investments in DG is a key question that will be explored in this chapter. The fourth group that sees opportunity consists of large-load customers (who started the EUIR movement) who look forward to deeper discounts than what they could achieve under a regulated structure, lower rates, or the freedom to self-generate and sell excess capacity to the grid.

Although the new entrants and investor-owned utilities have a common enthusiasm for EUIR, they have not created a common vision on how all of this is going to play out. This could be a classic case of fight them or join them, and the fighting or joining may well have its day in a PUC hearing room. The industrial and large commercial customers are almost universally favorable to EUIR because they see opportunity with either deeper utility discounts, less expensive self-generation, or utility investment in DG.

The EUIR change poses potential problems for inelastic customers left to the tender mercies of the market once the guaranteed rate reductions in the grand deal expire after a few years. These perils are particularly acute in states with one or more of the following characteristics: low rates, large rural areas, higher numbers of low income populations, or severe transmission constraints that make effective competition far more difficult. Given the magnitude of the rate and social equity issues at stake, EUIR skeptics may consider an analysis of the impact of EUIR on DG to be an academic exercise at best. Innovation, it is argued, can flourish under EUIR. The majority view sees the benefit of utility investment shifting away from generation and transmission, so Discos can concentrate on their core business, where DG resides.

What else might EUIR do for DG? Everyone would like to think that the future belongs to the efficient, and with utilities required to de-integrate and prices becoming transparent, customers will have a new-found opportunity to make choices that match their specific needs. Most optimistic is the view that EUIR will cause utilities to embrace DG as a smart business proposition. If Discos (who, at present, have few workable ideas on how to operationally cope with the complexities associated with thousands of distributed generators in their system) can find the financial incentives and answers to the safety, interconnection, and dispatch problems of DG, then change is possible. If the answers are there, then the underlying concern that the vast majority of residential and commercial customers have — they do not want to own, operate, maintain, and indemnify their own on-site generators — goes away.

The less popular view, perhaps considered contrarian by DG proponents, sees the ramifications of EUIR leading in a very different direction. If the supposed benefit of EUIR, lower rates, is achieved, then DG, which is already price-challenged, will have a difficult time competing with even lower-cost grid-supplied power. A more likely problem is the following: under traditional regulation, utilities frequently provide discounts to industrial and large commercial customers. However, under open markets, utilities are free, and will be even more inclined to offer discounted rates to retain these large loads. This would have the result of placing upward pressure on the inelastic residential and commercial customers. This may not necessarily work to the detriment of DG, but will be a factor considered by regulators. Of greater concern to distributed resource developers is that the more often the utilities provide discounts to large loads, the more they will erode the competitive position of the DG developers who want to make inroads in those markets.

9.4   Investment in DG

The comparative economics of self-generation and utility supplied power have been calculated and recalculated for well over 100 years. At the dawn of the electric age and continuing for about fifty years, it was commonplace for factories to self-generate. Gradually over time, the utilities' economies of scope and scale made grid-supplied power more attractive to large customers. Only the largest customers found it to be more economical to continue to self-generate. State laws typically provide complete freedom for companies to self-generate "within their fence." Most states have generally defined a public utility as any entity that provides electricity to others. States also often assign a certain geographic area as a legally enforceable exclusive franchised service territory granted to a utility in exchange for accepting public regulatory control over the rates charged. Under traditional electric regulatory law, this has confined industrial generation to within the fence.

The 1980s witnessed the combined impact of higher utility costs, PURPA incentives, and development and improvements of combustion turbine engines. These and other factors have caused more large loads to continually consider whether to self-generate or continue to buy power from the grid. With the advent of EUIR, there is pressure to remove laws permitting monopoly franchised service territories and prohibitions on selling outside the fence. All of this bodes well for DG, but how well is a function of the comparative economics of utility-supplied power and DG.

DG may develop more slowly in low-cost states than in high-cost states. Industrial and commercial facilities owners and managers typically prefer not to have to assume the burdens of buying equipment, procuring fuel in an uncertain fuel market, as well as maintaining and operating their own plants. Since there is a role here for third party providers, energy service companies (ESCOs)* are determining new ways to integrate traditional demand-side management energy services with DG. The ESCOs see the winners in this market to be those companies that position distributed generation not as a new device, but as a solution that gives customers more choice and control over their energy use, quality, and costs.

---

* See National Association of Energy Service Companies.

As a necessary reality check, it is interesting to note the relative lack of venture capital interest in DG compared to the other sectors of the economy.* The cause of this could be the unfortunate history of premature and exaggerated claims by energy technology developers. Another cause could be that there are less risky opportunities for venture capitalists. Developers have a vision of DG playing a highly productive role in the $250 billion per year U.S. electricity market, and that will always attract investors.

However, to date, DG suppliers may not have cost-effective technology, so cautious investors may hedge their bets until they see a clear answer to cost effectiveness and a signal as to where the utilities are going to go with DG. Private investors know that there is plenty of domestic and foreign** utility money eventually available to invest, through utility affiliates' structures, in DG. Could the serious money be holding back until they see which utilities are moving into DG, determine how successful they fare (given utilities' spotty record on affiliated interests), and then decide whether to invest? It is a classic chicken-or-egg situation, because a significant DG industry is unlikely to exist without venture capital.

Utility investment in DG will be competing with utility investment in traditional generation, transmission, and traditional distribution. Utilities have formed generation affiliates financing merchant plants, and the capacity additions announced have been extraordinary.*** As utilities invest in central generation, they are acutely aware of the painful lessons learned from excess capacity, coupled with knowing they are arriving at a point when they may not be able to ask regulators for recovery for their generation investments. Utility investment in transmission will continue to be problematic due to the underlying changed economics favoring the long-distance movement of energy through gas pipelines rather than across transmission lines,† the questionable profitability of transmission in the wake of open access, and siting problems. So, for the Disco aspect of the de-integrated utility, the question is whether they will invest in DG. Again, the answer depends on how cost-effective the technology is and how well the regulatory structure can incorporate DG into the new pricing paradigm.

9.5 Reliability, Restructuring, and DG

One important factor conditioning the development of DG is the impact of EUIR on electric reliability. As a general matter, and acknowledging that

---

* See the October 15, 1998 issue of *Public Utilities Fortnightly* (pur_info@pur.com).
** EPAct made it possible for the globalization that now characterizes the electric industry. Foreign companies, notably from England and Scotland, are buying major American utilities.
*** See Resources Date International, Boulder, Colorado.
† The author has prepared a PowerPoint comparison that reinforces what has become clear to energy analysts over the past several years.

there are many exceptions to this view, the broad base of customers, especially residential and commercial customers, probably prefer not to risk reliability problems resulting from a trade-off for lower rates. However, EUIR advocates have stated that restructuring will not only lower rates, but reliability will be enhanced with restructuring. Yet, if EUIR advocates are right about reliability getting better, higher utility-delivered electric reliability could make DG less attractive. This is because the motivation to have higher reliability* that results in installing DG, particularly in a growing number of high-tech companies where high reliability is of critical importance, could be achieved via a more reliable utility. It is also possible for competitive suppliers to use DG to achieve this increased reliability.

The new conventional wisdom suggests that the apparently more costly traditional engineering and regulatory approach of ensured reserve margins is being displaced by a market approach to valuing reliability, among other ancillary services, through pricing mechanisms. This sounds fine, and everyone involved hopes that the displacement works as projected. But if reliability suffers, and if the problem is tied, or perceived to be tied, to cost-cutting utilities (preparing for competition) who have not hired, retained, or adequately trained the necessary personnel to ensure reliability, the reaction could be swift. It could take the form of municipal lawsuits (in response to reliability problems in New York City, San Francisco, and Chicago in 1998 and 1999). The reaction could also be an attempt to reject the market approach and go back to the engineering/regulatory approach. This, of course, will be very difficult to do.

As is often the case, when system reliability problems are a result of inadequate generation during the peak hours, the deployment of DG could come into play. When coupled with sophisticated electronics that dispatch the distributed generation's output to the grid, DG can enhance reliability, and possibly do so at a price less than investing in peaking units used only during the few hours when the system peaks.

One manufacturer of electronics equipment** states:

> By upgrading and expanding current — and very familiar — interruptible-rate programs, utilities can add 100 to 400+ megawatts to their power supplies in each major metropolitan area in just a few months. When the customer's new control system receives a "dispatch" command from the utility, it automatically starts the generator, allows it to warm up properly, brings it up to speed and precisely synchronizes the generator's frequency, phase and voltage with the grid, closes a breaker to "lock on" to the

---

* A benchmark market assessment, performed by RKS Research & Counseling, a nationwide polling and market research firm, finds one-quarter of American businesses troubled by sporadic power interruptions (blackouts) and fluctuations (brownouts). They claim that close to a quarter of national accounts — America's largest and most energy-intensive national franchises and chains — are currently exploring options for generating their own electricity. The survey states that more than two-thirds of these customers have already taken some actions on their own to mitigate power interruptions.
** See *www.encorp.com*.

grid (referred to as "parallel" operation), and then gradually ramps up the generator output to transfer load off the grid and onto the generator. If this is technically sound, and the regulator (or market) knows this is a better deal than investing in peaking units, we should expect it to happen. The 100 to 400 megawatts referenced is an estimate of the number of existing back-up generators that could be aggregated into a "virtual power plant" distributed throughout an approximate 5000 megawatt peak load metropolitan area.

9.6 Fuel, Environmental Considerations, and DG

Always a key consideration brought into the calculus of electricity choices is the question of future natural gas prices. Gas prices spike and dip with great regularity as they move along the historic and projected future trend line. Gas prices were so high in the 1970s that the federal government barred gas as a boiler fuel for electric power generation. Yet, today, gas is nearly indisputably the fuel of choice for new electric capacity additions. Gas prices obviously vary according to supply and demand, and it is well known that very strong growth in natural gas is projected in electricity generation. There is also very strong growth projected in electricity generation due to the ever-increasing electrification of the economy, particularly driven by the information economy.* Since the predominance of new generation is gas-fired, it is now conventional wisdom that this will cause upward pressure on natural gas prices. This could affect DG technologies in very measurable ways. If the "dash for gas" does result in high gas prices, it could work to the disadvantage of less efficient DG technologies such as microturbines, and have less of an effect on more efficient DG technologies such as fuel cells.**

If markets behave as many expect, and EUIR really ends up favoring low-cost suppliers, the coal-fired generating stations that supply 55% of electricity in the U.S. have little to fear (except for public concern translated into political action surrounding visibility and global warming). Despite the contingent liability connected to environmental concerns, the market is responding favorably to coal-fired generating stations, reflected by the multiples over book value that coal plants are receiving when sold as part of the grand deal that requires utilities to divest their fossil generating assets. If the fuel, auto, and rural lobbies prevail on Congress to continue to thwart environmental initiatives, this will keep coal-derived electric power at a substantial competitive advantage over DG. The prospect of political success by the

* See the work of Mark Mills, technology strategist, president of Mills & McCarthy Associates Inc. of Chevy Chase, Maryland (*www.Mark_P_Mills@hotmail.com*).
** According to Mike Kujawa, Allied Business Intelligence, Inc., there are now 58 million U.S. households connected to the national gas grid and every one of them has the potential of having a household fuel cell for its electricity, hot water, and home heat.

coal, petroleum, and auto interests makes it all the more imperative that the cost of DG drop substantially.

9.7 Regulation of Air Pollution from Small Generators

Since DG spans the spectrum of technologies, the DG plants produce varying degrees of environmental impact. Some technologies, such as fuel cells or photovoltaics, can receive a blanket exemption from air regulation, while others, such as a diesel-powered backup generator pressed into service as a system peaking resource, will have limitations on the number of operating hours in certain locations. At the high-size end of the DG spectrum, a 50 MW unit combustion turbine will naturally require a whole other level of regulatory permit than a 75 kW microturbine. From a societal perspective, it is important that neighborhood and urban air quality not be adversely impacted by the deployment of DG. However, as the environmental implications of small scale on-peak generation are analyzed, their contribution to air degradation should be compared to the incumbent's contribution. To assist market development, statewide, regional, or national standards would likely yield better results than a grab bag of different local standards. It will be necessary to have continuing and active involvement by DG developers with air regulators to achieve workable solutions to air permits.

9.8 Evaluation Perspectives: Customer, Utility, Regulatory

According to the U.S. Energy Information Administration, generation represents about 72% of the embedded investment in the U.S. electric power system. About 8% of the embedded investment is in transmission, and about 20% is in distribution. This being the case, it is no surprise that the primary attention in electricity policy discussions has been focused on generation. One key question is how utility-owned generation can co-exist in an effective competitive market. Depending on the state-by-state particulars, the evolving structure of the industry will heavily depend on whether utilities will be required to, or will voluntarily, structurally divest their generation. Major attention is also afforded to the questions related to transmission, particularly regarding investment concerns, structure, ownership, and operation.

DG is obviously focused on distribution, but it must compete with the economics and utility investment in conventional generation, transmission, and distribution. Although regulators have tended to focus most on the generation and transmission side of the business through IRP and the issuance of certificates of public convenience and necessity, regulators are likely to start shifting their concentration towards distribution. This is primarily due to the widespread agreement that there will be a continuing role for regulators to oversee the post-restructuring monopoly Discos.

Five nodes in an electric utility have been described by Joe Ianucci:*

- Utility system (e.g., ~5000 megawatts)
- Distribution planning areas (e.g., ~150 megawatts)
- Distribution substations (~50 megawatts)
- Distribution feeders (~10 megawatts)
- Customers (one-third industrial at ~1 megawatt or larger; one-third commercial at ~100 kilowatts; and one-third residential at ~5 kilowatts)

These five nodes provide a basis for a discussion of DG evaluation perspectives. Ianucci states that there are two primary distributed generation evaluation perspectives: customer-centered and utility-centered. He is also of the opinion that the criteria used by the customer to evaluate DG contain the following: bill analysis/reduction, reliability enhancement, power quality improvement, cogeneration possibilities, distributed generation costs, capabilities, and operational difficulties. Added to this would be factors suggested by the Electric Power Research Institute: the cost of electricity that may be avoided by running the unit, fuel cost, installed cost of the system, and anticipated utilization of the unit (annual operating hours with an allowance for maintenance and outages) which would determine overall availability. Other important economic factors are net fuel rate of the equipment and maintenance costs (including spare parts, overhaul costs, and field service support). Labor costs to run the unit and utility standby fees also should be considered. Local utility tariffs may provide incentive to run or not run the unit depending on utility capacity requirements and market conditions.

The criteria used by the utility to evaluate DG contains the following: generation, transmission and distribution capital deferral/avoidance, system improvements (e.g., reliability) and problem mitigation (e.g., power quality), business implications (e.g., capital recovery), risk avoidance/managing uncertainty, customer partnerships (e.g., retention), DG costs, capabilities, and operational difficulties.

The criteria used by regulatory bodies will differ somewhat from state to state, but given the relatively common history and purpose of PUCs, certain expectations might be considered. The Regulatory Assistance Project has prepared a report** on distributed Resources for the Committee on Energy

---

* In his presentation to the Colorado Public Utilities Commission, November 14, 1999.
** Draft Report, "Profits and Progress Through Distributed Resources," July 15, 1999 (*Davidmosk@aol.com*).

Resources and the Environment of the National Association of Regulatory Utility Commissioners. The key conclusions include the following:

> The financial incentives favoring or disfavoring distributed resources deployment are generally unaffected by corporate structure. They are affected by the relationship between a utility's cost and price for distribution services. The location of the distributed resource is critical. Distributed resources installed on the utility side of the meter do not jeopardize profitability. The primary, and negative, impact on utility profitability of distributed resources deployment occurs when they are installed on the customer side of the meter. This is true for both demand-side and supply-side resources. From the utilities' perspective, demand- or supply-side resources installed on the customer side of the meter produce the same effect: sales go down and as a result revenues and profits go down.

Locating distributed resources in high-cost areas has significant potential benefits. The significant distribution cost savings resulting from distributed resources located in high-cost areas can reduce utility financial losses or even add to profits if the distributed resources are deployed only in high-cost areas.

The form of regulation also matters greatly, particularly whether the utility is subject to performance-based regulation (PBR) and, more importantly, whether the PBR is price- or revenue-based. Price regulation generally discourages distributed resources. Revenue regulation does not.

The deployment of distributed resources is affected by whether the utility has a fuel clause or similar regulatory provision; the nature of stranded cost recovery provisions, including the level of stranded costs, and stranded costs recovery mechanism (volumetric charges, exit fees, or other mechanisms that affect behavior); and whether there are balancing accounts for stranded costs.

> Regulators have a number of policies available to align utility profitability with the deployment of cost-effective distributed resources. Some, such as revenue-based PBR, go directly to the heart of the problem and fix the way regulation works. Others, such as distributed Resource Credits, distributed Resources Development Zones, and placing restrictions on pricing flexibility, aim at making distributed resources profitable to utilities by trying to direct distributed resources deployment to high cost parts of the utility's system.

> Getting utility profitability aligned with the deployment of cost-effective distributed resources is an important step, but it does not guarantee success. Even if regulation is able to completely align utility profits in the deployment of distributed resources, there may be other factors that overwhelm the power of any incentives. Such diversionary factors may include rate impacts, competitive and other risks, and issues of control or the lack thereof, each of which can undermine the incentives created in a PBR regulatory mechanism.

9.9 Participation at the PUC

Advocates of distributed generation (including Discos, manufacturers, customers, and others) will strategize on how best to seek the rules and rates from PUCs. As mentioned before, the regulatory focus is concentrated on the ~5000 megawatt level utility system. Of course, regulators do focus on customers, but primarily as a matter of rate design, not so much as a matter of engineering design. There is a threshold, often at a level of 10 megawatts or less, where PUCs allow utilities to make ordinary course-of-business investments without prior regulatory permission that may become subject to review in the context of a rate case or through tracking in a PBR review.

A change of state regulators' focus to distribution will take place first and foremost as the economic and engineering challenges of DG are resolved, and as a result of the outcome of EUIR. An important question for DG developers is: when is the time right for regulators to begin grappling with the DG issues?

Although the answer primarily may have to do with the state of the technological and economic development of DG, it may also have a lot to do with the particulars of each individual state and the role that Congress and others may take in setting standards and mandating a state response. There are national efforts, for example, to establish interconnection and net metering standards. If these succeed, it will make state proceedings move along much more easily. Speaking to a Congressional audience, Bev Jones, Vice President of External Affairs and Policy Development for Consolidated Natural Gas Company, said:

> No matter how enthusiastic [state legislatures] are about distributed power, the states can only travel part of the way to creating an environment which will foster optimum development of dispersed generation. Only the federal government can assure that the benefits of distributed power — greater flexibility, lower costs, lower emissions, enhanced reliability, greater customer choice — reach all corners of the marketplace.

Some states are not waiting for federal action to materialize. PUCs in some populous states, or some states with high electric rates, along with some states facing severe capacity adequacy problems, have initiated proceedings regarding interconnection standards, codes, net metering, and other regulatory actions to incorporate DG into the electric grid. The traditional IEEE standard of activity is underway, and new standards should be developed by 2001, assuming the present schedule is adhered to. Most notably, California, Texas, and New York have conducted detailed work on DG. Net metering legislation, often aimed at encouraging renewable energy development, is becoming state law in more states each year. Since the competitive position of DG is better in these states, it is natural to see regulatory activity take place first in high cost states. States that have experienced generation adequacy problems, such as Texas,* have produced valuable work that other states can draw upon.

One can expect that the less populous and lower cost states may take more of a wait-and-see attitude before initiating procedures to assist DG. When one recognizes the intense pressures on PUCs as they struggle to implement provisions of the Telecommunications Regulation Act of 1996,** it may not be realistic to expect most PUCs to initiate regulatory procedures on DG on their own accord. The question of who will carry the burden of proof is important, and PUCs and their staffs (particularly in smaller states) are not accustomed to volunteering to take on this burden. Utilities often carry the burden, volunteering to do so when it is in their interest, or when required to do so by the PUC.

Once DG developers are reconciled to expect neither regulators nor Discos to voluntarily carry the burden, they may have two choices. The first choice is to move forward on their own and see what happens. Unless done carefully and deliberately, this carries the risk of frustration, possibly failure. The second choice, arguably the more effective if conditions are right, is to work closely with the Disco and PUC staff to arrive at a mutually agreeable procedural and substantive arrangement that provides incentives to the Disco to be a partner while meeting the needs of DG developers. If the Disco and DG constituents are in agreement, they should encourage residential and environmental stakeholders to review the arrangement to consider the consequences to their interests. Once confident that the arrangement will not precipitate controversy, the arrangements should be brought forward to the PUC. As expected, participants in PUC proceedings are typically better off coming forward with all of their "ducks in a row" if they want to improve their chances of success.

9.10   Regulatory and Market Barriers

The following statement capsulizes what many consider to be the main regulatory and market barriers. Thomas R. Casten, CEO of Trigen,*** made the following remarks to Congress on August 1, 1999:

---

* Texas Public Utilities Commission (Nat Treadway, Office of Policy Development, and Ed Ethridge, Electric Industry Analysis, Project No. 19827, *Investigation into the Adequacy of Capacity for 1999 and 2000 Peak Periods*, Report on interconnection for distributed resources) *nat.treadway@puc.state.tx.us*.
** A key implementation requirement for PUCs is to ensure that interconnection agreements exist between incumbent local exchange carriers and competitive local exchange carriers. This has been an arduous task for regulators, but it has also created an experience base that may prove to be valuable when regulators consider interconnection agreements between distributed generators and Discos.
*** See *www.trigen.com*.

Under present rules, the emergency generators in hospitals, hotels, apartment and office buildings, and most government facilities are not allowed to provide power during high use periods. Old rules, based on the limits of 1940s technology, prevent emergency generators from being connected to the electric grid. These generators can operate only by interrupting power to the entire facility, then restarting and supplying power to priority loads. New modernized rules should allow emergency generators to routinely carry part of the electric load, thus easing pressure on overworked cables and transformers. Only simple wiring and control changes are necessary. These generators are already concentrated in the urban areas where peaking demands are the highest, and where construction of new transmission and distribution facilities is most problematic.

Congress allowed qualifying cogeneration facilities to operate in parallel with the electric grid in 1978, but did not modify the rules governing emergency generators. Our country has a huge fleet of distributed generation assets, but isn't able to take advantage of them even during times where society would be very well served by their operation. By immediately changing the rules that block use of these valuable assets, Congress and FERC can assure that high demand period power outages will not recur. Please enact immediate legislation mandating interconnection standards by an impartial federal agency (such as FERC) and encouraging state regulatory commissions to offer lower rates to users who can generate on demand and shave life threatening electric system overloads. Each of the state regulatory commissions should establish impartial governance over interconnection and offer lower rates to those who can generate on demand.

Another valuable perspective is that of Jay Hakes, head of the U.S. Energy Information Administration, Department of Energy, who testified to Congress in April, 1999:

Predicting which technologies will be successful is highly speculative. A direct link cannot be established between levels of funding for research and development and specific improvements in the characteristics and availability of energy technologies. In addition, successful development of new technologies may not lead to immediate penetration in the marketplace. Low prices for fossil energy and conventional technologies; unfamiliarity with the benefits, use, and maintenance of new products; and uncertainties concerning the reliability and further development of new technologies are all factors that may slow technology penetration and are barriers that the tax credits are intended to address.

However, these limitations do not mean that the impacts of the research, development, and deployment programs could not be substantial over time. There are a number of barriers to technology penetration that may account for seemingly slow penetration of technologies that appear cost-effective. Lack of information about new technologies is one barrier that may be overcome with information programs.* Subsidies or regulated prices may hold energy prices artificially low and hamper the penetration of technologies. Builders and homeowners or tenants may have different incentives for energy efficiency. It may be difficult for the builder or landlord to recover the additional costs for more expensive, energy-efficient equipment from a buyer or tenant who may not value energy efficiency highly. Conversely, the buyer or tenant who will be paying the energy bills may not readily have the option of making the equipment choices. Even if energy consumers are aware of potential cost savings from a more efficient technology, they may have preferences for other equipment characteristics, for example, valuing vehicle size over efficiency. Also, consumers may prefer a relatively short payback period for investments in energy-consuming technologies. Technology penetration can also be slowed by uncertainties about reliability, installation and maintenance, availability of the next generation of the technology, and necessary infrastructure.

Some of these barriers can be addressed by information programs, collaborative efforts for development and diffusion, research and development to improve technologies and reduce costs, and incentives to enhance the cost effectiveness of new technologies. All these initiatives may help to encourage earlier penetration of technologies. Subsequently, the initial penetration may have the additional impact of reducing costs through learning, establishing the infrastructure, and increasing familiarity with new technologies. Finally, equipment standards and other mandates such as renewable portfolio standards can also lead to earlier penetration of new, more advanced technologies; however, standards may not be the most cost-effective methods for encouraging improvements in energy efficiency.

A presenter at a conference on DG offered what he considered to be the seven "sins" that represent barriers to distributed generation. They are offered here for consideration, without attempting to address each individually:

- Zonal or non-locational energy pricing
- Absence of real time energy pricing
- Standard-offer type energy pricing for default customers
- Rolled in pricing of new transmission investments
- Market rules that artificially suppress the energy clearing price through price discrimination

---

* RKS Research & Consulting states that American business is already familiar with distributed technology options. More than half of large businesses are aware of cogeneration and on-site generation, for example, while one-third of these firms recognize fuel cells and small turbines. But these respondents are less conversant with the features, environmental performance, and costs of these systems.

- Restrictions on access to the spot market
- Restrictions on access to the system operators' economic dispatch

There may be very real social equity, economic, and other reasons why regulators would permit at least some of these sins to remain. Regulators have a balancing act to perform when crafting workable compromises between many competing values and goals. It may not be realistic to expect even the most enthusiastic pro-distributed generation regulator to address all perceived barriers to DG without giving real consideration to the impacts that removing barriers may have on other valid regulatory goals.

Regulators can be expected to assume their traditional role as surrogates for competition when regulators know that the alternative would be unmitigated market power, when anti-competitive behavior would be at risk of being exercised. However, where regulators see the opportunity for effective competition to prevail, they will likely view competition as preferable to retaining traditional regulation. With the prospect for effective competition, they will gladly shift to an impartial arbiter of resolving disputes among competitors. They see their evolving role as the referee that ensures that the competitive market is played in a fair and equitable manner.

9.11 Legislative and Regulatory Opportunities

The operational framework for DG requires that statutes and rules define the rules of the road; otherwise, DG will have a disadvantage in the marketplace. Therefore, legislative action at the state and federal level is required and, as referenced earlier, is taking place. Once the legislative framework is in place, a policy decision by legislators is made. This provides the legal framework that allows the administrative agency, such as the PUC, to establish rules and regulations, tackling the details that the legislation did not, as a practical matter, address. The Distributed Power Coalition of America states that the benefits of DG to utilities include saving money on transmission and distribution substation costs by deferring more costly traditional investments. They also state that there are environmental benefits to utilities through emission credits, including SO_2 emission offsets, NO_x emissions-avoided control costs, or emissions offsets. The coalition also quantifies new business opportunities in the areas of cogeneration and enhanced reliability. Additionally, the coalition identifies the opportunity for utilities using distributed power to add small capacity increments instead of large ones, by requiring only low capital investments and by enjoying short lead times to construct a facility. All of these proposed benefits, if they can be quantified to exist as stated, are sure to be of keen interest to electric regulators.

9.12   Research and Development

Energy is the force that drives the economy, and energy choices shape our lifestyles and environmental quality. Research and development (R&D) is part of the continuum leading to demonstration and commercialization. It is important that the commitment of private R&D, federal R&D, and R&D sponsored by industry groups such as the Electric Power Research Institute and the Gas Research Institute for DG be continued and expanded. Obviously, if DG expects to compete with grid-supplied power, with or without regulatory assistance and with or without utility affiliate buy-in, it will do so primarily on the basis of its own economic viability. That viability will be materially assisted by a strong commitment to R&D. Research and development has played a vital role in virtually every technology that is now part of the electric infrastructure. When considering energy R&D budget priorities, policymakers should not short-change DG. However, there are disturbing trends indicating that this is starting to happen for a variety of reasons. Continued vigilance by proponents of DG is in order.*

References

Kuttner, R., *Everything for Sale: the Virtues and Limits of Markets*, Century Foundation, 1997.

Hirsch, R. G., *Power Loss: The Origins of Deregulation and Restructuring in the American Electric Utility System*, MIT Press, Cambridge, MA, 1999.

---

* An interesting point/counterpoint on EPRI treatment of distributed generation is found in the April 1, 1999, issue of *Public Utilities Fortnightly* (*pur_info@pur.com*).

10

Combined Heat and Power (CHP)

Bruce Hedman and Tina Kaarsberg

CONTENTS

0-8493-0074-6/01/$0.00+$1.50
© 2001 by CRC Press LLC

Power generation systems create large amounts of heat in the process of converting fuel into electricity. For the average utility-sized power plant, more than two-thirds of the energy content of the input fuel is converted to heat. Conventional power plants discard this waste heat, and by the time electricity reaches the average American outlet, only 30% of the energy remains. Distributed generation (DG), due to its load-appropriate size and siting, enables the economic recovery of this heat. An end user can generate both thermal and electrical energy in a single combined heat and power (CHP) system located at or near its facility. CHP systems can deliver energy with efficiencies exceeding 90% (Casten, 1998).

CHP systems have been used by energy intensive industries (e.g., pulp and paper, petroleum) to meet their steam and power needs for more than 100 years. They can be deployed in a wide variety of sizes and configurations for industrial, commercial, and institutional users. CHP strategies can even be used with utility-sized generation, usually in conjunction with a district energy system (Spurr, 1999). CHP systems can also involve nonelectric or shaft power, or the electricity can be used only internally.* In the U.S., however, most CHP applications have been cogeneration (medium-sized CHP for electricity and steam).** This chapter discusses only CHP applications that generate electricity using DG prime movers.

10.1   CHP Definition and Overview

Combined heat and power (CHP) systems capture the heat energy from electric generation for a wide variety of thermal needs, including hot water, steam, and process heating or cooling. Figure 10.1 gives an example of the efficiency difference between separate and combined heat and power. A typical U.S. CHP system converts 80 out of 100 units of input fuel to useful energy — 30 to electricity and 50 to heat. By contrast, traditional separated heat and power components require 163 units of energy to provide the same

---

* The U.S. Department of Energy's CHP Web site, *www.oit.doe.gov/chpchallenge*, and the U.S. CHPA Web site, *www.nemw.org/uschpa*, contain numerous references.
** For an excellent review of small-scale CHP worldwide, see Major, G., *Small Scale Cogeneration*, CADDET Energy Efficiency Analysis Series, 1995.

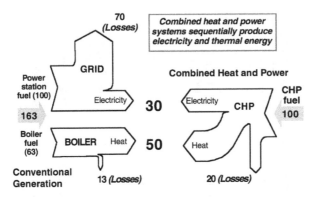

FIGURE 10.1
CHP versus separate power.

amount of heat and power. Thus, with today's technologies, CHP can cut fuel use nearly 40%.*

10.2 Available Technologies

Commercially available CHP technologies for DG include diesel engines, natural gas engines, steam turbines, gas turbines, microturbines and phosphoric acid fuel cells.** Table 10.1 summarizes the characteristics of commercial CHP prime movers. The table shows the wide range in CHP capacity — from 1 kW Stirling engine CHP systems to 250 MW gas turbines. All are projected to have lower costs and emissions and higher efficiencies due to incremental technology advances.

10.2.1 Reciprocating Engines

For CHP applications, the two principle types of combustion engines are four-cycle spark-ignited (Otto cycle) and compression-ignited (diesel cycle) engines. CHP projects using reciprocating engines are typically installed for $800 to $1500/kW. The high end of this range is typical for small capacity projects that are sensitive to other costs associated with constructing a facility, such as fuel supply, engine enclosures, engineering costs, and permitting fees.

---

* Based on a paper by Roop, J. M. and Kaarsberg, T. M., Combined heat and power: a closer look, *Proceedings of the 21st National Industrial Energy Technology Conference*, Houston, TX, May 1999. These are national averages for existing installed boilers and central generating plants.
** Nearly all the 171 PAFCs installed in the U.S. obtained a federal government subsidy of as much as $1000/kW; thus, in what follows, they are not included in cost estimates.

TABLE 10.1

Comparison of DG/CHP Technologies

	Diesel Engine	Natural Gas Engine	Gas Turbine	Microturbine	Fuel Cells	Stirling Engine[a]
Electric efficiency (LHV)	30–50%	25–45%	25–40% (simple) 40–60% (combined)	20–30%	40–70%	25–40%
Part load	Best	Ok	Poor	Poor	—	Ok
Size (MW)	0.05–5	0.05–5	3–200	0.025–0.25	0.2–2	0.001–0.1
CHP installed cost ($/kW)	800–1500	800–1500	700–900	500–1300	> 3000	> 1000
Start-up time	10 sec	10 sec	10 min–1 hr	60 sec	3–48 hrs	60 sec
Fuel pressure (psi)	< 5	1–45	120–500	40–100	0.5–45	—
Fuels	Diesel, residual oil	Natural gas, biogas, propane	Natural gas, biogas, propane, distillate oil	Natural gas, biogas, propane, distillate oil	H_2, natural gas, propane	All
Uses for heat recovery	Hot water, LP steam, district heating	Hot water, LP steam, district heating	Heat, hot water, LP-HP steam, district heating	Heat, hot water, LP steam	Hot water, LP-HP steam	Direct heat, hot water, LP steam
CHP output (Btu/kWh)	3400	1000–5000	3400–12,000	4000–15,000	500–3700	3000–6000
Usable temp. for CHP (°F)	180–900	300–500	500–1100	400–650	140–700	500–1000

[a] Expected to be commercial by 2005.
Source: ONSITE SYCOM Energy Corporation, Market Assessment of CHP in the State of California, draft report to the California Energy Commission, September, 1999 (except for Stirling data).

10.2.1.1 Heat Recovery

Energy in the fuel is released during combustion and converted to shaft work and heat. Shaft work drives the generator while heat is released from the engine through coolant, exhaust gas, and surface radiation. Approximately 60 to 70% of the total energy input is converted to heat that can be recovered from the engine exhaust and jacket coolant. Smaller amounts of heat are also available from the lube oil cooler and, if available, the turbocharger's intercooler and aftercooler. Steam or hot water can be generated from recovered heat that is typically used for space heating, reheat, domestic hot water, and absorption cooling.

By recovering heat from the jacket water and exhaust, approximately 70 to 80% of the fuel's energy can be effectively used. Heat in the engine jacket coolant accounts for up to 30% of the energy input and is capable of producing 200°F hot water. Some engines, such as those with high pressure or ebullient cooling systems, can operate with water jacket temperatures of up to 265°. Engine exhaust heat is 10 to 30% of the fuel input energy. Exhaust temperatures of 850 to 1200°F are typical. Only a portion of the exhaust heat can be recovered since exhaust gas temperatures are generally kept above condensation thresholds. Most heat recovery units are designed for a 300 to 350°F exhaust outlet temperature to avoid the corrosive effects of condensation in the exhaust piping. Exhaust heat can be used for thermal applications ranging from hot water to about 230°F or low-pressure steam (15 psig).

10.2.1.2 Closed-Loop Hot Water Cooling Systems

The most common method of recovering engine heat is the closed-loop cooling system, as shown in Figure 10.2. These systems are designed to cool the engine by forced circulation of a coolant through engine passages and an external heat exchanger. A heat exchanger transfers excess engine heat to a cooling tower or radiator. Closed-loop water cooling systems can operate at coolant temperatures of 190 to 250°F.

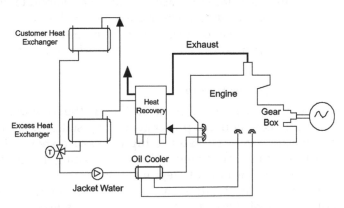

FIGURE 10.2
Closed-loop heat recovery system.

10.2.1.3 Ebullient Cooling Systems

Ebullient cooling systems cool the engine by natural circulation of a boiling coolant through the engine. This maintains the temperature throughout the coolant circuit. The uniform temperature extends engine life, improves combustion efficiency, and reduces friction in the engine. It is typically used in conjunction with exhaust heat recovery for production of low-pressure steam. Cooling water is introduced at the bottom of the engine where the transferred heat begins to boil the coolant, generating two-phase flow. The formation of bubbles lowers the density of the coolant, causing a natural circulation to the top of the engine. The coolant at the engine outlet is maintained at saturated steam conditions and is usually limited to 250°F and a maximum of 15 psig. Inlet cooling water is also near saturation and is generally 2 to 3°F below the outlet temperature.

10.2.2 Steam Turbines

Steam turbine technology is one of the most versatile and oldest prime mover technologies used to drive generators or mechanical machinery. Steam turbines are widely used for CHP applications in the U.S. and Europe where special designs have been developed to maximize efficient steam utilization. Most of the electricity (>80%) in the U.S. is generated by conventional steam turbine power plants. The capacity of steam turbines can range from fractional horsepower to more than 1300 MW for large utility power plants.

A steam turbine does not directly convert a fuel source to electric energy but requires a source of high-pressure steam. The steam is usually produced in a boiler or heat recovery steam generator (HRSG). Boiler fuels can include fossil fuels such as coal, oil, and natural gas, or renewable fuels such as wood or municipal solid waste. Steam turbines offer a wide array of designs and complexity to match the desired application and/or performance specifications. Traditional utility applications maximize efficiency of electric power production with multiple pressure casings and other elaborate design features. For industrial applications, steam turbines are less complicated to increase reliability and reduce cost (e.g., generally single or dual casing). Both utility and industrial steam turbine designs can be adapted for CHP.

10.2.2.1 Technology Description

The thermodynamic cycle for the steam turbine is the Rankine cycle. The cycle is the basis for conventional power generating stations. In this cycle, a heat source (boiler) converts water to high-pressure steam. The steam expands in a turbine to produce power. The steam exiting the turbine is condensed and returned to the boiler to repeat the process.

A steam turbine consists of a stationary set of blades (called nozzles) and a moving set of adjacent blades (called buckets or rotor blades) installed within a casing. The steam pushing the blades turns the shaft of the turbine and the connected load. A steam turbine converts pressure energy into velocity energy as it passes through the blades. The energy in high-pressure steam from a boiler* or other source** spins a steam turbine, and the turbine spins the shaft of a generator. All "steam-electric" generating plants — whether coal, natural gas, oil, nuclear, or geothermal energy powered — use this basic process.

The primary type of turbine used for central power generation is the *condensing* turbine. Steam exhausts from the turbine at subatmospheric pressures, maximizing the heat extracted from the steam to produce useful work. The *non-condensing turbine* (also referred to as a back-pressure turbine) exhausts steam at atmospheric pressures and above. In these turbines, a downstream process actually does the condensing to drive the cycle. An innovative application for steam turbines is as a replacement for pressure-reducing values. This application is generally quite small-scale, i.e., less than one megawatt. Manufacturers have lowered costs dramatically on such modular, load-following, back-pressure steam turbine generators.*** The discharge pressure is established by the specific CHP application.

The *extraction turbine* has opening(s) in its casing for extraction of steam either for process or feedwater heating. The extraction pressure may or may not be automatically regulated depending on the turbine design. Regulated extraction permits more steam to flow through the turbine to generate additional electricity during periods of low thermal demand by the CHP system. In utility type steam turbines, there may be several extraction points, each at a different pressure.

Modern large condensing steam turbine plants have efficiencies approaching 40 to 45%; however, efficiencies of smaller industrial or back-pressure turbines can range from 15 to 35%. Boiler/steam turbine installation costs are between $800 and $1000/kW or greater depending on environmental requirements. The incremental cost of adding a steam turbine to an existing boiler system or to a combined cycle plant is approximately $400 to $800/kW.

10.2.2.2   Heat Recovery

Heat recovery from a steam turbine is somewhat misleading since a steam turbine can be defined as a heat recovery device. Producing electricity in a

---

* Boilers convert thermal energy, produced in the combustion of a wide range of fuels (or waste heat from industrial processes), into steam or hot water. They come in various sizes, and boiler fuels can include fossil fuels such as coal, oil, natural gas, and refining crude, or renewable fuels like wood or municipal waste. Boilers are often used with byproduct fuels. Recent developments include improved tubing material that increases durability and modular designs that require less operating expertise.
** The most typical source, other than boilers, is a heat recovery steam generator (HRSG). Other sources of pressurized steam (> 400°F or so) include industrial waste heat, garbage incinerators, nuclear reactions, or geothermal energy.
*** The capital cost of a Trigen-Ewing backpressure steam turbine varies from about $700/kW for very small systems (50 kW) to $200/kW for large systems (over two megawatts). These costs do not include the expense of the steam system.

steam turbine from the exhaust heat of a gas turbine (combined cycle) is a form of heat recovery. Heat recovery methods from a steam turbine use exhaust or extraction steam with an economizer or air preheater. A steam turbine can also be used as a mechanical drive for a centrifugal chiller.

The amount and quality of the recovered heat is a function of the entering steam conditions and the design of the steam turbine. Exhaust steam from the turbine can be used directly in a process or for district heating. Alternatively, it can be converted to other forms of thermal energy including hot water or chilled water. Steam discharged or extracted from a steam turbine can be used in a single- or double-effect absorption chiller.

10.2.2.3   *Industrial, Commercial, and Institutional Applications*

In industrial applications, steam turbines may drive an electric generator or equipment such as boiler feedwater pumps, process pumps, air compressors, or refrigeration chillers. Turbines used as industrial drivers are almost always a single casing machine, either single stage or multistage. They can be either condensing or noncondensing depending on steam conditions and the value of the steam. Steam turbines can operate at a single speed to drive an electric generator or operate over a speed range to drive a refrigeration compressor.

For noncondensing applications and load-following applications, steam is exhausted from the turbine at a pressure and temperature sufficient for the CHP heating application. Although back-pressure turbines are less efficient than condensing turbines, they also are less expensive and do not require a surface condenser. In these turbines, a downstream process actually does the condensing to drive the cycle. These turbines can operate over a wide pressure range (typically between 5 and 150 psig) depending on the process requirements and exhaust steam. Small turbines can be used to replace pressure-reducing values (PRV) — a very cost-effective application which converts normally wasted energy into valuable electricity. The PRV replacement is applied mainly in institutional or industrial settings where high-pressure steam is available and low-pressure steam is needed for process or space heating.

10.2.3   Gas Turbines

Gas turbines are a cost-effective CHP alternative for commercial and industrial end users with a base-load electric demand greater than about 5 MW. Although gas turbines can operate satisfactorily at part-load, they perform best at full power in base-load operation. Gas turbines are frequently used in U.S. district steam heating systems since their high quality thermal output can be used for most medium pressure steam systems.

Gas turbines for CHP can be in either a simple-cycle or a combined-cycle configuration. Simple-cycle applications are most prevalent in smaller installations of typically less than 25 MW. Waste heat is recovered in an HRSG to generate high- or low-pressure steam or hot water. The thermal product can be used directly or converted to chilled water with single- or double-effect absorption chillers.

The simple-cycle gas turbine has the lowest electric efficiency and power to heat ratio since there is no recovery of heat in the exhaust gas. Hot exhaust gas can be used directly in a process, or, by adding an HRSG, exhaust heat can generate steam or hot water. Combined cycles involving a steam turbine become economical for larger installations. The most advanced utility-class gas turbines achieve up to 60% electric generation efficiency — but achieving this high efficiency means that only very low-grade waste heat is available for CHP.

More energy can be extracted from the turbine by burning the oxygen-rich exhaust gas (supplemental firing). A duct burner is usually fitted within the HRSG to increase the exhaust gas temperature at efficiencies of 90% and greater.

10.2.3.1 *Absorption Chilling*

Absorption chilling systems can be provided to produce chilled water directly from the gas turbine exhaust (Figure 10.3). The most common application of absorption chilling, however, is to use low, 2 to 4 bar (~30 to 60 psig), or medium, 10 bar (~150 psig), pressure-saturated steam. Absorption chillers generate chilled water using a working fluid operating between high-temperature gas turbine exhaust and a lower-temperature sink (Figure 10.4). Most common absorption chillers in industrial applications use a lithium bromide (LiBr) and water solution to provide chilled water at 7°C (44°F). Lower water temperatures can be achieved with ammonia and water systems.

In a lithium bromide absorption chiller, chilled water is produced through the evaporation of water in the evaporator section. Water evaporation is induced by a concentrated solution of lithium bromide that has a high affinity for water vapor. As water vapor is absorbed in the absorber section, additional water evaporation chills the refrigerant by boiling at low pressure and temperature. The absorbent solution becomes diluted and is pumped to the regenerator, where steam is used to boil out excess water. The water vapor is condensed in the condenser section by a cooling water system and returned to the evaporator section. In a separate flow path, the strong lithium bromide returns to the absorber section.

Low pressure steam, 2 to 4 bar (~30 to 60 psig), is used in single-stage lithium bromide absorption chillers at a rate per refrigeration ton (RT) of 7.7 kg/RT (17 lbm/RT). Medium pressure steam, 10 bar (~150 psig), is used in two-stage absorption chillers at a rate of 4.5 kg/RT (10 lbm/RT).

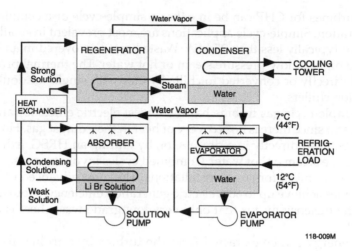

FIGURE 10.3
Absorption cooling cycle.

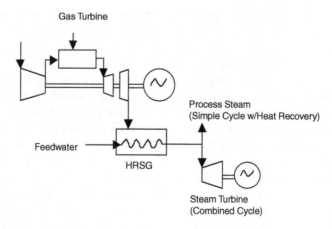

FIGURE 10.4
Gas turbine heat recovery.

10.2.4   Microturbines

Hot exhaust gas from the turbine section is available for CHP applications. Recovered heat can be used for hot water heating or low-pressure steam applications, although most designs incorporate a recuperator that limits the amount of heat available for CHP.

10.2.5   Fuel Cells

Phosphoric acid electrolyte fuel cells (PAFC) are the only commercial fuel cells sold in any quantity to date. The PAFC has been installed in more than 200 locations in the U.S., Europe, and Japan — almost all in CHP mode. PAFCs were the first practical application because, unlike other aqueous acids, PAFCs can be operated above the boiling point of water — typically around 200°C. PAFCs, therefore, can use waste heat from the fuel cell stack to directly reform methane into a hydrogen-rich gas for use as a fuel. They are produced in 200 kW modules that can easily be combined. The heat is used for space heating or hot water, but is not of a high enough quality to be used in other cogeneration applications. The Japanese are particularly advanced in PAFC research and design.

Molten carbonate fuel cells (MCFC) operate at higher temperatures and are more efficient than the PAFC, with estimated efficiencies up to 55% lower heating valve (LHV). The carbonate electrolyte is solid at room temperature but liquid at the operating temperatures of 650 to 800°C. The high exhaust temperature of an MCFC can generate additional electricity in a steam turbine or in a gas turbine combined cycle. The MCFC is expected to target 1 to 20 MW stationary power applications and should be well suited for industrial CHP.

Solid oxide fuel cells (SOFC) are the newest type of fuel cell and are still in laboratory testing for various configurations. Like MCFCs, the high-grade waste heat produced by SOFCs can be used for internal reforming and many other applications, including steam that a steam turbine can use to generate extra electricity in a combined (bottoming) cycle. Even after this process, the heat remaining is sufficient for cogeneration applications. Hybrid systems using gas turbines or microturbines could increase electric efficiencies to 60%.

Proton exchange membrane fuel cells (PEM) operate at relatively low temperatures (80°C) because the polymer membrane melts at higher temperatures. The fuel cell's low-temperature heat is not adequate for traditional cogeneration,* but does allow a quick startup time.

The type of fuel cell determines the temperature of the heat liberated during the process and its suitability for CHP applications. Low-temperature fuel cells generate a thermal product suitable for low-pressure steam and hot water CHP applications. High-temperature fuel cells produce high-pressure steam that can be used in combined cycles and other CHP process applications. Although some fuel cells can operate at part-load, other designs do not permit on/off cycling and can only operate under continuous base-load conditions.

---

* Though hydronic heading and desiccant regeneration are possible.

10.2.6   Stirling Engines

The Stirling engine — so named because it is based on the Stirling thermo-dynamic cycle — was conceived more than a century ago. Stirling engines produce power not by explosive internal combustion, but by an external heat source — usually a continuous-combustion burner. Until recently, reliability problems have limited their use to hobbyists. It is only in the past generation that a viable free-piston Stirling was developed. All Stirling engines can be operated with a wide variety of fuels, including fossil fuels, biomass,* solar, geothermal, and nuclear energy. When used with fossil and biomass fuel, the continuous-combustion heater head avoids temperature spikes, which makes emissions very low and easy to control. The Stirling engine is a heat recovery device, like the steam turbine. Several European utilities are demonstrating this technology for residential micro-CHP applications. Even at these very small sizes, electric efficiencies of more than 30% have been achieved.

10.3   Typical CHP Applications

CHP systems can provide cost savings as well as substantial emissions reductions for industrial, institutional, and commercial users. This section reviews some of the primary issues faced by the design engineer in selecting and designing an optimized CHP system. Selecting the right CHP technology for a specific application depends on many factors, including amount of power needed, the duty cycle, space constraints, thermal needs, emission regulations, fuel availability, utility prices, and interconnection issues.

Designing a technically and economically feasible CHP system for a specific application requires detailed engineering and site data. Engineering information should include electric and thermal load profiles, capacity factor, fuel type, and performance characteristics of the prime mover. Site-specific criteria such as maximum noise levels and footprint constraints must be taken into account.

10.3.1   Electric and Thermal Load Profiles

CHP by definition implies the simultaneous generation of two or more energy products that function as a system. One of the first and most important elements in the analysis of CHP feasibility is obtaining accurate representations of electric and thermal loads. This is particularly true in situations where CHP systems are not allowed to export to the grid. Such applications

---

* Biomass can be used in several ways, including direct combustion, two-stage combustion, and (the cleanest alternative) with a gasifier.

are usually load-following applications where the prime mover must adjust its electric output to match the demand of the end user while maintaining zero output to the grid. A 30-minute or hourly load profile provides the best results for such an analysis. Thermal load profiles can consist of hot water use, low- and high-pressure steam consumption, and cooling loads. The shape of the electric load profile and the spread between minimum and maximum values will largely dictate the number, size, and type of prime mover. It is recommended that electric and thermal loads be monitored if such information is not available.

For base-load CHP applications that export power to the grid and meet a minimum thermal load required under the Public Utility Regulatory Policies Act (PURPA), sizing a CHP facility is largely dictated by capacity requirements in the wholesale energy market. Rather than meeting the demand of an end user, such plants are dispatched to the grid along with other generating systems as a function of cost of generation.

Capacity factor is a key indicator of how the capacity of the prime mover is utilized during operation. Capacity factor is a useful means of indicating the overall economics of the CHP system. It indicates the facility's proximity to baseload operation. Capacity factor is defined as follows:

$$Capacity\ Factor = \frac{Actual\ Energy\ Consumption}{Peak\ Capacity\ of\ Prime\ Mover \times 8760\ hours}$$

A low capacity factor is indicative of peaking applications that derive economic benefits generally through the avoidance of high demand charges. A high capacity factor is desirable for most CHP applications to obtain the greatest economic benefit. A high capacity factor effectively reduces the fixed unit costs of the system ($/kWh) and helps to maintain its competitiveness with grid-supplied power.

Gas turbines are typically selected for applications with relatively constant electric load profiles so as to minimize cycling the turbine or operating the turbine for a large percentage of hours at part-load conditions where efficiency declines rapidly. Gas turbines are ideal for industrial or institutional end users with 24-hour operations or where export to the grid is intended.

Most commercial end users have varying electric load profiles, i.e., high peak loads during the day and low loads after business hours at night. Natural gas reciprocating engines are a popular choice for commercial CHP due to good part-load operation, ability to obtain an air quality permit, and availability of size ranges that match the load of many commercial and institutional end users. Reciprocating engines exhibit high electric efficiencies, meaning that there is less available rejected heat. This is often compatible with the thermal requirements of the end user.

Thermal demand of a commercial or institutional end user often consists of hot water or low-pressure steam demand in the winter and a cooling demand in the summer. Heat from the prime mover is often used in a single-stage steam or hot water absorption chiller. This option allows the CHP system to operate continuously throughout the year while maintaining a good thermal load without the need to reject heat to the environment.

10.3.1.1   Quality of Recoverable Heat

The thermal requirements of the end user may dictate the feasibility of a CHP system or the selection of the prime mover. Gas turbines offer the highest quality heat that is often used to generate power in a steam turbine. Gas turbines reject heat almost exclusively in their exhaust gas streams. The high temperature of this exhaust can be used to generate high-pressure steam or lower-temperature applications such as low-pressure steam or hot water. Larger gas turbines (typically above 25 MW) are frequently used in combined cycles where high-pressure steam is produced in the HRSG and is used in a steam turbine to generate additional electricity. The high levels of oxygen present in the exhaust stream allow for supplemental fuel addition to generate additional steam at high efficiency.

Some of the developing fuel cell technologies, including MCFC and SOFC, will also provide high quality rejected heat comparable to a gas turbine. Reciprocating engines and the commercially available PAFC produce a lower grade of rejected heat. Heating applications that require low-pressure steam (15 psig) or hot water are most suitable, although the exhaust from a reciprocating engine can generate steam up to 100 psig.

Reciprocating engines typically have higher efficiency than most gas turbines in the same output range and are a good fit where the thermal load is low relative to electric demand. Reciprocating engines can produce low- and high-pressure steam from their exhaust gas, although low-pressure steam or hot water is generally specified. Jacket water temperatures are typically limited to 210°F so that jacket heat is usually recovered in the form of hot water. All the jacket heat can be recovered if there is sufficient demand; however, only 40 to 60% of the exhaust heat can be recovered to prevent condensation of corrosive exhaust products in the stack that will limit equipment life.

10.3.1.2   Industrial Heat Recovery

Industrial sites that produce excess heat or steam from a process may offer a CHP opportunity. If the excess thermal energy is continuously available or at a high-load factor and is of sufficient quality, this heat can be used in a bottoming cycle to generate electricity in a steam turbine. In addition to electrical generation, steam turbines are often used to drive rotating equipment like air compressors or refrigeration compressors. Through a variety of turbine designs, the steam exhausted from the turbine can be used for lower-grade heating applications or cooling in a CHP configuration. Excess steam could also be used for reforming natural gas for a fuel cell.

10.3.1.3 Noise

Although fuel cells are relatively expensive to install, they are being tested in a number of sites, typically where the cost of a power outage is significant to lost revenues or lost productivity and uninterrupted power is mandatory. Stirling engines should also do well in these markets. Their relatively quiet operation has appeal and, thus, these units are being installed in congested commercial areas. Locating a turbine or engine in a residential area usually requires special consideration and design modifications.

Engine and turbine installations are often installed in building enclosures to attenuate noise to surrounding communities. Special exhaust silencers or mufflers are typically required on exhaust stacks. Gas turbines require a high volume of combustion air, causing high velocities and associated noise. Inlet air filters can be fitted with silencers to substantially reduce noise levels. Gas turbines are more easily confined within a factory-supplied enclosure than reciprocating engines. Reciprocating engines require greater ventilation due to radiated heat that makes their installation in a sound-attenuating building often the most practical solution. Gas turbines require much less ventilation and can be concealed within a compact steel enclosure.

10.3.1.4 Footprint

Three technologies in particular offer compact packaging and have an appeal to end users seeking an unobtrusive CHP system. Stirling engines are the smallest, followed by fuel cells and microturbines. Larger steam turbines, gas turbines, and reciprocating engines are generally isolated in either a factory enclosure or a separate building along with ancillary equipment. Table 10.2 shows equivalent footprint size for several different CHP types.

TABLE 10.2

Equivalent Footprint Size for Different CHP Types

Technology	Equivalent CHP Dimensions (for indicated KWe)
Steam turbine	Refrigerator (100)/Garage (100,000)
Gas turbine	Ryder Truck (3000)
Reciprocating engines	2 Refrigerators (30)
Micro turbine	Refrigerator (30)
Fuel cell[a]	Ryder Truck (200)
Stirling engine	Oven (3)

[a] Includes the fuel cell stack and the balance-of-plant (BOP).

10.3.1.5 Fuel Considerations

Since not all fuels can be used with every technology, fuel can also dictate fixed costs as shown in Table 10.3.

TABLE 10.3

Cost Comparison for CHP Fuels

Fuel Type	Installed Cost	Fuel Cost	O&M Costs
Coal	Medium — new Very low — old	Low	Medium
Natural gas	Low	Low	Low
Petroleum	Medium	Medium	High
Waste heat	Medium	Zero	Medium
Biomass	Medium-high	Low	High

A potential system issue for gas turbines is the supply pressure of the natural gas distribution system at the end user's property line. Gas turbines need minimum gas pressures of about 120 psig for small turbines, with substantially higher pressures for larger turbines. Assuming there is no high-pressure gas service, the local gas distribution company would have to construct a high-pressure gas line or the end user purchase a gas compressor. The economics of constructing a new line must consider the volume of gas sales over the life of the project.

Gas compressors may have reliability problems, especially in the smaller size ranges. If black start capability is required, then a reciprocating engine may be needed to turn the gas compressor, adding cost and complexity. Reciprocating engines and fuel cells are more accommodating to the fuel pressure issue, generally requiring under 50 psig. Reciprocating engines operating on diesel fuel storage do not have fuel pressure as an issue; however, there may be special permit requirements for on-site fuel storage.

Diesel engines should be considered where natural gas is not available or very expensive. Diesel engines have excellent part-load operating characteristics and high power densities. In most localities, environmental regulations have largely restricted their use for CHP. In California and elsewhere in the U.S., diesel engines are almost exclusively used for emergency power or where uninterrupted power supply is needed, such as in hospitals and critical data operating centers. As emergency generators, diesel engines can be started and achieve full power in a relatively short period of time.

10.4 CHP Economics

The economic competitiveness of CHP is site specific and varies according to size and load applications. With new market rules and new technologies, past guidelines and rules of thumb may no longer apply. In this section, the competitive position of CHP is evaluated in terms of future electric and gas prices, CHP technology cost and performance, and a set of plausible demand patterns of customers by size and application. Forecasters typically provide only average prices. Because CHP significantly alters a customer's load profile, an analysis based on average prices is of limited value. Typically, analysts divide rates into base-load, intermediate, and peaking (in order of increasing costs).

10.4.1   CHP Technology Cost and Performance Characteristics

Table 10.4 shows characteristics for application sizes from as small as 50 kW to as large as 25 MW. The heat rates and recoverable thermal energy factors are based on commercial product specifications, with the exception of the microturbine, for which performance factors are estimated. Microturbine cost factors were estimated based on assessment of early market entry economics and not manufacturers' projections for high volume production.

Package costs, heat recovery equipment costs, and balance-of-plant costs can vary widely by application and the degree of competition. The costs in the table reflect realistic estimates for costs for these technologies. The bottom two rows of Table 10.4 are examples of generated power costs. These are effective average power costs achievable by base-load operation of these technologies at the assumed costs for both power-only and CHP applications. The small engine and microturbine technologies are assumed to have an economic life of 10 years; the remaining technologies are assumed to have an economic life of 15 years. The CHP costs differ from the power-only costs by the addition of the heat recovery capital costs and the assumption that the heat recovered replaces that produced by an 80% efficient gas-fired boiler. The gas cost for the analysis was assumed to be $2.50/MM Btu.

Operating costs include both fuel and nonfuel expenses (such as replacement of spark plugs for engines, and replacement of stacks for fuel cells). As discussed above, many of the most efficient technologies can operate on only very pure (expensive) fuels. Per Btu, the cheapest fuel is coal, which can be used only with boiler/steam turbine and Stirling engine CHP applications. The primary economic driver for CHP is production of power at rates that are lower than the utility's delivered price. Figure 10.5 demonstrates graphically how CHP compares with traditional central station generation combined with the necessary transmission and distribution (T&D) to move the power to the load.

By comparison, the cost to produce electricity from a CHP system using an industrial-sized gas turbine, including fuel, capital, and operation and maintenance (O&M) expenses, is less than $0.04/kWh for base-load purposes. This cost compares favorably with a base-load central-station combined-cycle plant at the busbar, even before T&D charges are added in. As shown in Figure 10.5, CHP can also compete against large simple-cycle gas turbine plants for intermediate-load purposes and peaking power once T&D costs are factored in. The T&D charges represented in this exhibit include 7% line losses and a $150/kW investment.

The cost of CHP varies, of course, by application, technology, and grid circumstances, but, as this example illustrates, the economic fundamentals will

TABLE 10.4

CHP Technology Cost and Performance Estimates

Cost/Performance Characteristics	Microturbine	Gas Engine	Gas Engine	Gas Turbine	Gas Turbine
Performance					
Size kW	50	100	800	5000	25,000
Heat rate (Btu/kWh HHV)	13,306	13,127	10,605	11,779	10,311
Exhaust heat (Btu/kWh)	4498	1786	1443	5193	4522
Coolant (Btu/kWh)		3404	2750		
Cost					
Package cost ($/kW)	$500	$650	$350	$400	$300
Heat recovery	$150	$100	$75	$75	$75
Emission controls	$0	$70	$29	$102	$100
Project management	$25	$33	$18	$20	$15
Site & construction	$35	$46	$25	$28	$21
Engineering	$20	$26	$14	$16	$12
Civil	$50	$75	$38	$15	$13
Labor/installation	$100	$130	$44	$60	$45
CEMS	$0	$0	$0	$30	$20
Fuel supply-compressor	$40	$0	$0	$20	$15
Interconnect/switchgear	$150	$150	$63	$20	$8
Contingency	$25	$33	$18	$20	$15
General contractor markup	$164	$197	$101	$81	$64
Bonding/performance	$44	$39	$20	$24	$19
Constr. carry charges	$83	$99	$51	$87	$69
Basic turnkey cost ($/kW)	$1375	$1647	$842	$998	$789
Maintenance cost ($/kWh)	$0.010	$0.014	$0.011	$0.003	$0.003
Power Cost ($/kWh)					
No heat recovery	$0.075	$0.085	$0.053	$0.051	$0.040
With heat recovery	$0.067	$0.075	$0.042	$0.037	$0.027

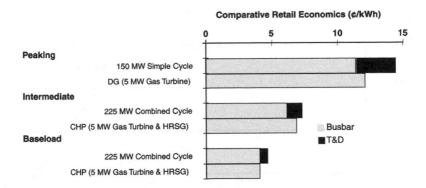

FIGURE 10.5
Cost of power from on-site CHP versus delivered price. From ONSITE SYCOM Energy Corporation, Market Assessment of CHP in the State of California, draft report to the California Energy Commission, September 1999.

frequently favor CHP. In a restructured environment, users may also begin to place significant economic value on the standby capability and increased power reliability that CHP can provide, further enhancing the potential economic benefits of on-site CHP.

Figure 10.6 shows the convergence of first cost of many CHP technologies. While it is true that the costs of all the technologies have fallen steadily, Figure 10.6 (which shows the average capital cost of each technology) reveals that some have declined more quickly than others. Technologies just becoming commercial, such as fuel cells and Stirling engines, are much more expensive, but have faster falling costs than those of established technologies.

While smaller technologies were more expensive in the past, with manufacturing advances and material and sensor enhancements this is no longer the case. The ability to use volume manufacturing is now the variable that drives the cost.

10.4.2   CHP Financing

CHP projects generally are financed with a mixture of internal funds and debt financing. Because of this, CHP decision makers use conservative metrics (such as very short payback) before they are willing to invest. This conservatism may be due to their limited ability to borrow to finance CHP projects, or because they have limited investment funds. It also allows for risk that the future savings will not be realized.

Sometimes, leasing arrangements are developed that eliminate the need for the site customer to come up with initial investment funds for the project, relying instead on third parties to own the system and take a portion of the benefits. For these types of projects, it is common for some kind of discounted cash flow analysis to be conducted. Here again, decision makers may protect themselves from risk by setting high hurdle rates for

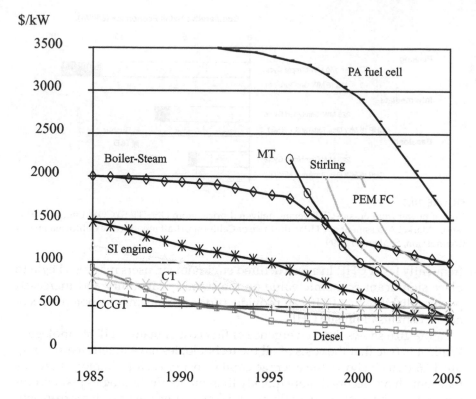

FIGURE 10.6
Cost trajectories for DG technologies applicable to CHP systems.

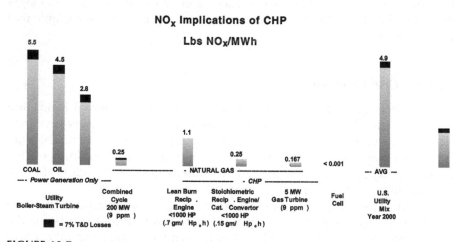

FIGURE 10.7
NO_x-related emissions reductions with CHP. (*Source:* (1) GRI Report Light Duty Vehicle Full Fuel Cycle Emissions Analysis, 1994; (2) Gas Turbine Environmental Analysis and Policy Considerations, Onsite Energy, 1997; (3) Sierra Energy and Risk Assessment, SoCalGas UEG Customer End-Use Specific Avoided Energy GT&D Costs and Emissions, 1997.)

the analysis. Thus, decision makers' perceptions of risk are an important element of what is considered an economic investment.

10.4.2.1 Payback Analysis

Payback analysis is a rule-of-thumb method often used for preliminary evaluation of potential energy projects; it is the time it takes (usually measured in years) for the initial investment to be recovered from the savings. For example, a simple payback analysis of a CHP project would divide the first year's savings into the initial capital cost to estimate the number of years required to return the initial investment. Often, decision makers require paybacks of two to three years or less, even though far longer paybacks might be excellent investments.

10.4.2.2 Discounted Cash Flow Analysis

Another way to determine whether a project is economically feasible is to use net present value (NPV) or internal rate of return (IRR). If the NPV is positive or if the IRR is greater than the decision maker's cost of money, then the project is considered economically feasible. For this analysis, the customer's acceptance of CHP was based on the projected IRR. A "myopic" IRR was used, i.e., the decision maker was assumed to value his yearly savings based on prevailing energy rates when the investment decision was made rather than on perfect knowledge of all future prices. Any project with an IRR above 10% provides customers with economic benefits, but acceptance levels drop off as the IRR declines to the economic floor. Simple paybacks were also calculated for comparison purposes, though these were not used in the market acceptance calculations.

10.5 Benefits of CHP

In the fall of 1998, the U.S. Department of Energy (DOE) and the Environmental Protection Agency (EPA) initiated the "CHP Challenge" to double the amount of CHP generating power in the U.S. by 2010.* The benefits estimated by DOE analysts (U.S. Department of Energy, 1998) include net additional energy savings of 1.3 Quads, carbon reductions of 40 Mtc, SO_2 reductions of 0.94 million tons, NO_x reductions of 0.42 million tons, and economic savings of $5.5 billion.

---

* For an update on the progress of the CHP Challenge, see *http:\\www.oit.doe.gov/chpchallenge* or *http:\\www.nemw.org\uschpa*.

10.5.1   CHP Efficiency

Power generation systems create large amounts of heat in the process of converting fuel into electricity. Over two-thirds of the energy content of the input fuel is converted to heat and wasted in many older central generating plants. As an alternative, an end user with significant thermal and power needs can generate both its thermal and electrical energy in a single combined heat and power system located at or near its facility. Figure 10.1 shows how a well-balanced CHP system outperforms a traditional remote electricity supply and on-site boiler combination. The chart illustrates that out of 100 units of input fuel, CHP converts 80 to useful output, 30 to electricity, and 50 to steam or some other useful thermal output; traditional separate heat and power components require 163 units of energy to accomplish the same end-use tasks. While future central station plants will be able to generate electricity more efficiently than the 30% average rate used in developing the chart, CHP installations with proper thermal/electric balance have design efficiencies of 80 to 90% and will still result in significant overall energy savings. On-site use of CHP also reduces transmission and distribution system line losses to zero from typical central unit line losses of 4 to 7%.

10.5.2   Emissions Reductions

By increasing the efficiency of energy use, CHP can significantly reduce emissions of criteria pollutants, such as NO_x and SO_2, and noncriteria greenhouse gases, such as CO_2.

Figures 10.7 and 10.8 show NO_x and CO_2 emissions comparisons respectively by power generation technology and fuel type. While reductions in both NO_x and CO_2 result from moving from solid and liquid fuels to natural gas, the figures show the added reductions that efficiency can provide. CHP technologies can significantly reduce emissions and compare favorably to advanced low-emission central station technologies such as gas-fired combined cycle.

10.5.3   Ancillary Benefits

In a restructured electric industry, CHP and other DG options can offer grid support to the distribution utility. They also give energy service providers (ESPs) or users the ability to offer ancillary services to the system, including:

- Voltage and frequency support to enhance reliability and power quality
- Avoidance or deferral of high cost; long lead time T&D upgrades

- Bulk power risk management
- Reduced line losses; reactive power control
- Outage cost savings
- Reduced central station generating reserve requirements
- Transmission capacity release

ESPs are working now to determine the quantity and value of benefits derived from grid-support and ancillary services that accrue from installing CHP and other DG systems.

CHP offers a customer enhanced reliability, operational and load management flexibility (when also connected to the grid), ability to arbitrage electric and gas prices, and energy management (including peak shaving and possibilities for enhanced thermal energy storage). The value of these benefits will depend on the characteristics of the facility, the form and amount of energy it uses, load profile, rate tariffs, prices of electricity and gas, and other factors. A facility making a CHP purchase decision will have to consider the ancillary benefits, including the revenue stream possible from sale of the transmission and distribution benefits to the ISO and reduced operating costs, along with the other costs and benefits of the project.

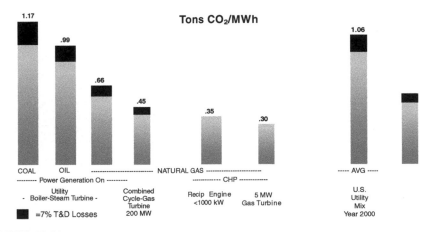

FIGURE 10.8
Comparison of CO_2 emissions from electricity generation. (*Source:* (1) GRI Report Light Duty Vehicle Full Fuel Cycle Emissions Analysis, 1994; (2) Gas Turbine Environmental Analysis and Policy Considerations, Onsite Energy, 1997; (3) Sierra Energy and Risk Assessment, SoCalGas UEG Customer End-Use Specific Avoided Energy GT&D Costs and Emissions, 1997.)

References

Casten, T., CHP — Policy Implications for Climate Change and Electric Deregulation, 2, 1998.

Spurr, M., *District Energy Systems Integrated with Combined Heat and Power*, International District Energy Association, October 1999.

U.S. Department of Energy, Energy department initiative aims at recovering industry power losses, press release PRL-98-046, Washington, D.C., 1998.

11

Electric Power Distribution Systems

Lawrence A. Schienbein and Jeffrey E. Dagle

CONTENTS

This chapter provides a basic introduction to *electric power distribution systems*. When considering the application of DG, it is important to consider issues that apply to the environment in which these systems are deployed. An overview of key issues that should be considered is provided in this context. While this section is not intended to provide details of specific interconnection standards that may apply to the interconnection of distributed resources with the electric power system, it is intended to provide insight into the technical basis behind those standards. A brief introduction of the electric power grid is followed by a discussion of physical and operational characteristics of typical electric power transmission and distribution systems. Economic and distributed resource interconnection considerations are discussed in later sections.

11.1 Transmission and Distribution System Characteristics

11.1.1 Physical Characteristics

The *transmission* system connects the generating stations and loads together through nodes called *substations* (see Figure 11.1). The substations contain switches and circuit breakers, transformers to connected different voltage levels, and other ancillary substation equipment (voltage control capacitor banks, reactors, metering and control equipment, etc.). Substation layout and complexity vary widely depending on the application.

Typical transmission voltages in the U.S. include extra-high voltages of 765 kV, 500 kV, and 345 kV. Other common voltages include 230 kV, 161 kV, 138 kV, and 115 kV. Lower voltages, such as 115 kV and 69 kV, are sometimes called *subtransmission* voltages. The difference between transmission and subtransmission has little to do with actual voltages — subtransmission refers to a lower-level grid hierarchy that interfaces with the bulk transmission backbone.

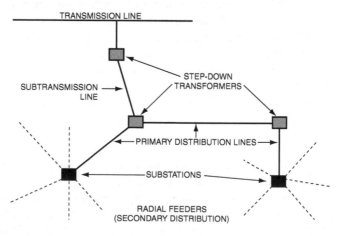

FIGURE 11.1
Typical transmission and distribution system (courtesy of Nicoara Graphics).

The *distribution* system provides the infrastructure to deliver power from the substations to the loads. Figure 11.2 shows the most common designs for distribution feeders. Typically radial in nature, the distribution system includes *feeders* and *laterals*. Typical voltages are 34.5 kV, 14.4 kV, 13.8 kV, 13.2 kV, 12.5 kV, 12 kV, and sometimes lower voltages. The distribution voltages in a specific service territory are likely similar because it is easier and more cost effective to stock spare parts when the system voltages are consistent.

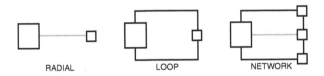

FIGURE 11.2
Typical distribution feeder configurations. The radial design is the most common in the United States (courtesy of Nicoara Graphics).

Power flow on a line is a function of voltage and current. Because the current itself is inherently bidirectional, power can typically readily flow in either direction. However, other operational constraints such as circuit breakers and other control devices may not be able to accommodate reversal of power flow without replacement or modification.

The electric power grid operates as a *three-phase network* down to the level of the service point to residential and small commercial loads. Feeders are usually three-phase overhead pole line or underground cables. As one gets closer to the loads (many of which are single phase), three-phase or single-phase laterals provide spurs to the various customer connections.

Three-phase electricity refers to voltage waveforms (and corresponding current) 120° out of phase with each other. This provides advantageous characteristics for rotating machines (both generators and motors) by inducing a smooth rotating magnetic field with which the rotating magnetic field can be coupled. Also, this offers a significant advantage for electric power transmission and distribution because each of the three phases will cancel each other out when combined. This makes it possible to string three conductors carrying voltage and rely on the mathematical cancellation of this power to provide a virtual neutral. Without a metallic return wire, significant cost savings can be achieved. Thus, nearly all transmission and distribution power lines have only three conductors.

To ensure efficient operation, it is important to balance the phases so that they are approximately equal. This is achieved through load balance, and also obtained through transformer configurations (see Figure 11.3) depending on whether the transformer terminals are configured with a three-conductor delta, a three-conductor wye, or a four-conductor wye (with this fourth connection optionally grounded). It is also common for the *primary and secondary terminals* of a transformer to have different configurations (hence a delta–wye transformer). These transformer connections are important for

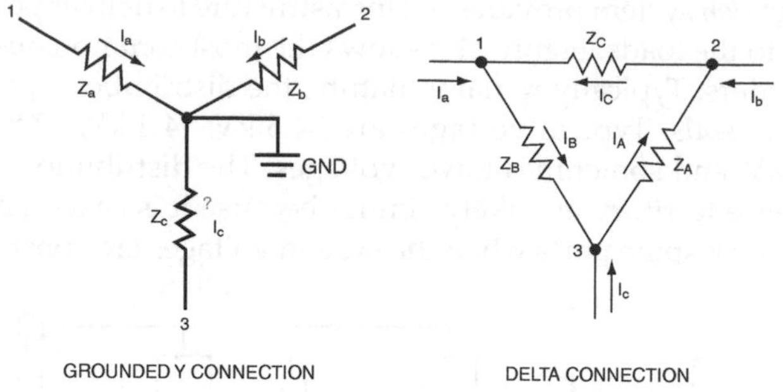

GROUNDED Y CONNECTION   DELTA CONNECTION

FIGURE 11.3
Single-phase transformer configurations for delivering three-phase power (courtesy of Nicoara
Graphics).

balancing phases. These considerations are also important for how ground
fault current will flow.

11.1.2   Load Characteristics

Depending on the needs of the customer, the voltage supplied can be as low
as 120 V single-phase or 120/240 V single-phase, where the 240 V secondary
of the distribution transformer has a center tap that also provides two 120 V
single-phase circuits. Larger customers may utilize three-phase power, with
120/208 or 277/480 V service.* Very large, industrial customers, for example,
can include higher three-phase voltages, such as 2400 V, 4160 V, or greater
(see above). Depending on the voltage at which it acquires power, the customer is usually responsible for providing the transformers and other infrastructure to serve all of the lower voltage requirements needed by its facility
or site.

A typical *service transformer* (ground mounted or pole mounted), supplying
five to ten residences, converts the three-phase power at a distribution voltage of 13.8 kV (transformer primary), for example, to power at 120/240 V.
Each of the 240V, single-phase transformer secondaries has a center tap that
provides two 120 V single-phase circuits. Therefore, an individual residence
can be supplied with both 120 and 240 V single-phase service. The higher
voltage is necessary for appliances such as clothes dryers.

11.1.3   System Protection

To protect distribution system equipment from damaging overloads, that
equipment contains *protective devices*. This protection can be manifest in a

---

* A common way to denote three-phase service is to list the line-to-neutral followed by the line-
to-line voltage. Three-phase circuits are defined by the line-to-line voltage, with its line-to-neutral voltage multiplied by $\sqrt{3}$.

variety of forms, but usually through fuses or circuit breakers. The devices sometimes employed to provide the logic to decide when to trip a circuit breaker are referred to as *protective relays*, or sometimes simply *relays*. Relays can be programmed for a variety of functions, including (but certainly not limited to) overcurrent, over- and under-voltage, over- and underfrequency, differential current, and reverse current.

The primary objective of protective devices is to de-energize equipment when conditions warrant. To protect equipment or maintain safety, these devices typically respond to *faults* (e.g., a short circuit) by *isolating* appropriate equipment. The goal of good protective coordination is to limit the isolation to as small a portion of the system as possible. This contains the disruption and prevents adjacent portions of the system from being affected.

Also, protection schemes need to account for the failure in isolating devices (e.g., fuses, circuit breakers, etc.) by providing redundant protection capability. This is usually accomplished through protective zones whereby protective relays operate quickly for faults within their primary zones and more slowly for faults that are farther away in secondary or tertiary zones. Therefore, if a relay (or its associated circuit breaker) fails, another relay and its circuit breaker will operate as a backup, but, because this secondary system is farther away, it will disrupt a larger portion of the system. A simple example illustrates this principle. A short in a residential circuit will trip the circuit breaker associated with that circuit. Should that breaker fail to clear the fault in time, the main circuit breaker will trip and interrupt power to the entire panel. Similarly, there are other overlapping zones of protection upstream in the distribution system.

Because many faults in overhead transmission and distribution lines are transient in nature (e.g., lightning and small animal contact), the fault persists as long as the circuit is energized and the arc is sustained. When the fault is *cleared* (circuit is de-energized), the arc is extinguished. In those cases, it is safe to *reclose* (energize the circuit again). Therefore, most overhead circuits have *reclosers* that automatically re-energize the line after clearing a fault. If the line continues to trip, that means that a *persistent* fault exists, and the line goes into *lockout* (remains de-energized until a line crew inspects the line and manually resets the circuit breaker).

11.2 Operational Concerns

The overriding consideration regarding the interconnection of distributed resources to an electric utility system is ensuring the safety of the maintenance crews that must work on the distribution system. There must be absolute assurance that these systems cannot pose a risk to the safety of the utility linemen at any time. The most robust situation is where the generation device is physically prevented from backflow because it is connected through a transfer switch. A transfer switch connects the load to a normal source of power or an emergency source of power, but because of the construction of the switch itself, these two sources of power can never be connected to each other, even inadvertently. Use of a transfer switch also provides assurance that two sources that have not been synchronized cannot be inadvertently connected. With such a transfer switch configuration, there is no possibility that the generation can backfeed into the power system, and thus there are no adverse impacts to the power system.

However, operation of the on-site generation in parallel with the utility source is often desired. This can be a vitally important means of ensuring that adequate resources are available to meet peak load requirements in excess of generating capacity, or fulfilling a desire to maximize generation by generating either base-load (constant) or peaking power requirements. This could be driven by economies associated with the specific application or by resource availability, as is the case with certain renewable resources. In these cases, the generator may provide only a portion of the total on-site load, in excess of the on-site load, or combinations of the two at various times. The consequence of inadvertently connecting two sources out of sync is immediate and severe. Therefore, any time that it is possible to operate in parallel with the grid, appropriate synchronizing and sync check relays are absolutely necessary.

It is also imperative that the generator does not inadvertently feed the utility during a utility outage. This can be as simple as a directional relay that prevents any backflow. However, in many instances the design may call for the generator to feed back into the utility under normal conditions. This issue can be complicated if there are multiple generators or alternate sources of power generation (e.g., wind or solar generation with battery backup). In those cases, other types of protection such as under-frequency or undervoltage will be necessary to determine when to isolate from the utility source. The specific design requirements can vary greatly because they depend on many factors.

11.2.1   Planning Considerations

An electric power system is the ultimate just-in-time delivery system; the consumption is being simultaneously produced. Maintaining this balance between generation and load for various time scales is one of the primary objectives of the electric utility industry. Deregulation has significantly compromised the ability of industry to maintain this balance throughout the system.

At the longest time scale, a source of adequate supply requires multi-year planning horizons. Where to build generation capacity, how much, and what type of fuel is to be used are questions that often require years of lead time in order to be adequately prepared to meet demand. Planning studies to determine adequate transmission capacity and the process of citing and constructing this infrastructure also take several years. These assets are very expensive (a large coal-fired plant can have a capital cost of over two billion dollars; large transmission lines can be as much as one million dollars per mile). Thus, if wrong choices are made, unnecessary investment or *stranded assets* can result.

Historically vertically integrated, utility companies, with the approval of their associated regulatory oversight bodies, made these decisions. By virtue of their monopoly status, investor-owned utilities (IOU) are heavily regulated by appropriate state and federal regulators. A regulatory body, typically a state public utility commission (PUC), approves these investments by allowing the capital outlays for capacity expansion to be added to the utilities' rate base. The rate for electrical service is then determined to be that necessary to recover this investment, including operating expenses and a specified rate of return for the shareholders.

Deregulation is intended to change the way these long-term investment decisions are made. Removing energy generation from the domain of the monopoly utility companies and allowing open competition for electrical production, will enable energy generators to make market-based decisions regarding where and how much generation should be built. The theory is that the free market will be more efficient, encourage more innovation, and result in fewer stranded assets when too much or the wrong type of generation is built.

To ensure that incumbent utility players will not exert excessive market power, many regions that have begun the deregulation process have instituted independent system operators (ISO) to operate the transmission and distribution grid. The nature of the ISO varies dramatically throughout the U.S. It can either be a wholly new organization, as is the case in California, or a slight modification of an existing power pool, such as the Pennsylvania–New Jersey–Maryland (PJM) interconnection.

At the next time scale, arranging for the operation of existing generating plants requires that fuel contracts be arranged, maintenance outages scheduled and coordinated, and long-term contracts to either buy or sell power with neighboring systems or between regions determined. It is common for utilities to plan fuel purchases years in advance. Sometimes, however, these decisions may be made in relatively short time scales (e.g., spot market purchases). The primary consideration is economic optimization — how to best meet the power-generating obligations with a minimum fuel investment.

Unit commitment is the next time scale whereby generating plants are committed to operation. Which units are committed to be on-line, and which units are taken off-line and shut down? This decision generally takes into account near-term load requirements as well as short-term load forecasts. The basic question here is how to optimize the available generation assets in order to minimize the cost of starting plants and shutting them down as the load changes throughout the day.

11.2.2   Real-Time Operations

Unit dispatch is the near real-time matching of available generation capacity to the load. The control center is responsible for determining how much generation must be dispatched to meet load requirements. The systems used for this function are referred to as the energy management system (EMS) and supervisory control and data acquisition (SCADA).

In North America there are about 140 control centers, each responsible for controlling a portion of the interconnected grid. These *control areas* together form an interconnected system. The three interconnections in the US. are comprised of the eastern interconnection, the western interconnection, and Texas.

Traditionally, the control areas were the service territories of major electric utility companies. However, there is a trend toward control areas defined by an ISO service area that provides control area functions but is not part of the traditional vertically integrated utility company.

Because the interconnected grid involves interstate commerce, the Federal Energy Regulatory Commission (FERC) provides regulatory oversight and approves wholesale transmission tariffs. The North American Electric Reliability Council (NERC) was established following the 1965 northeast blackout. Chartered with ensuring the reliability of interconnected systems, NERC develops standards of operations and encourages interaction among transmission grid operators to ensure smooth interconnected operation. Unlike FERC, which is a federal regulatory body, NERC is an industry association by voluntary membership.

NERC has recently established regional security coordinators to enhance communication among the control area operators. It has become apparent that control areas, through information assets such as SCADA and EMS, have a relatively complete and comprehensive view of their own service territories, but not necessarily enough information about the unfolding situation in neighboring service territories. Because the integrated system behaves as one cohesive system, regardless of ownership boundaries, an impending problem that develops remotely can have a cascading effect and, thus, create problems in adjacent portions of the grid. The motivation for regional security coordinators is to monitor information from multiple control areas, share information with the other regional security coordinators, and develop a real-time situational awareness of the interconnected system that transcends individual geographic domains of responsibility. Specific actions of these regional coordinators includes approving maintenance scheduling that might have an impact on reliability, providing indications and warnings of impeding problems, and facilitating coordination among control center operators who may have incomplete information on grid health and current operational status.

Other actions taken by NERC include adopting transmission loading relief (TLR) procedures for backing off transactions on overloaded transmission corridors. Various databases and analysis tools have been developed to support this, and a formalized process for tagging energy transactions has been implemented.

11.3 Distribution System Economics

While this section is not intended to be a comprehensive treatment of distribution system economics, it may be instructive to the reader to understand some fundamental economic issues associated with distribution system planning and operations. Transmission and distribution system equipment must be sized to accommodate the maximum load to which the equipment will be subjected. Because average demand is usually much less than maximum demand (as shown in Figure 11.4), there is normally a lot of unused capacity in the system. Because of load diversity characteristics, this is usually more prevalent closer to the load.

When capacity needs to be added, expansion must come in increments. This incremental expansion can be quite large compared to the rate of growth. For example, if there is a 30 MVA transformer serving a community that becomes overloaded, it is a very large investment to replace it with a 45 MVA transformer. Other options include adding a second transformer in parallel, or other methods of reconfiguring load. However, all of these options can be quite expensive.

Planning how to accommodate this capacity expansion requires accurate load forecasting. Often, where load growth occurs has little correlation with existing load. For example, if a utility predicts annual growth of 4%, most of this will likely occur in outlying areas with new construction rather than in the areas in which load is currently served. Therefore, siting and sizing substation capacity, and the feeders and other distribution infrastructure, becomes a difficult planning exercise. It is anticipated that distributed resources will increasingly enhance the ability of planners to cope with this growth without investing in infrastructure upgrades that have a likelihood of becoming stranded assets if the planning assumptions are wrong.

11.3.1 Ancillary Services

Ancillary services are functions performed by the transmission grid control area operator to ensure grid reliability. Electricity is more than a commodity, with many aspects of ensuring a continuous supply of high-quality power also necessary. FERC has identified specific ancillary services that must be unbundled from the overall transmission tariff (see Table 11.1). These have been traditionally provided as part of the overall rate a customer pays for electricity (or the rate a marketer would pay for transmission service in the wholesale market). Unbundling these ancillary services

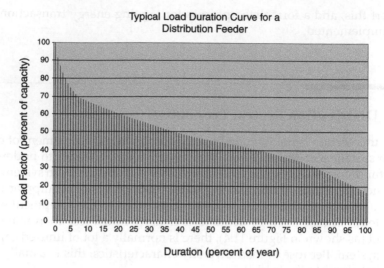

FIGURE 11.4

Duration of distribution system utilization.

TABLE 11.1

Key Ancillary Services and Their Definitions

Service	Description	Time Scale
Services FERC Requires Transmission Providers to Offer and Customers to Take from the Transmission Provider		
System control	The control area operator functions that schedule generation and transactions before the fact and that control some generation in real-time to maintain generation/load balance; interconnected operations services working group definition more restricted, with a focus on reliability, not commercial, activities, including generation/load balance, transmission security, and emergency preparedness	Seconds to hours
Reactive supply and voltage control from generation	The injection or absorption of reactive power from generators to maintain transmission-system voltages within required ranges	Seconds
Services FERC Requires Transmission Providers to Offer but which Customers Can Take from the Transmission Provider or Third Parties or Self-Provide		
Regulation	The use of generation equipped with governors and automatic-generation control to maintain minute-to-minute generation/load balance within the control area to meet NERC control-performance standards	~1 minute

TABLE 11.1 (CONTINUED)

Key Ancillary Services and Their Definitions

Service	Description	Time Scale
Operating reserve — spinning	The provision of generating capacity (usually with governors and automatic-generation control) that is synchronized to the grid and unloaded and that can respond immediately to correct for generation/load imbalances caused by generation and transmission outages and that is fully available within 10 minutes	Seconds to < 10 minutes
Operating reserve — supplemental	The provision of generating capacity and curtailable load used to correct for generation/load imbalances caused by generation and transmission outages and that is fully available within 10 minutes (unlike spinning reserve, supplemental reserve is not required to begin responding immediately)	< 10 minutes
Energy imbalance	The use of generation to correct for hourly mismatches between actual and scheduled transactions between suppliers and their customers	Hourly
Services the FERC Does Not Require Transmission Providers to Offer		
Load following	The use of generation to meet the hour-to-hour and daily variations in system load	Hours
Backup supply	Generating capacity that can be made fully available within one hour; used to back up operating reserves and for commercial purposes	30–60 minutes
Real power loss replacement	The use of generation to compensate for the transmission system losses from generators to loads	Hourly
Dynamic scheduling	Real-time metering, telemetering, and computer software and hardware to electronically transfer some or all of a generator's output or a customer's load from one control area to another	Seconds
System black-start capability	The ability of a generating unit to go from a shutdown condition to an operating condition without assistance from the electrical grid and to then energize the grid to help other units start after a blackout occurs	When outages occur
Network stability services	Maintenance and use of special equipment (e.g., power-system stabilizers and dynamic-braking resistors) to maintain a secure transmission system	Cycles

Source: Hirst, E. and Kirby, B., Creating Competitive Markets for Ancillary Services. ORNC/CON-448, Oak Ridge National Laboratory, Oak Ridge, TN, 1997 (with permission).

individually and adopting market-based pricing should encourage competition and eventually lead to innovation and alternate lower-cost means for obtaining these functions.

The first two ancillary services defined by FERC, system control and reactive power supply from generation (for grid voltage control), are obtained from the transmission providers. System control refers to those functions that are provided by the control area operators. Maintaining the 60 Hz frequency on the grid (by maintaining a real-time balance between generation and load), maintaining reliability of the bulk transmission grid, and other functions of operating the interconnected power grid are part of this ancillary service.

Reactive power control from generators is necessary to maintain an appropriate voltage profile of the transmission grid. The control area operator dispatches the terminal voltage of key generating units. The generating unit consequently produces reactive power commensurate with that needed to maintain the desired voltage. This control is necessary to maintain appropriate voltages on the transmission grid, which is also an important function of controlling power flows through the network. Figure 11.5 shows the basic idea of reactive power.

The next ancillary services defined by FERC and described below may be obtained from the transmission provider, purchased from other third-party providers, or self-provided. Regulation is the process whereby individual generating units respond to signals from the control area operator to change generation output in real-time to assist with the balance of load and generation (and, consequently, contribute to the regulation of the 60 Hz grid frequency). This process, also called *automatic generation control*, determines the total generation needed to maintain the balance with load. Derived from measuring the total flow of power with adjacent control areas and subtracting the scheduled power flow, *inadvertent energy exchange* is added to the frequency error multiplied by the *frequency bias*. Frequency bias is an empirically derived constant that recognizes the natural regional frequency response characteristics. Relatively large frequency differences and transients cause *governors* on individual generators to respond in real-time (increasing power output when measured frequency decreases and vice versa). Recognizing that not all of the generation needs to be controlled in this manner, only a portion of the total on-line generation is included in this automatic generation control. This ancillary service compensates those generators that must change their real power output in response to these control signals.

Spinning reserve is on-line generating capacity that is immediately available to compensate for unexpected changes in the network, such as generator units or transmission lines tripping out of service. This spinning reserve absorbs these changes to maintain a reliable and stable grid. While the exact amount of spinning reserve varies depending on peculiarities of regional reliability standards, it is usually 5 to 7% of total on-line capacity or equivalent to the single largest on-line generator unit, whichever is less. This ancillary service compensates those generators that are on-line but must operate below their full rated output to provide spinning reserve.

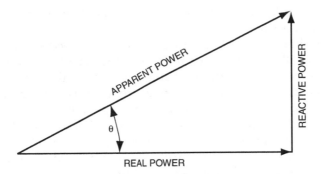

FIGURE 11.5
Power factor (ratio of real to apparent power) impacts every aspect of electrical distribution system management. Reactive power, which does no "work," consumes up to 15% of all electrical transmission capacity (courtesy of Nicoara Graphics).

Supplemental reserve is similar to spinning reserve but is not necessarily an immediate response to changing network conditions. Supplemental reserves might be called into action as a result of events that deplete spinning reserves or errors in short-term load forecasts that result in the need for additional capacity, or they may be used to accommodate other changes in generator or load requirements. This ancillary service maintains generating capacity in ready reserve or compensates for load curtailment action.

In order to accommodate mismatches between scheduled and actual deliveries of power, hourly reconciliation is necessary. The energy imbalance ancillary service can be procured from the transmission provider or a third-party provider, or can be self-provided. The remaining ancillary services are not mandated by FERC and can be arranged by separate agreement with transmission providers as necessary, depending on local or regional requirements. Load following is the ability to accommodate expected variations in customer loads that may not be easy for power markets or generators who are providing a fixed resource. This ancillary service is provided by those who have access to variable sources of generation that can complement that available from fixed sources. Backup supply is necessary when the primary source of generation must be taken off line due to equipment failure or unexpected maintenance. This ancillary service provides an alternate means of supplying customers with other sources of power that can be brought into the market.

The power loss in the transmission network between the point of delivery and the point of supply can be replaced with the next ancillary service, real power loss replacement. By separating this as an ancillary service, alternative means for providing this difference (also recognizing that many transactions will have many different points of supply and delivery that may vary throughout the loading cycle) may be easier to accommodate by the transmission provider or a third party. Also, because the total power loss on the transmission network is a function of its total utilization (rather than a linear summation of losses from individual transactions), accounting may be simplified if handled as an ancillary service rather than a direct overhead on individual transactions.

When real-time changes to existing schedules are needed, dynamic scheduling is the ancillary service that recognizes the need for an automated system to be able to handle these changes in near real time. Allowing the transmission provider or a third party to provide this infrastructure relieves the burden on power marketers to have this capability in-house.

If the grid fails, restoration requires power generation facilities that can start (without any off-site power) and feed the grid (thereby starting other power generating units that require off-site power in order to start). The ability to provide this service may require additional capital and maintenance expense that is compensated through the black-start ancillary service.

The final ancillary service is network stability services. How the generator interacts with the grid, and whether it creates or alleviates network stability problems, is vitally important to maintaining a reliable grid. The proper design, installation, maintenance, and operation of key control equipment associated with the generating facility are imperative to maintaining grid stability. Examples of these devices include power system stabilizers and other supplementary controls associated with exciters or governors. This ancillary service reimburses the cost of maintaining these controls in proper working order and may include modeling and verification activities to ensure that proper contribution to network stability is maintained.

11.4   Distributed Resource Interconnection Considerations

11.4.1   Introduction

This section discusses protection and control, the power quality issues, and other operating concerns (such as possible intermittent operation) of distributed resource interconnection with the power distribution network. The Electric Power Research Institute (EPRI) defines distributed generation (DG) as "the integrated or stand-alone use of small modular resources by utilities, utility customers, and third parties in applications that benefit the electric system, specific customers, or both." The term is synonymous with other commonly used phrases like self-generation, on-site generation, cogeneration, and "inside the fence generation." This chapter defines distributed resources to include distributed, smaller, power generating units and energy storage devices such as battery banks.

Although the proliferation of energy storage units as distributed resources today greatly lags behind the interconnection of distributed generators, the unique requirements of energy storage devices with respect to interconnection should be considered in planning for future distributed resource growth.

For example, pumped hydroelectric facilities can often be considered as distributed energy storage where the return of the stored energy to the network requires the use of water-powered turbines and conventional electric generators. By comparison, a conventional battery bank returns power to the network through a power electronic-based inverter that changes direct current to alternating current. While both are energy storage systems, the protection and control requirements for each will differ greatly.

Distributed resources which feed electricity directly to the utility power grid are sometimes referred to as *utility interactive* power systems. It is precisely this interaction with the utility that mandates the need for comprehensive and reliable control of the distributed resource unit and protection of the network, distributed resource and customer load assets, and personnel.

As Figure 11.6 demonstrates, a typical distribution feeder moves electricity from higher to lower voltage in a single direction. In Figure 11.7, the DG unit does not backfeed more electrical voltage than the line is designed for, and therefore does not disrupt the existing protection devices.

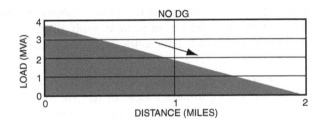

FIGURE 11.6
Typical load flow along a distribution feeder without DG unit (courtesy of Nicoara Graphics).

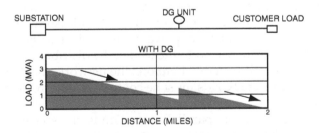

FIGURE 11.7
Distribution feeder with DG unit sized for the line, no backfeeding (courtesy of Nicoara Graphics).

Distributed resource configurations and operating modes can range from the fairly straightforward and occasional start-up, synchronization, and connection of a small, on-site, backup diesel generator unit for peak shaving purposes to the nearly continuous operation of several large diesel generator units or fuel cell systems in parallel, supplying power both to the on-site load and to the utility feeder (a form of cogeneration). The latter system would include fully automatic controls, transfer switches, and protective relays to facilitate seamless transfer (that is, power transfer that is completely unnoticed by the power user) between grid-connected operation and grid-independent operation and to ensure protection of the network and user assets. (In the grid-independent, or off-grid, configuration, the on-site load is disconnected from the feeder and supplied only by the on-site generation. This can also be termed islanded operation). The system might also be dispatchable by the network control center through SCADA.

11.4.2   Distributed Generation and Distribution System Control and Protection

11.4.2.1   Parallel and Grid-Independent Operation

Control and protection requirements can be discussed more readily by dividing the distributed resources into two main configuration groups:

- Distributed resources configured to operate mostly *in parallel with the utility network* (and, hence, in parallel with all other generators connected to the network). This configuration is sometimes termed *"utility interactive."*

- Distributed resources configured to operate primarily *independent of the utility network* (*grid-independent* or *off-grid* mode) but which may be occasionally interconnected with the network. This mode is sometimes also called the intentional islanded mode. However, this must be clearly differentiated from accidental or unintentional islanded mode, which can have severe consequences.

Parallel (or grid-interactive) distributed resources are defined as those that can be and are routinely connected to the network (sometimes called the *common bus*). Here, the transfer of power between the two systems is desired and expected. When connected to the feeder, the distributed generator is, in fact, part of the overall system. Therefore, the network, the distributed generator and the connected loads must be controlled and protected as an integrated system.

Grid-independent (or nonparallel) operation is of lesser importance here because the distributed resources in this case are designed to supply power only to an isolated load, and there may be no (or very infrequent) interaction with the network. Distributed resources of this type can be broadly classified as backup power systems and uninterruptible power supplies (UPS). An example of a grid-independent system is a battery bank and inverter (the distributed resource in this case) that supplies AC power to loads in a home following the interruption of power from the utility feeder/distribution transformer. The battery and inverter system is connected to the home distribution panel only after the panel is disconnected from the feeder. When utility service is restored, the inverter/battery system is disconnected from the loads before the distribution panel is reconnected to the power supply from the feeder. The battery bank charger is reconnected to the distribution panel, and the batteries are recharged while they await the next outage.*

The fundamental utility requirement for nonparallel distributed generators is that the load is transferred from the network supply to the backup generator in an open transition as described — in short, a switching sequence where the load is disconnected positively from the network before it is switched to the distributed resource (or *islanded* with the distributed resource). Of course, in addition, the user is concerned about the protection of the connected loads, and the distributed generator is concerned about the protection of the generating unit. However, the power distribution network operators and other power users on that network are no longer concerned about the operation of that distributed generator.

11.4.2.2 *Discussion of Control and Protection*

The objective of the protection and control system is to enable the distributed resource generators and/or storage devices to deliver the intended services to the users and the distribution system reliably, safely, and cost effectively. *Protection*, in the context of distributed resources connected to the utility network (i.e., operating in parallel), includes protection of network physical assets (lines, breakers, disconnects, transformers, etc.) and the utility line personnel, protection of the distributed resource assets and personnel, and protection of the loads served by the distributed resource in combination with the utility supply as it pertains to the flow of electrical power.

The *distributed resource control* function is focused on starting, stopping, paralleling, and disconnecting the generators and/or storage systems in an orderly and reliable manner. In addition, the controller monitors the health of the distributed resource subsystems such as generator over temperature, shutting down the units in an orderly manner should any fault or combination of fault conditions not related to the network connection, and delivery of power to or from the network occur.

Network protection and control systems have, until now, been designed for the central generation type of utility system, as described earlier. The introduction of distributed resources into these systems demands changes in the protection and control schemes of both the network and the distributed resource systems. In addition, the scope of the changes increases as the penetration of distributed resources, in terms of total number of units and total capacity, increases.

The need for additional protective devices and control logic is true even at the most fundamental level of DG. For example, a very small power-generating unit, such as a rooftop photovoltaic (PV) panel and inverter (perhaps a

---

* It is worth noting that even this very basic form of grid independent system can be very useful to the operation of the distribution system because, in principal, this specific load could be "shed" by remote control whenever necessary without any significant reduction in the power supplied to the load (of course, this depends on the duration of the load shedding).

few hundred W up to perhaps 3000 W peak capacity), is connected on the load side of a home's main distribution panel. Power flows both to and from the feeder on a net-metering basis. Such an installation requires anti-islanding protection logic and a transfer switch arrangement and associated control logic to ensure that there will be positive, immediate, and reliable disconnection from the feeder in the event of a feeder outage or an inverter/PV array fault.

While protection logic is part of the overall control logic, it is focused on solving two important problems:

1. Reducing potentially damaging transients when connecting and synchronizing the distributed generator units to the network and when disconnecting those units from the network

2. Protecting the utility feeder, the loads, and the utility personnel by ensuring that there is no possibility of one or more distributed generators continuing to supply a utility feeder and its loads following the disconnection of that feeder from the utility network (or, in the case of a general utility network outage, the occurrence of this unwanted continuing connection and supply of power to the feeder is termed islanding or "run on")

Transients are mitigated by the proper selection, installation, maintenance, and control of the transfer switch subsystem and the auto synchronizer. Both are highly standardized and mature products that can be readily selected for the specific distributed generator system.

As stated, the objective of the protection and control system is to enable the distributed resource generators and/or storage devices to deliver the intended services to the users and the distribution network reliably, safely, and cost effectively. The requirements (and, therefore, the complexity and cost) of protection and control systems for distributed resource systems, beyond the requirements of various standards, codes, and required certifications, depend primarily on:

- The size of the DG system with respect to the minimum total customer load on the feeder
- The number, size, and location of other DG units on the feeder
- The purpose of the DG — grid-connected or primarily grid-independent operating mode
- The type of DG — diesel generator, gas turbine generator, fuel cell, etc.
- The specific configuration of the feeder system (including laterals to the loads), including the size, location, operating mode, type of relays, breakers, and fuses, the feeder voltage, and the location, size, and configuration of all transformers
- Network operator requirements specific to that network (possibly as a result of experience with unique and unusual loads) and any additional safety requirements of local jurisdictions

Safety, in the broadest sense, pertains to (1) the safety of physical assets, e.g., the network, the distributed resource systems, and the loads served, and (2) the safety of humans, e.g., utility line personnel, the distributed resource operators and maintainers, and the operators of the end user's electrically powered equipment and infrastructure. The safety of personnel is mostly related to the potential for accidental or unintentional operation of a feeder, isolated from the network and powered by the distributed resources. Anti-islanding control logic and associated hardware devices designed specifically for the feeder characteristics are absolutely essential for distributed resource systems. The safety of utility line personnel is greatly jeopardized whenever anti-islanding cannot be ensured when the feeder is disconnected from the network.

11.4.2.3 Utility Protection and Control Guidelines and Requirements

Most utilities have published guidelines and requirements for operating, metering, and protective relaying for the interconnection of small power generators to the utility network. The fundamental requirements of utilities for small generators (less than about 200 kVA), assuming that the ratio of installed generation capacity to minimum load requirements on the feeder is 10% or less and including protection and control requirements, can be summarized as follows:

- The power supplied must be 60Hz alternating current.
- Basic designs must meet applicable minimum electrical standards as adopted by, but not limited to, national, state, and local governing bodies. This includes the National Electrical Code (NEC) and others.
- A manual and lockable disconnecting device must be installed at the point of interconnection in series with a protective fuse and a fused disconnect.
- A line voltage relay/contactors must be installed to disconnect the generator from the de-energized feeder and to prevent its reconnection until the line is re-energized by the utility. Undervoltage, overvoltage, underfrequency, and overfrequency sensors must be installed and connected to the relay. This is the anti-islanding requirement.
- All reactive power requirements for induction generators or power inverters are supplied by the utility. This is intended to reduce the possibility of self-excited operation in the event that the feeder is de-energized.

Many utility guidelines were developed with the expectation that few interconnected generators would be installed on a given feeder and that the penetration, in terms of the ratio of the capacity of the generators to the minimum load demand on the feeder, would be very small (typically less than 10%). Therefore, it was assumed that there would be no appreciable effect on relays and breakers designed and set up for current flow in one direction only, and also no effect on the overall stability of the feeder. Furthermore, where penetration was small it was assumed that, should the feeder be isolated from the network for any reason, the distributed resources could not, under any circumstance, continue to supply the load, the voltage would collapse, and the distributed generators would shut down automatically (on an undervoltage trip).

However, the most comprehensive utility guidelines recognize the need for more protective devices and other requirements as the penetration increases. These guidelines, along with standards already in place and being developed, form a very good basis for protection and control specifications for distributed resources. See, for example, the proposed New York State standardized interconnection requirement (N.Y. State Department of Public Service, 1999) and the Oklahoma Gas and Electric interconnection guidelines (Oklahoma Electric Company).

For many utilities, distributed resource installations, particularly small PV and small wind power generators, are already relatively common, and substantial operating experience has been gained. Nevertheless, many owners of these smaller distributed generators have found that obtaining the necessary interconnection approvals can be costly. In general, this cost barrier still exists for smaller generators. Furthermore, standardization of requirements is not yet an accomplished fact. The wind power industry has been a leader in establishing standards for the interconnection and operation of distributed wind power generators (National Standard of Canada, 1991). Many of the standards were already in place in the early 1980s.

11.4.3   Islanded Operation

In its broadest sense, islanding can be defined as the operation of the distributed resource and the intended user's loads in complete isolation from the utility grid — the feeder line. This definition does not discriminate between intentional (safe) islanding, where there is positive disconnection from the feeder, and unintentional or accidental islanding, where power flows not only to the direct user's load but also to the feeder (and possibly to all or some of the other loads on that feeder). Therefore, this definition must be refined when referring to protection issues.

With respect to protection (for the most part, the protection of the utility's power grid), islanding is defined as the unintended supply of power from one or more distributed power plants to a portion of the utility network (for example, a feeder line) following the separation of the feeder from the distribution network. Islanding is possible if the distributed resource controller misinterprets or does not detect the opening of the utility's feeder breaker so that the distributed power unit continues to feed power to the intended customer and to the dead feeder line. It is possible that some or all of the other loads may remain connected to the feeder. They will be fed power with possibly poor voltage and frequency regulation (that could damage some loads) until the distributed generators and/or loads are disconnected (trip) as a result of the detection of over- or undervoltage or over- or underfrequency by conventional, passive protection systems.

The quality of the voltage and frequency delivered depends on the rating of the DG that remains connected and the size of the loads being supported when the feeder breaker opened. For example, it is possible that the voltage will collapse almost immediately if the connected load demand far exceeds the rating of the connected DG. In this case, undervoltage will be detected. However, if the connected DG rating is such that it can meet the demand of the loads, fairly stable operation within the over- or underfrequency and over- or undervoltage acceptable operating limits is possible. Utility personnel dispatched to service the apparently disconnected feeder may be placed in great danger. Therefore, the distributed generator system controller must include a means to continuously and reliably test the status of the feeder line and instantly disconnect the distributed generator in the event of a tripped feeder.

The basic protection philosophy followed by utilities until now has been to limit the total capacity of interconnected capacity on a given feeder to less than about 10% of the minimum expected load on the feeder. Therefore, for any fault on the feeder it is expected that the feeder protective relays will operate, isolating the feeder and leaving the loads connected to a much smaller amount of DG. The voltage is therefore expected to collapse, resulting in the automatic shutdown of the distributed generators due to undervoltage. This basic protection approach has been successfully implemented and practiced by small, interconnected wind and PV power generators. In particular, wind turbines driving induction generators have demonstrated that basic over- or undervoltage and over- or underfrequency sensors, relays, and control logic are quite adequate to prevent islanding.

The islanding hazard can be exacerbated when more than one DG source is connected to the feeder and the ratio of the total distributed generator capacity to the instantaneous demand load is relatively large. The results of tests conducted by Sandia National Laboratories on PV inverters in 1997 showed that when several inverters (with different anti-islanding techniques) were operating on a single 120 V circuit, the inverters frequently continued to feed power from the PV arrays to the circuit loads for more than two seconds (times greater than 30 seconds were observed) following the disconnection of the circuit from the utility network (Sandia National Laboratories, 1998). In these cases, the ratio of power being generated to power being used by the loads on the circuit was in the range of 0.8 to 1.2. Interestingly, it was observed that the presence of a transformer in the circuit resulted in much shorter disconnect times, less than 0.5 seconds. This was a result of the fact that most inverters cannot supply the nonlinear magnetizing current required by the transformers. Sandia has proposed a method for designing an anti-islanding inverter that includes both active and passive techniques. The active methods are termed SFS (Sandia frequency shift) and SVS (Sandia voltage shift). The passive methods are over- or underfrequency and over- or undervoltage.

11.4.4 Power Quality

Distributed resources must provide the intended services to the users and electrical power networks. Fundamentally, these services are simply to provide high quality electric power in a safe manner when it is needed (i.e., reliability of supply) and in the quantity demanded. (Reliability — in the narrowest sense, the minimization of outages of any duration — is now considered by many to be a component of power quality.) New high technology equipment requires higher quality power. Furthermore, the rapid spread of computers and automated equipment has made customers aware of the effects of poor power quality. It is important to keep in mind that the user (possibly, but not necessarily, also the owner of the distributed resource) is buying services and not hardware, software, and operation and maintenance contracts. With the deregulation of the power industry, retail and commercial, electric power providers can and will differentiate their services through power quality. Therefore, power quality has moved to the forefront. At the same time, the advent of distributed generation has brought new problems that have the potential to degrade the power quality that existed before the installation of DG.

In more specific technical terms, power quality describes how closely the actual electrical signal at various points in the network (including at the terminals of the user's load device) follows the ideal stable, sinusoidal waveform that we associate with utility grade power. The term distortion is used to describe any deviation from this perfect sinusoid. Several important distortions or components of power quality are power interruption (complete loss of the waveform), voltage sag (decrease in amplitude of the waveform), voltage "flicker" (momentary voltage swings), and the presence of harmonics (higher frequency waves combined with the fundamental 60 Hz sinusoid). (See Figure 11.8.)

It is argued that distributed resource interconnection and operation will have both negative and positive impacts on the quality of power delivered to

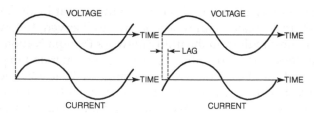

FIGURE 11.8
Effect of lagging power factor (courtesy of Nicoara Graphics).

the user. On the positive side, a decrease in service interruptions is ensured for the end users that can count on operating their on-site generation in the off-grid mode when required. On the negative side, it is possible for stability to be compromised when numerous distributed resources are operated on the same feeder. Because stability margins are specific to the configuration of the feeder and its loads, the distributed resources connected to the feeder, and the load being served at a given moment, general guidelines for determining whether there may be a stability problem cannot be defined.

Distributed resources have the potential to mitigate the effects of utility interruptions in service. Many end users would want a seamless transfer of their loads from the utility to the distributed resource should a utility power outage occur. Later, when utility service is restored, a seamless reconnection and synchronization of the two systems is desirable. In practice, a rather broad spectrum of configurations is available because system cost increases rapidly with decreasing transition time, and many users do not require virtually seamless transition. Some configurations are similar to conventional UPS systems, where, for example, an energy storage unit (typically a battery bank) immediately picks up the end user's load following the utility outage. The power is delivered to the load through an inverter that converts direct current (DC) to alternating current (AC) power. Seconds or minutes later, after the engine–generator set has reached rated speed, it is connected to the load and the battery/inverter is disconnected. Fuel cells are ideally suited for this application since they must already include energy storage to operate successfully when operating in the grid-independent mode. In addition, fuel cell power is delivered as DC power. Ideally, one power inverter would service both the fuel cell and the energy storage unit.

Simpler and less costly configurations use only the diesel generator unit, for example. Power will be interrupted for a few minutes while the diesel starts and reaches operating speed. In these simple systems, there is no transfer switch and no synchronizing gear, so when the utility power is restored, the engine–generator unit is disconnected from the load before the utility service is reconnected. The end user experiences a very short interruption in service in return for a much less expensive system. At present, the penetration of distributed resources into the electric distribution systems is very small; therefore, few reports exist on direct operating experience, particularly with respect to power quality.

11.4.5 Operational Concerns

As discussed previously, operational concerns are minor whenever just a few small distributed generator units are installed on a distribution feeder and the penetration, in terms of the ratio of the capacity of the generators to the minimum load demand on the feeder, is very small (typically less than 10%). This scenario is already commonplace in the U.S. and Canada on some rural distribution feeders where small wind and PV systems are in operation at widely-spaced farms and ranches. DG operational concerns are expected to move to the forefront on highly-loaded urban feeders, where the scenario of many distributed generators (high density) and much higher penetration is expected to become the norm.

It is argued that without centralized control of the distributed resources (and possibly some of the loads), network stability and reliability will be greatly compromised. Ironically, this will require the design and implementation of control and dispatch approaches on a small scale that, on a larger scale, are now a necessary and vital part of the of the conventional centralized generation paradigm. This system of many distributed generators on a given feeder or feeders, along with the necessary monitoring and control system able to dispatch the generators individually, has been termed the *virtual utility*. The concept includes not only the ability to dispatch specific generators to supply the network, but also the ability, through remote and centralized control, to dispatch specific generators to serve loads that are being disconnected from the network feeder (i.e., shed) to mitigate peaks, for example. The disconnected loads go into intentional islanded operation but are returned to the network supply when conditions warrant. Inherent in this approach is the assumption that individual generators will start, connect to the network (if required), and operate reliably when called upon. In addition, the transfer switches and protective relays must also operate reliably. To some degree, each generator will operate intermittently.

While renewable power sources such as wind turbine generators and PV power generators can be connected and disconnected from the network at will, their power output varies continuously (albeit less so for PV). Therefore, more careful planning is required for the use of these resources in the high density, virtual utility generating mix. Nevertheless, their value can be substantial in situations where the wind and solar flux are substantial and predictable.

Most utilities require performance verification testing of larger distributed generating units. The rationale is quite simple — the larger the generating unit, the greater the impact it can have on network reliability and stability. Verification testing is, of necessity, site specific. To ensure reliable performance, the tests must be repeated periodically over the life of the installation. Until now, test procedures for interconnected generators have focused on the components and subsystems critical to protecting the network. In the new paradigm of DG, where effective utilization of the distributed resources is vital, the reliability of the distributed generators and their control and protection systems is of equal importance and they must also be periodically tested.

All critical interconnection, control, and protection components and sub-systems should have type approval based on type testing carried out by independent testing laboratories. Typically, these components and sub-systems are required to meet the requirements of relevant UL, CSA, IEEE, ANSI, and other established and relevant standards as an integral part of the approval process for interconnecting and operating the distributed power plant.

11.4.6 Institute of Electrical and Electronic Engineers (IEEE) Standards Coordinating Committee 21

When parallel operation is considered, issues of concern are raised by both the electric power system (EPS) and distributed resource (DR) owner. Among these issues are the continued quality and reliability of power and the safety of personnel and equipment. In addressing these concerns, EPS owners have established protective relaying schemes that must be met by the DR owners. DR owners claim that these schemes are restrictive, cost prohibitive, and unnecessary. In between these diverse interests sits the State Public Utility Commission given the task of establishing the real requirements. Consequently, there is a need to establish a single standard defining the requirements for proper protection of both the EPS and the DR at the point of interconnection. The IEEE has taken on the task of creating and publishing such a standard.

The IEEE Standards Coordinating Committee 21 (IEEE SCC-21) is attempting to develop this standard. This is probably the most qualified organization to develop such a standard. For decades, the IEEE has developed recommended standards addressing the safe and proper configuration of equipment for generation, transmission, distribution, and utilization of electric power. Through their many societies and sponsored conferences in various disciplines, IEEE members are kept current with emerging technologies and trends in electric power. Clearly, the IEEE is preeminently qualified to draft a standard defining the appropriate configuration of electrical equipment at the point of interconnection. At this writing, a working draft of the standard, Draft 02, has been created and is circulating for comments and input. This draft is still missing some important sections. However, there is a significant body of text and content to indicate the specification's intent. Bear in mind that anything discussed in this commentary may well be superceded or overridden by the final draft of the standard.

Those unschooled in the science and art of electric power production, distribution, and utilization would prefer to have a standard that clearly defines all of the requirements for every combination of electric power equipment. Unfortunately, any such standard would be too large to carry. The standard must provide for the infinite permutations of EPS and DR ratings and capacities at the point of interconnection and common coupling. Accordingly, as found in the current draft of the standard, several factors are considered in the determination of the protective and control schemes. While some attention is still paid to size and ratings, this standard defers to the more important determinants of interconnect protection. These are defined as:

- Stiffness ratio — a term developed for this study that compares the utility system's available fault current at the planned interconnection point of the DR unit with rated output of the DR unit (looking at the DR)
- Contribution ratio — a term developed for this study that compares the DR's fault contribution to the power system at the planned interconnection point of the DR unit to that available from the utility system (looking at the system)

From the standpoint of the system and equipment, available energy must be determined at every point to ensure that the response to anomalies is correct and adequate. In selecting switching devices, if the device is to interrupt current, it must be capable of interrupting the maximum potential current available at the location of the device in the circuit. Similarly, if the device is to remain closed during a fault so that another device may interrupt the fault, it must be capable of withstanding the maximum potential current available at the location of the device in the circuit. For example, at any given fault location, circuits downstream of the DR must be capable of handling the maximum potential current of the EPS plus the DR. Circuits upstream of the DR need be capable of handling the maximum potential current of the DR. The significance of this approach is that consideration is given to the contribution capacity of the DR, as is the case with every other generator on the system. When this distinction is made, the nature of the generation process becomes transparent and proper protection and coordination are achieved.

Distributed resource generation will be comprised of a number of technologies. Among these will be PVs, fuel cells, wind turbines, and induction and synchronous generators. Those processes providing power through inverters will not have the same transient performance as synchronous generators. Since the transient performance determines the protection profile, differing protection packages are required for different generation processes. Protection packages are intended to respond to the abnormal condition. Normal conditions are handled by the process controls. It is not difficult to design a system to operate under normal conditions. The difficulty comes in designing the system to provide for normal operation and respond to the transient situations that arise. For example, when the DR operates in parallel with the EPS, controlling its output flow of power is readily achieved. Adjusting for increases and decreases in power flow as a function of normal variations in available fuel, (e.g., as the angle of the sun's rays varies the output of a PV system) is relatively easy. Determining what happened and initiating the appropriate response to a transient condition is another issue.

When a fault occurs on the distribution grid to which the DR is connected, the protection scheme is required to identify the nature, location, and proper action to take. Taking the case of a fault on a distribution circuit, if it is a single-phase fault (a power pole knocked down by a car, a power line broken due to ice loading, a tree blown over in a storm knocking the lines down, etc.), the response must be to isolate the affected circuit to provide continuity of service to the surviving circuits. A fault on a circuit appears as both a voltage reduction and current increase. The distinguishing features that differentiate the event as a transient or normal occurrence are the magnitude and quantity of changes. The protection scheme must be able to make the distinction.

The various generation processes will have different responses to transient conditions. The output of an inverter will not provide as much initial fault current as it would were it equipped with a battery on its DC bus. The output of an induction generator will not provide as high a contribution to fault current magnitude and duration as will that of a synchronous generator. As the present draft of the proposed standard indicates, it is necessary to take the measure of the DR in stiffness and contribution at the point of connection to the EPS to develop an appropriate protective scheme. The outcome of the final design is to leave the EPS system no less reliable nor lower in power quality than what existed before the DR was integrated. While the draft as it now stands is still only a work-in-progress, it is headed towards becoming a standard that will be a useful guide in defining interconnect requirements for DR.

References

Hirst, E. and Kirby, B., Creating Competitive Markets for Ancillary Services. ORNC/CON-448, Oak Ridge National Laboratory, Oak Ridge, TN, 1997.

Oklahoma Electric Company, Guidelines — Operating, Metering and Protective Relaying for Interconnection of Cogenerator, Small Power Producers, and Other Non-Company Sources of Generation to the OG&E System, 1993.

National Standard of Canada, Wind Energy Conversion systems (WECS) — Interconnection to the Electric Utility, CAN/CSA — F418-M91, January 1991.

NY State Department of Public Service, New York State Standardized Interconnection Requirements, Application Process, and Contract for New Distributed Generators, 300 kVA or Less, Connected in Parallel with Radial Distribution Lines, staff proposal, July 1999.

Sandia National Laboratories, *Quarterly Highlights of Sandia's Photovoltaics Program*, vol. 3, 1998.

12

Installation and Interconnection

James M. Daley and Anne-Marie Borbely

CONTENTS

Installing new electricity-producing equipment near, on, or within buildings requires the same permit evaluation process as any other modification to the

0-8493-0074-6/01/$0.00+$1.50
© 2001 by CRC Press LLC

site, with one telling exception: the existing code structure was never designed for wide-scale deployment of electric generators outside the ownership and control of electric utilities.

As of the year 2000, the competitive position leveled somewhat for both utility and nonutility owned generation units. Any equipment not located on utility property and used expressly for the operation of that utility became subject to the same local code requirements as all other owner-operators. Therefore, utilities attempting to enter the DG marketplace by installing generators at customers' sites will undergo the same permitting process as energy service companies, local distributors, manufacturers, or the energy customer.

The essential roadblock still exists, however, for all new DG technologies, regardless of ownership. The sourcebooks for local code officials — the National Electrical Code, the International Fuel Gas, Plumbing, Mechanical, Building, and Fire Codes, and the National Life-Safety Code — contain no reference for microturbines, Stirling engines, or, until recently, fuel cells.

Although standards exist for the installation of traditional on-site generators, their interpretation — and the building codes they must interact with — vary between state and local jurisdictions. For this reason, this section merely presents the suite of issues that may be encountered; it is not a definitive guide to the codes or standards.

12.1 The Cost of Ignorance

Figure 12.1 shows the cost to a developer for failing to adequately account for code requirements in his product development. By year 8, the product in code compliance is no longer generating sunk costs, and by year 15 it has returned all previous investment. In contrast, the product out of compliance fails to establish a revenue stream, and by year 16 the cost of investment capital is beyond recovery. As shown in later sections, it can take three years to develop a new consensus standard and another two years to have it referenced in the model codes. For entirely new technologies that may impact life safety issues and building construction, a new standard must be initiated five years prior to their commercial introduction, or manufacturers will continue to sink capital into businesses physically incapable of generating substantial revenue.

Absent any explicit definition of what comprises safe installation and operation, each building code official in the 44,000 state and local code authorities across the United States must independently determine the appropriate requirements one site at a time. Each unit will be evaluated under an "alternative methods and materials" clause that does not imply approval. A code official can require any number of design, test, and documentation reviews

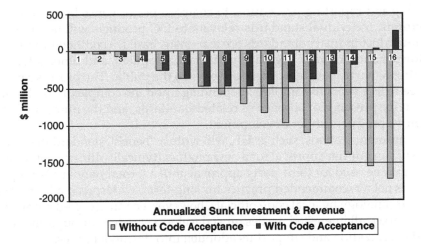

FIGURE 12.1
Revenue comparison of code-compliant versus non-compliant technology.

before ruling on the installation. In many cases, this may result in the unit being denied permission to operate, and there is no appellate process to fall back upon.

12.2 Codes and Standards

Standards are documents that outline the agreed-upon design and performance of a given technology, while model codes address the design, installation, and operation of materials and equipment as it relates to public health and life safety. Although standards are not automatically accepted into model codes, thousands of standards are referenced as required practice for a given installation. For example, the National Electrical Code references NFPA 110, Standard for Emergency and Standby Power Systems, for buildings required to have on-site power in the event of a grid failure. A key difference between model codes and standards is that standards, and any modifications to them, are approved through a vote that involves any and all interested parties, while model code modifications are voted on solely by code officials.

12.2.1 Standards

Several types of standards have been developed for industry, but the two most relevant to DG technologies are (1) product testing standards that outline the criteria for safe operation, and (2) rating standards, which define

326 Distributed Generation: The Power Paradigm for the New Millennium

testing and reporting procedures to compare performance between manufacturers. Individual standards relevant to DG products will be discussed below. Most industry standards are consensus based, a codified process that allows manufacturers, regulatory officials, suppliers, and other interested parties to participate in a procedure open to the public. The process can take three to five years to complete, depending upon the complexity of the standard, the presence or absence of related standards, and the number of competitive products under development.

Some organizations, such as UL, will write a "bench" standard that can be used as an interim approval for a new product (typically during prototyping) and may be used for third-party approval until a formal standard is adopted. This is not a recommended practice for long-term, widespread adoption of a new technology by local code officials, however. For an accepted pathway to the market, all interested parties must develop a standard through an accredited organization and support its adoption into the model codes.

12.2.2   Model Codes

Model codes can be adopted (although not required; state or local agencies can write their own codes) to address the design, construction, and operation of buildings and facilities. Model codes are not designed to exclude a given technology and generally focus on prescriptive solutions that allow performance-based alternatives. That said, approval of non-referenced designs is dependent upon the local code official and his comfort level with how a new product respects life safety concerns.

Model building codes were historically developed by three regional organizations that covered the northeastern (Building Code Officials and Code Administrators, or BOCA), southeastern (Southern Building Code Congress International, or SBCCI), midwestern, and western U.S. (International Conference of Building Officials, or ICBO), respectively. This regionalization led to contradictory building requirements between even adjacent local jurisdictions, driving up the cost of equipment, supplies, and building designs. In the early 1990s, the three code organizations agreed to cooperate within a single framework for national guidelines, resulting in the International Code Council (ICC).

The ICC is responsible for separate international energy conservation, fuel gas, mechanical, one- and two-family dwelling, plumbing, building, fire, and residential model codes. The three regional model code groups continue to operate, competing as service organizations (education services, plan review, etc.) to the building and code communities.

The National Fire Protection Association (NFPA) also develops documents in model code language that, because they are developed by a standards organization, are not considered true model codes. NFPA 70, the National Electric Code, and NFPA 54, the National Fuel Gas Code, are relied upon heavily by local code authorities, however, and are referenced extensively in the model codes.

12.3 Installation of a DG Unit

This portion of the code approval process addresses electrical safety, fuel supply and storage, and access (by the fire department or other public safety officials).

UL 2200 is the most commonly cited reference for combustion engines and gas turbines in stationary power applications and can be considered to cover microturbines, although the product is not currently referenced. The requirements cover engine–generator assemblies rated 600 volts or less and intended for use in ordinary locations in accordance with the National Electrical Code, NFPA 37 (Standard for the Installation and Use of Stationary Combustion Engines and Gas Turbines), NFPA 99 (Standard for Health Care Facilities), and NFPA 110 (Standard for Emergency and Standby Power Systems). UL 2200 does not cover hazardous (classified) locations or uninterruptible power source (UPS) equipment. This is not a performance standard.

NFPA 853, Standard for the Installation of Fuel Cells, provides for the design, construction, and installation of both prepackaged and field-constructed power plants above 50 kW gross electrical output. Unlike the previously approved ANSI standard Z21.83, NFPA 853 covers a variety of fuel sources (Z21.83 is for natural-gas supplied systems only).

NFPA 37, Standard for the Installation and Use of Stationary Combustion Engines and Gas Turbines, historically covered units to 7500 horsepower output. That language was eliminated in the current draft (1998) and can now be considered to cover microturbines as well.

12.4 Operation

The code official will assess how the unit, when operating, interacts with other systems in the building. As mentioned previously, most code officials have not encountered on-site generators designed for full-time operation. Absent any national education program for local code officials, this portion of the approval process requires the inspector to speculate on potential hazards that may arise within or outside the building in question. For new power generation technologies not referenced in any installation standard or code, the code official now has total responsibility for all potential impacts and no codified methodology to fall back on. Figure 12.2 presents the most common issues that may arise during the code approval process.

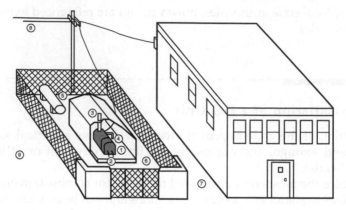

FIGURE 12.2
Selected code requirements for DG units.

12.4.1   Notes on Figure 12.2

1. In Figure 12.2, the "box" itself will require some form of third-party testing and certification, such as UL or CSA. Design, testing, and listing of fuel cell power plants up to 600 volts AC and 1 MW output fall back on ANSI Z21.83, American National Standard for Fuel Cell Power Plants.

2. As a general rule, all electrical wiring must be stranded annealed copper, regardless of prime mover.

3. Although neither NFPA 37 nor NFPA 853 specifies enclosure requirements beyond a reasonable level of protection against unauthorized access and general protection against hazardous conditions, the local code official may require that the cabinet meet NEMA standards for explosion-proof enclosures (no one has reported this yet, however) and that any unit installed outdoors be protected from natural elements and vehicular impact.

4. All fuel gas systems utilizing service pressures under 125 psig must be installed and operated in accordance with NFPA 54, the National Fuel Gas Code. Fuel piping must be steel or other metal and in compliance with NFPA 30, Flammable and Combustible Liquids Code. Additionally, all pressure-boosting equipment must be certified for design, construction, and testing according to ASME Boiler and Pressure Vessel Code. Fuel cell systems utilizing compressed natural gas must also meet NFPA 52 requirements and the Compressed Natural Gas Vehicular Fuel Systems Code; hydrogen piping falls under ASME B31.3, Process Piping.

5. All liquid petroleum gas systems (liquid or vapor phase) must be installed in accordance with NFPA 58, Standard for the Storage and Handling of Liquid Petroleum Gases. The Flammable and Combustible Liquids Code and API 620 (American Petroleum Institute), Design and Construction of Large Welded Low-Pressure Storage Tanks, may also apply. On-site hydrogen — gaseous or liquefied — storage falls under NFPA 50A and NFPA 50B, respectively. Liquid fuels such as diesel, ethanol, and methanol must be installed as prescribed in NFPA 30, Flammable and Combustible Liquids Code.

6. Outdoor and rooftop installations generally require a cement foundation for the integrated package.

7. The distance between the unit and buildings, ventilation systems, or access ways may be clearly defined, such as a minimum of five feet, or it may be left to the local code official to determine reasonable access.

8. Interconnection to the local electric distribution system will fall under IEEE 1547, Standard for Distributed Resources Interconnected with Electric Power Systems, expected to be completed by 2002 (see Section 12.5, Interconnection, for further information).

9. Local zoning ordinances (definition of hazardous materials and relation to residential zones, distance to property line and rights-of-way, access by local fire and safety authorities, etc.) may need to be consulted in some areas. Additionally, local building inspectors may require that a fire risk evaluation be performed for each installation with respect to design, layout, and operating conditions of the unit. The inspector may then require any or several fire protection systems (portable versus fixed, foam or gaseous extinguishers, automatic sprinklers or dry chemical fire suppression systems).

12.5 Interconnection

Distributed power generation sources will generally be sited within the electric distribution grid. Historically, this system has been designed to accommodate a one-way electron flow from the transmission system to the load being served. Grid connection equipment for feeding into the system, therefore, was designed for megawatt-sized power plants that fed into the primary transmission lines. Such utility-grade equipment is vastly over scaled for kilowatt-sized generators and effectively impedes the introduction of grid-connected DG resources. Additionally, electric utilities have little or no incentive to allow even small-scale generators to randomly inter-connect (reducing their own revenue potential and introducing uncertainty to the system).

The Institute of Electrical and Electronic Engineers (IEEE) has formed a working group, SCC-21, to develop an interconnection standard (P-1547) for all DG technologies connecting to radial distribution feeders. This ambitious undertaking, headed by the National Renewable Energy Laboratory (NREL), is expected to be complete by 2002. The issues surrounding integrated elec-trical power generation and distribution are outlined below.

12.5.1   Power Sources

The interaction of paralleled power sources is largely a function of the dispar-ity in size of the sources. Additionally, this interaction can be separated into the categories of normal operation, fluctuations in parameters, and transient occurrences. The circuit shown in Figure 12.3 is a simplified schematic of a distribution grid, important in understanding the differences in these occur-rences. Figure 12.3 will be explored as a power system from the point of gen-eration to the point of utilization.

Utility-scale generation is quite large; the mix will contain base generators rated at several hundreds of megawatts up to 1100 MW. Intermediate gener-ation plants range from 100 to 400 MW. Generators of 25 to 100 MW are used for peaking purposes and are typically disbursed around the grid system at load concentrations to relieve overloading on the transmission system. Since this text concentrates on DG with ratings of 10 MW or less, attention will be focused on utility circuits likely to have generation of this size or less.

The grid is fed by distributed utility generation rated 25 MW and higher. Thus, the grid will have capacity of at least 2.5 times that of the DG unit con-sidered herein. This generation, however, will most likely be connected to the radial distribution feeders. An important consideration in siting this unit is the stiffness of the radial at the point of the connection. This chapter

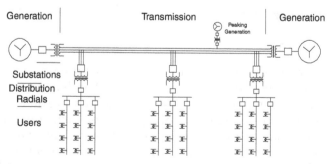

FIGURE 12.3
Simplified schematic of the power grid from generation to utilization.

defines stiffness as a measure of the capacity of a radial with respect to its ability to handle load with minimum fluctuation in voltage. A properly sited DG unit below 10 MW will probably have no noticeable impact on the frequency of the grid system. The grid system, however, will control the frequency of the unit.

Assuming a DG unit in parallel operation with the grid, the controls of the prime mover will set fuel flow and excitation to enable the unit to take on real and reactive load up to the desired level for the steady-state condition. Fuel control adjusts the amount of real load (kW) the unit will carry. Excitation control adjusts the amount of reactive load (kVAR) the unit will carry. The response times of these controls (to changes of state on the grid) have time constants between several tenths of a second to a few seconds; these units will be comparatively slow in response to sudden changes in grid voltage at the point of interface. Thus, responses to transients will differ as a function of the type of prime mover and excitation control.

This discussion assumes that the unit is connected to the grid at a distribution radial. Depending on the distance from the substation, conductor size, and other users on the radial, the available mega-volt-amp (MVA) at the point of connection will vary. Generally, the higher the available MVA at the point of connection, the stiffer the circuit. What does all of this mean? When electrical engineers design power distribution systems for buildings, they must determine the available MVA at the point of connection so that circuit protection devices can be adequately sized to interrupt a fault (short circuit) in the building. While a value of available MVA cannot be assumed across the board, for commercial and industrial buildings with service demands higher than 500 kW, one can expect to have 50 MVA available at the point of connection. At residences, one can expect to have approximately 50 times the rating of the pole transformer available. For facilities between these two levels, one can expect about 25 times the transformer rating. Engineers count on this high available MVA to provide for voltage stability in the facility, and they rely on sufficient fault current to permit coordination of circuit protective devices. The objective of this coordination is to clear the fault with the protective device closest to it. This allows unaffected circuits to remain in service.

12.5.2   Perspectives

DG units will be deployed for a suite of applications. Some will be designed and installed for DG service. Some will come from installed capacity, such as emergency and standby power systems. Some will use fuel, others will use renewable resources. As a general rule, those installed for commercial, industrial, and institutional facilities will have ratings less than the full load demand of the host facility. Those used for residential facilities may be rated higher than the host facility but typically will not exceed the rating of the transformer between the facility and the distribution circuit. Retail facilities may have generation capacity near the facility full load demand. However, for the range of DG units considered in this work, the most likely scenario will be the distribution circuit having an available MVA at the point of interconnect considerably higher than that of the unit.

Figure 12. 4 illustrates a typical distribution radial feeder that would originate at a substation. Several users and user classifications are fed from the same feeder. If one of these users were particularly large with respect to the feeder's capacity, a separate feeder would be run from the substation to that facility. For example, a real-life scenario may be that residences (single phase) comprise 5 kW loads, commercial users 25 to 200 kW loads, the industrials 500 to 3000 kW loads, and one large user, for example, requires a 10 MW load to be connected to the feeder. In a worst-case scenario, the large user would be located furthest from the substation. With a 10,000 kW load, the likelihood is that this user would have a few very large unit loads (generally motors). With every starting, the motor inrush current would cause a noticeable voltage drop along the distribution feeder. Proximity would dictate how much voltage drop smaller facilities would experience. For this reason, the connection to this large user would either be at the start of the feeder or on a separate feeder entirely.

12.5.3   System Faults

For this discussion, a fault is a short circuit imposed on the power system. The short circuit can occur at any point in the circuit from the substation transformer to the light switch on the wall. The response of loads and DG units will differ as a function of fault location in the system. An understanding of Ohm's law is required to understand system response to a fault. Simply,

$$E = I \times Z$$

where,

$$E = \text{Voltage}$$
$$I = \text{Current}$$
$$Z = \text{Impedance}$$

Simply stated, the voltage at any point in a circuit is equal to the product of current flow in the circuit and the impedance to that point in the circuit. For example, if there were a source of 100 volts having an impedance of 1 ohm and a current flow of 10 amps, the total impedance of the circuit would be 10 ohms. The voltage at the terminals of the source would be 90 volts because the voltage drop across the source impedance is 10 A × 1 ohm = 10 volts.

If a fault occurred at a point in the circuit such that the total impedance reduced to 5 ohms, current in the circuit would increase to 20 A and source terminal voltage would drop to 80 volts. If the fault occurred at the terminals of the source, terminal voltage would reduce to zero volts and current would

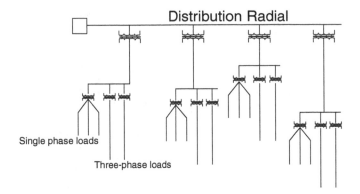

FIGURE 12.4
Simplified schematic of a radial distribution feeder.

increase to 100 A. Putting this into context and looking at Figure 12.4, a fault occurring at the substation would produce higher energy into the fault than one occurring at the far end of the feeder. Protective devices would clear this fault. Should the fault occur in the distribution feeder, the breaker (represented by the square at the beginning of the feeder) would open.

Typically, distribution feeder circuits are protected by high-speed reclosers. When a fault occurs, the recloser disconnects the feeder and then immediately recloses it. This operation is used because faults on a distribution circuit can have many causes. If the cause is temporary, when the recloser closes, the fault will have disappeared. If the fault were the result of an automobile accident that knocked a pole down, causing wires to come into contact with each other, upon reclosure, the fault would still be present and the recloser would trip again. Typically, a recloser will reenergize a feeder circuit one or two times before it locks the circuit out and requires resetting. This operation is used to ensure maximum continuity of power to the feeder.

The operation of the recloser on initial opening occurs within six cycles of fault initiation. Reclosing occurs in about another six cycles. Accordingly, the total outage would last 12 cycles or more. When the lights blink, it is typically because a fault has occurred in the power system and a fault protective device has operated to clear the fault. When the lights go out and stay out, it is likely the result of the fault remaining on the feeder after the reclosure.

The occurrence of a fault in any user's system can produce a flicker in the lights of his neighbors for as long as it takes his overcurrent device to operate. The impedance of the circuit to the point of fault in the user's system would typically be too high to allow sufficient current flow to cause the recloser to operate. Figure 12.5 is an example of a typical larger user facility. Such a system would be found in facilities with maximum demand of 3 MW. Facilities with larger demands would have several systems like that shown.

Evaluating the system for response to faults, four locations have been selected. A fault at the secondary side of T2, location 1, would produce the highest fault current available on the load side of this transformer. The value

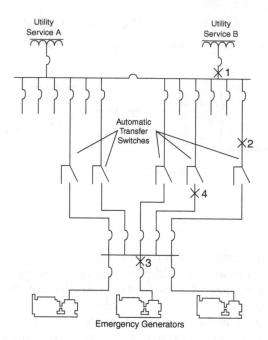

FIGURE 12.5
Typical single line power diagram of a commercial or institutional user.

of this fault would be equivalent to the secondary voltage of the transformer divided by the impedance of the transformer where the distribution feeder can be considered an infinite source. The available MVA at this point is reduced when the impedance of the feeder and its source cannot be ignored. For the sake of discussion, assume that the transformer is rated 2.5 MVA and its impedance is the normal value, 5.75%; where the impedance of the feeder source can be ignored, the available fault current at location 1 would be 3000/.0575 = 52 kA, 17.4 times the full load rating. At location 2, the fault current is reduced by the additional impedance in the circuit path to the point of fault.

It is necessary to make a similar evaluation of the on-site generation fault current capability. To understand this, the various generation schemes must be evaluated. For DG, the power will be generated either by a static or dynamic source. The static source will be an inverter because the prime movers of these sources will be generating DC power. These sources will be a fuel cell, microturbine, or photovoltaic system. In the case of the photovoltaic, the prime mover is the sun. As the sun's incident radiation on photo cells increases, the output current of the cell increases to some maximum value. This current is passed through an inverter that converts the DC to AC.

Similarly, a fuel cell produces DC output, except its output current is derived from the processing of a fuel to release electrons that form the current. The microturbine is an extremely high-speed generator producing AC current. Because of its high speed, it cannot be directly connected to the grid.

Its output is rectified to produce DC and then inverted to produce AC at the grid frequency. In some cases, storage batteries will be included in these inverter power systems. Where there are no storage batteries, the transient output capability of these systems is quite limited. When batteries are included, the transient output can be higher than the full load rating. In most cases, the magnitude of transient output current of an inverter is limited to protect the solid state switching devices that produce the AC output. Typically, operating on the inverter alone, these power sources cannot be used to start motors of equivalent size. One can assume that transient output current will be limited to less than six times the full load current rating.

The dynamic sources will have rotating generators producing AC outputs. The prime movers of these sources will be combustion, steam, wind or hydraulic turbines, or reciprocating engines. These machines will drive either induction or synchronous generators. The induction generator will depend upon the grid for excitation. The synchronous generator will have its own excitation system. For the case of the induction generator, the prime mover is started, and, as it approaches synchronous speed, it is connected to the grid. The terminal voltage of the induction generator is determined by the grid voltage. When the speed of this generator exceeds synchronous speed, it begins to generate power. The higher the speed, the higher the output power will be. To avoid overloading the induction generator, its speed is closely controlled by metering the power stream (fuel, steam, wind, or water). The synchronous generator produces terminal voltage when it is running.

A voltage regulator is included in the excitation system to control the terminal voltage. Synchronous generator terminal voltage is determined by the generated voltage minus the voltage drop due to winding resistance in the stator. This drop increases as the current flow out of the generator increases. Consequently, the voltage regulator increases excitation to produce constant voltage for increases in load. Controlling the energy source flow to the input of these prime movers controls the amount of power they will produce. Again, controlling input energy flow controls real load, and controlling excitation controls reactive load. In the case of the induction machine, since its excitation is taken from the grid, its reactive loading is determined by the real load being supplied. As is the case with an induction motor whose power factor improves with loading, so is the case with the induction generator — its proportionate reactive loading varies with real load.

A characteristic of the dynamic generator is that it stores electric energy in its air gap. This energy can help the generator respond to fluctuations in load. This stored energy also contributes current flow into a fault. Recall that when conducting a short circuit study of an AC power system, the designer must add the contribution from all running motors. In practice, this can be about four times the full load current of the motor. A motor having a 100 A full load current can contribute up to 400 A to a fault on its power system. The actual value will be a function of the motor impedance and circuit impedance between the motor and the fault. Similarly, the synchronous generator will also produce a significant contribution to a fault on its power system. In each case, the magnitude and duration of this contribution will be a function of machine and control characteristics.

Because the induction generator excitation is derived from the grid, when the voltage reduces, the energy is drained from the machine. This will happen in a very short period of time for a close in fault. For example, if the fault causes the grid voltage to go to zero at the location of the induction generator, one can expect the fault current contribution to decay from its initial value of 4 times full load current to 1.5 times full load current in about 20 to 30 milliseconds, 1.25 to 1.9 cycles when the generator was operating at full load at the time of the fault. Were the generator operating at less than full load, the contribution to the fault current would be less.

In the case of the synchronous generator, because it has an excitation system, the duration of the fault current contribution would be somewhat longer. Excitation systems for synchronous generators are typically designed to produce up to 300% full load current for up to 10 seconds. To accomplish this, generator excitation is supported. However, initial fault current flow will be limited by the load on the generator at the time of the fault and the generator's subtransient reactance. Generators rated up to 12.5 MVA* can have a subtransient reactance of 10% or more. This translates to a maximum contribution to fault current flow of ten times full-load rating if the generator is operating at full load when the fault occurs. The subtransient time constant is the most significant determining factor in initial fault current magnitude. Generators of the sizes discussed herein will have subtransient time constants of 20 to 30 milliseconds. At the end of the second cycle after initiation of the fault current flow, the generator contribution will typically be less than half its initial value.

Given the preceding conditions, going back to Figure 12.3 and determining the fault current contribution from a single generator at 2 MW and 10% subtransient reactance, the initial maximum fault current contribution from a single generator would be 30 kA. With zero impedance to the paralleling bus, the available fault at point 3 would be 90 kA. At point 4, this fault current would be reduced by the impedance of the path from the paralleling bus to the location of the fault in the feeder.

12.5.4   DG Considerations

The opportunity for DG is driven by one of two factors: research or economics. Where research is the objective, actual cost is less important than operating strategy. Where economics is the objective, risk management is the important consideration. DG resources of the photovoltaic, wind, microturbine, and fuel cell type are the subjects of many research programs. The objective of this

---

* Generating sets are typically rated in kW and power factor. The industry standard is to rate the set at some kW and 0.8 pf. Accordingly, a 10 MW generator set would have a 12.5 MVA generator.

research is to find an economical environment that would encourage investment of user capital in these types of DG. Those that find economic justification will quickly promulgate throughout the DG environment. Prime movers that burn hydrocarbon or other fuels are in broad-scale use throughout the electrical environment today. To the extent that these prime movers will meet siting EPA requirements, they can be implemented. Accordingly, this discussion will ignore considerations of the prime movers and concentrate on the electrical environment with which the outputs will interface.

There are two ways to implement DG: operation in parallel with the grid or operation independent of the grid (islanded). In the islanded mode, electric load will be transferred from the grid to the DG unit. A subset of the islanded mode is closed transition transfer to and from the grid. In this mode, the two power sources are momentarily paralleled. This operation minimizes load disturbances. Islanded power plant operation is in wide-scale use throughout the world. It is typically called emergency and standby power. There are pros and cons for each mode. In the parallel mode, protective relaying is required to protect the DG unit and facility power system from adverse effects introduced by the grid and to protect the grid from adverse effects introduced by the DG unit.

12.5.5 Protective Relaying

Protective relay schemes are designed to respond to changes in the parameters they monitor to initiate an appropriate control sequence when any of those parameters exceed a preset limit. While such relay schemes are referred to as protective, they are also permissive. That is, the relays will prevent a control sequence until the monitored parameters satisfy preset conditions. The IEEE and ANSI have established a shorthand method of defining protective relay functions. The method assigns unique numbers to relay functions. Table 12.1 is a partial list of the assigned numbers and their functions for the population of protective relays likely to be found in the DG environment.

The relay functions are selected as functions of their location in a circuit and the characteristics of the equipment at that location. For example, many grid operators will require a directional power relay at the point of interconnection of the DG unit and the grid. The relay will likely be set to trip at a specific power level flowing back into the grid. The device usually called for is Dev. 32. In this scenario, the relay would initiate a designed control response whenever power flow from the DG unit into the grid exceeds a preset value. Thus, it would operate as a protective device. Similarly, grid operators may require Dev. 27 at the point of interconnect. This would act as a permissive device. It could also be used as a protective device in a backup role to Dev. 32. It is useful to provide some commentary on each of these devices and the roles they would likely play in the DG environment.

TABLE 12.1

ANSI and IEEE Device Protective Relay Functions

ANSI Device Number	Device Function
25	Automatic synchronizing or synchronism check
27	Undervoltage
32	Directional or overpower
40	Loss of excitation
46	Negative sequence and unbalanced current
47	Negative sequence voltage
50	Instantaneous overcurrent
51	Time-delayed overcurrent
52	Circuit breaker
59	Overvoltage
67	Directional overcurrent
81	Frequency, over- or underfrequency, or both
87	Differential overcurrent

12.5.5.1 Descriptions of the Devices

As an automatic synchronizing device, Dev. 25 will control the output of the DG unit to cause the frequency and voltage of the DG to match the grid and then reduce the phase angle between the DG and grid voltage wave. Minimizing this phase angle reduces synchronizing current and the associated torque at the instant of connection of the power sources. In the synchronizing check mode, this device would not control voltage and frequency.

Dev. 27 is an undervoltage relay. In the permissive role, this device would prevent an operation until the monitored voltage achieves some acceptable value. In the protective role, it can serve as a backup to other relays to initiate the disconnect of a power source. This device typically has adjustable trip and reset values and an adjustable operating time delay.

Dev. 32 acts in a protective mode. It monitors power flow in level and direction. It would typically operate to separate circuits when the power flow in a specific direction exceeds a preset value.

Dev. 40 monitors a synchronous generator. It acts to disconnect the generator from a bus if it should lose excitation.*

Dev. 46 provides protection to the generator. Current unbalance and negative sequence currents can result in excessive heating of the generator iron, resulting in damage. The unacceptable unbalance or negative sequence current can be the result of harmonics or single-phase faults on the power system. Since the generator can tolerate this condition for a short time, this device usually acts as a backup device to overcurrent relays.

---

* Because DG machines of the rating under consideration here are typically brushless generators, actual measurement of excitation is not practical. Therefore, excess reverse VARS is usually monitored to indicate loss of excitation. Accordingly, a directional power relay connected to measure VARS is typically used in the Dev. 40 role.

Dev. 47 provides protection against negative sequence voltages. Again, because the power system can tolerate negative sequence voltages for a short time, this relay can also act as a backup protective device.

Dev. 50 operates when the current in the circuit exceeds a preset value. This is an instantaneous operating relay typically used to protect circuits against fault currents. It is usually set to disconnect the protected circuit for currents in excess of six or more times normal full-load current in the circuit.

Dev. 51 serves to disconnect the protected circuit for currents in excess of a preset value. However, it usually operates on a defined time versus current curve. The higher the current flow, the faster the relay will operate. This device is set to operate at current levels in excess of 1.25 to 10 times normal full-load current in the circuit.

Dev. 52 is typically operated by the permissive and protective relays. It serves to connect or disconnect the protected circuit.

Dev. 59 is an overvoltage relay. In the permissive role, this device would prevent an operation until the monitored voltage achieves some acceptable value. In the protective role, it can serve as a backup to other relays to initiate the disconnect of a power source. This device typically has adjustable trip and reset values and an adjustable operating time delay.

Dev. 67 monitors current flow for magnitude and direction. When the current flow in a circuit exceeds a preset value in a defined direction, the relay operates to disconnect the circuit. Depending on the application, this device can be either instantaneous or time-versus-current delayed.

Dev. 81 is an over-, under-, or over- and underfrequency relay. In the permissive role, it acts to initiate an operation when the frequency of the monitored circuit achieves an acceptable value. In the protective role, it can serve as a backup to other relays to initiate the disconnect of a power source. This device typically has adjustable trip and reset values and an adjustable operating time delay.

Dev. 87 is a zone protection device used to compare the flow of current into a circuit zone with the current out of the zone. When these currents differ by a preset value, it can be concluded that an unacceptable condition exists and the protected zone needs to be disconnected. This can serve as a primary and/or backup protective device. It is typically used to limit damage to the windings of a generator resulting from part-winding faults.

12.5.6 Perspective

Up to this point, several issues concerning DG have been presented. It is necessary to understand the various aspects of these issues if a safe and acceptable operating environment is to be achieved. There are two governing codes that will come to bear on this environment: the National Electrical Code (NEC), NFPA 70, and the National Electrical Safety Code (NESC), IEEE/C2. The first of these defines the requirements for a safe and acceptable power system in a facility. The second defines the requirements for a safe and acceptable utility distribution system.

The purpose of these codes is twofold. The first consideration is personnel safety. The second consideration is property safety. From the personnel safety aspect, the installed circuit must provide assurance to any personnel that a circuit can be isolated and locked out to make it safe to work on. This is required on both sides of the point of connection, the grid and the facility. This is also an OSHA requirement. From the property safety aspect, the installed facilities must curtail the environment for creating a fire.

In designing the DG environment, the circuits on both sides of the point of common connection must be protected in accordance with the respective code, the equipment must be protected from the undesirable effects of operation, an economical operating strategy must be achieved, and risk must be curtailed. The host facility of the DG unit is likely not in the electric power generation business. In the case of coincidental use of installed units (emergency and standby power systems), those units exist to mitigate the impact of power outages. That they can be used for DG allows a return on the sunken capital by curtailing the facility operating cost of electricity. In the case of generation installed specifically for the DG environment, the equipment investment must be configured to provide the maximum return on invested capital.*

12.5.7   Power Source Control

In the DG application, the power source immediately connected to the grid will be an inverter or an induction or synchronous generator. (Wind turbine generators can use inverters, induction generators, or synchronous generators.) For inverters, electronic circuitry controls the firing of output semiconductors to create an electric waveshape that will cause power to flow from the source to the grid. Controlling these switching devices controls the rate of energy flow. Notwithstanding that, these controls must also respond quickly enough to prevent damage to the solid-state switching components.

---

* The owner of dispersed generation facilities is not likely to be in the business of electric power generation. Rather, electric power generation will be the coincidental result of controlling the cost of operations. In the past, this translated to the installation of optional standby power systems. With utility deregulation upon us, dispersed generation becomes a viable alternative in the energy source mix. Now, cost control can be a result of either business continuity during a loss of the normal power supply or avoidance of the high cost of electric energy during peak demand periods. Whatever purpose is served, both roles have the potential to enhance operating profit. the coincidental use of an optional standby power system for peaking provides a real and measurable return on the incremental equipment, as well as the initial investment in the system. If the peaking strategy is initially included in the optional standby power system at the design phase, its cost could be little or nothing. Thus, peaking could provide a real return on invested capital that otherwise could not be achieved. Businesses that have a large disparity in electric energy demand between day and night will have an energy cost structure supportive of dual use of the optional standby power system.

Fundamentally, the energy source feeding the inverter input is typically dependent upon a conversion technique that responds to an uncontrollable source for that energy. Photovoltaic and wind turbines are examples of this. The input to the inverter is dependent on the amount and quality of incident solar rays for photovoltaic systems. For wind turbines, the power input to an inverter and generator can be controlled by changing the pitch of the rotor blades for varying wind conditions. Therefore, wind generation can have its output controlled to some extent by controlling the rate of input energy. For these cases, the grid and local load accept all the power that the source is capable of delivering whenever the source delivers it. If an inverter has a battery source in parallel with its primary input source, such as might be found with photovoltaic and fuel cell systems, then the inverter is capable of delivering more power to its output terminals than might be available at any instant in time.

Power sources with fuel inputs (i.e., combustion turbines, reciprocating engines, fuel cells, etc.) will have their output controlled for both power and excitation. The fuel cell may feed the grid in parallel with a battery. Where there is no battery, an acceptable and operating grid is required to develop power through its inverter. The power output of a fuel cell is a function of the rate of fuel flow through it. In the chemical processing of the fuel in the cell, electrons are freed at the anode and caused to flow through some external path, the inverter, and back to the cathode of the cell. Metering the rate of fuel flow will control the output current flow and, consequently, the rate of power flow. For turbines and reciprocating prime movers, metering the rate of fuel flow to the combustion process controls developed shaft horsepower.

As can be seen from the simplified control strategy of Figure 12.6, fuel flow is controlled in response to various inputs. When the prime mover operates independently of the grid and there is no load connected to it, fuel control is a function of matching the speed feedback signal from the prime mover to the reference signal. If load is being carried, the speed feedback still controls prime mover speed. For any given fuel flow, the prime mover will develop a fixed level of power for constant speed. As load is applied to the prime

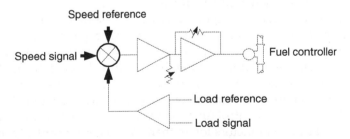

FIGURE 12.6
Simplified strategy for output power control of the generating system.

mover, speed reduces. In response to this, the fuel control increases flow to the prime mover until the speed feedback matches the reference. Fuel control for grid-connected generators functions in a similar manner.

For the induction generator, the load feedback signal is matched to the load reference signal to cause the prime mover to take on load. The speed of the prime mover must exceed synchronous speed for the generator to take on load. Recall that when operated as an induction motor, slip frequency increases as load is increased. Similarly, to increase load on an induction generator, the slip frequency must also be increased. The difference is that in motoring, shaft speed is slower than synchronous speed, and in generating, shaft speed is faster than synchronous speed. In the case of the synchronous generator, shaft speed operates at synchronous speed all of the time. Independent of the grid, speed reference and feedback signals are compared to control fuel for synchronous operation at any load. When operated grid-connected, load reference is compared to load feedback to bring the prime to a load level, and shaft speed remains constant.

The size disparity between the prime mover and grid capacity prevents the prime mover from changing the frequency of the grid; at whatever frequency the grid operates, the synchronous generator will operate at the same frequency. The prime mover is typically not capable of driving the generator beyond synchronous speed because it cannot develop sufficient horsepower to slip synchronous speed. However, the prime mover can lose its load production capability and become a load motor while grid-connected.

It is accepted that load flow control is readily and accurately controllable regardless of the initial source of power. Attention is now directed toward excitation control. As fuel control meters real load, kW loading, excitation control meters reactive load, kVAR loading. Since the induction machine is excited from the grid, control of kVAR loading is accomplished by adjusting real load. (Recall that the operating power factor of any induction motor is a function of shaft load at any time. Similarly, the kVAR loading of the induction generator is a function of how much power it is producing at any time.) kVAR loading of inverters can be controlled in a similar fashion to the control of their real load. Increasing the magnitude of the effective voltage will cause the inverter to carry increased kVAR load.

A synchronous generator requires an excitation source and control. Generators in the size range considered in this work are typically self-excited and, for the most part, brushless. The fuel flow controller controls kW loading. For the grid-connected generator, varying the excitation level of the field controls kVAR loading. As excitation is increased, kVAR increases. Following the example of load control, the voltage regulator controls excitation. When grid-connected, the VAR/pf control function varies excitation to achieve a desired power factor. For power factors lower than 0.7, excitation control can become unstable. It is recommended, therefore, that synchronous generators be operated in the power factor control mode rather than the VAR mode.

12.5.8   Time Constants

In evaluating device performance, it is useful to determine performance with respect to time. For example, in the case of a generator picking up load, it is useful to know how long it will take the engine and generator controls to bring the output back to desired voltage and frequency after a load change is experienced. It should be obvious that the larger the change, the longer it will take to return to desired values.

To apply some relative means of comparison, the time constant of the system is measured. By definition, one time constant in decay is the time it takes for the monitored parameter to fall to 37% of its initial value. This decay is typically exponential. In the context of fuel and regulation controls, it is important to know the time it takes for the device to respond to a change of state and restore original values. It is also important to keep in mind that not all generation processes can contribute substantially to a change in state at the operator's output terminals.

In the case of a fuel cell, if the load on its output increases, the voltage will decrease until the fuel control increases fuel flow and the chemical frees additional electrons. This can take considerable time, many seconds to minutes. In the case of the reciprocating prime mover and synchronous generator, voltage can be restored in under a second to a few seconds depending on the load change and its inrush requirements. The prime mover will be restored to frequency within one to a few seconds depending on the percent change in load. When connected to the grid, load changes will follow the control algorithm.

Voltage changes will be a function of the change in bus voltage. With the synchronous and induction generator (and, in some cases, inverters having batteries in parallel with their inputs), consideration must be given to transient response. Synchronous and induction generators have energy stored in the air gap of the running machine. When grid-connected, any transient decrease in grid voltage will cause the energy to be fed into the grid. Consider the case of a fault on a grid radial distribution feeder near a host facility with an induction generator DG unit on line. From engineering practices used in fault studies on power systems, the energy taken as a contribution to the fault from the induction machine is four times the full-load current rating and has a subtransient time constant of approximately 20 ms (0.020 s). Evaluation reveals that the initial contribution to the fault will be reduced to less than half its original value within the first two cycles. At six cycles, the contribution to the fault will be zero. Given a synchronous generator that has its own excitation system, the contribution to the fault will be a function of the subtransient reactance and the excitation support system. The minimum subtransient reactance to be expected in the generators considered herein is 10%. The subtransient time constant is typically 20 to 30 ms. The transient reactance is typically 0.3 to 0.4, and its time constant is typically 150 to 200 ms. At the end of 1.5 cycles, the contribution to a fault would be less than half of its original value. Given battery-supported inverters, their control circuits can detect the fault in less than half a cycle, and the turn of the output components in the next half-cycle. Therefore, these devices can limit fault current contribution as well. As previously mentioned, fault current capability of a synchronous generator can be 10 times full-load rating.

Putting this into perspective, in the overwhelming majority of installations that will economically support DG, the likelihood is that the rating of that generation will be less than 25% of the peak demand of the facility. It is as likely that the DG unit will only operate when peaking power units drive the rates for electricity from the grid. It is thus seen that the fault current-driving ability of the DG unit discussed herein can be well coordinated with the grid at the point of connection. The availability of fault driving kVA from the grid is likely to be 10 or more times that of the DG unit.

12.5.9　Power System Design

The design of the power system will vary as a function of the type of prime mover, mode of generation, its size in comparison to the grid at the point of connection, and mode of operation among other points of consideration. Recall that the DG unit can operate in either of two modes: grid-connected or islanded. A subset of the islanded mode is temporary grid connection to shift load from the grid to the DG unit and vice versa. Not all power generating devices can operate separate from the grid. Static generation systems can operate separate from the grid if the load is held constant and does not have an inrush requirement for starting. Where there is an inrush requirement or varying load, static generation systems employing batteries or other sources of power in parallel with the generation process can be operated separately from the grid. As an example, a photovoltaic system with a storage battery is commonly used in remote locations where grid-supplied power is not available. It is desirable to analyze the design requirements of a DG unit beginning with the simplest system, the islanded DG unit.

12.5.9.1　Islanded Operation

Brief mention was made in Chapter 10 of this type of DG system, which was referred to as load transfer. Operation typically begins with starting the DG prime mover and then transferring load to it. The typical protective/permissive relaying functions found in the islanded system will include Dev. 27 on the grid source and Devs. 27 and 81 on the DG source. Because these types of systems are typically installed for emergency and standby service, Dev. 27 on the grid source terminals of the transfer switch will initiate starting of the DG when the grid voltage falls below a preset value for a preset time. Because fault current clearing takes more than six cycles and the grid distribution circuit may have a recloser, it is undesirable to start the DG unless the outage is extended. Therefore, there will be a time delay on the operation of Dev. 27 to start the DG. This delay is approximately one to six seconds long. Devs. 27

and 81 are used to prevent load transfer to the DG until it has achieved acceptable voltage and frequency levels. In the protective role, these two devices will initiate load transfer to the grid (when it is acceptable) if the DG fails. Load transfer in this mode is open transition transfer.

12.5.9.2 Closed Transition Transfer

The issue of confidence in the control strategy of any power system that parallels with a utility-derived power source is a universal concern. The electric utility supplier in any such application must provide safety to its maintenance staff as well as continuity and reliability in electric service to its customers. For these reasons, utility companies are quite demanding of the synchronizing and paralleling control strategies that they will allow at the point of common coupling.

On the other side of this issue are deregulation and the increasing demand that automation of customer facilities places on continuity of electric service. Deregulation and increased automation, coupled with reductions in operations staffing, impose less intrusive load transfer strategies on electrical loads. Having suffered an interruption in the process upon loss of the utility power source, it is often costly to suffer a second disruption upon restoration to the utility service. Consequently, upon restoration of the critical load to the normal power source and for test transfer in either direction, closed transition transfer has become the preferred transfer control strategy. This transfer strategy is equally suitable for the DG scenario. The equipment is comprised of a power switching module and a control strategy.

The power switching module of the closed transition transfer switch is a dual operator type. The single operator type is the traditional open transition, double throw switch. When the operator is energized, both sets of contacts move simultaneously. In this operation, the closed contacts open before the open contacts can close. This is a binary device; in the steady-state condition, only one set of main contacts can be closed. In the dual operator switch, each set of main contacts has its own operator. This switch provides both closed and open transition transfer. Typical operation, when both power sources are available and synchronism exists, is for the open set of main contacts to close first, followed quickly by the opening of the (initially) closed main contacts. Because this transfer switch must provide for transfer from an inadequate source to an adequate source, it must also be capable of open transition transfer. In open transition transfer, the closed main contacts are opened before the open set is closed.

Regardless of the operating strategy, both switch designs should be compliant with the type test regimens of the CSA, IEC, and UL standards. The control module should be qualified to these same standards. Typically, control strategies must meet specified repetitive accuracies and stability over a temperature range of -5 to $40°C$. The control strategies would include voltage and frequency sensing, time delays, and various other control functions; of interest to this discussion is only that strategy committed to closed transition transfer. At a minimum, the repetitive accuracy of the control strategy should be ± 1% of nominal at ambient temperature, and the stability should be ± 0.5% of setting across the temperature range.

In addition to measuring acceptability of the two power sources, the control strategy should also measure the difference between the sources in voltage, frequency, and phase angle. A typical operation sequence for closed transition transfer is to first determine acceptability of both power sources. Only when both sources are determined acceptable and have remained acceptable for a preset time delay should the strategy determine if the differentials are met. Only when the voltage difference between the two sources is less than 5%, the frequency difference is less than 0.2 Hz, and the phase angle crosses 5° (electrical) will the strategy initiate a closed transition transfer. Upon initiation of transfer, the open contacts are closed. After closure, the initially closed mains are opened. Timing for this type of closed transition transfer should be arranged to limit parallel operation of the sources to less than 100 ms.

Figure 12.7 shows a simplified logic flow chart of a suitable control strategy. Because the closed transition transfer strategy must be capable of both open and closed transition, both logic paths are illustrated in the figure. The path on the left is for closed transition transfer, the other is for open transition. Because the strategy is required to determine when either strategy is to be implemented, it must be capable of both open and closed transition. As an overview, closed transition transfer is initiated by restoration of the preferred source of power to acceptable values for the time set in the respective delay functions or by initiation of either the unit-mounted test switch or remote initiating contact. Open transition transfer is automatically initiated when the source to which the load is connected becomes unacceptable and the other source is determined to be acceptable.

Looking at Figure 12.7 and following the left logic path, assume that a closed transition transfer is called for and initiated. The strategy continuously checks the two sources of power to determine their acceptability. As long as they remain acceptable, the strategy will make the differential determination of voltage, frequency, and phase angle. Only when the sources are acceptable, the voltage difference is less than 5%, the frequency difference is less than 0.2 Hz, and the relative phase angle difference between the two sources is less than 5° (electrical) will the strategy initiate the transfer operation. When transferring the load to the normal (preferred) source, the strategy will close the CN contacts and initiate a timing function to permit an overlap time of no more than 100 ms. At the end of this time, the strategy will initiate opening of the CE contacts. (CN are the operator and main contacts for the preferred source, and CE are the operator and main contacts for the alternate source.)

The control strategy discussed herein assures the power system operators that paralleling will only occur when both sources are adequate and within a very narrow window of synchronism. Additionally, the control strategy includes automatic recovery should a malfunction occur during the closed transition transfer operation. The stability and repetitive accuracies cited are

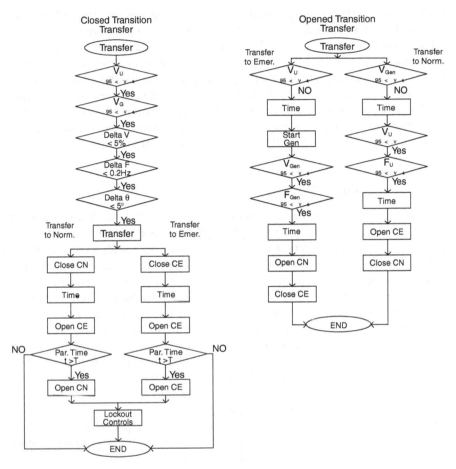

FIGURE 12.7
Logic flow diagram for closed transition transfer of electric load; (left) closed transfer, (right) open transfer.

recommended because these values meet both ANSI protective relay and the commercial standards mentioned.

12.5.9.3 Extended Parallel Operation

Because the closed transition transfer strategy has an overlap time limited to 100 ms, there is no need for protective relaying. The sources will be out of parallel before a protective relay can operate and before a recloser can reclose. When parallel operation extends beyond that very short period, the designer must evaluate how many and which protective relays will be needed and where they are to be applied. It has been suggested that the protective relaying scheme should be seen as two schemes. The first of these schemes is to prevent or permit an operating mode. The second is to terminate an operating mode. It has also been suggested that relays serving a permissive role can, and usually do, act as backup relays to those relays in the protective role.

The extent of the protective relay scheme, having met safety considerations, is affected somewhat by the length of time the sources operate in parallel. If the sources are to operate in parallel just long enough for the load to be shifted from one source to the other, thus eliminating loading transients, the sources need to be in parallel for less than two to three minutes. In theory, such operating strategies should not require as extensive a protective relaying scheme as would be required for continuous parallel operation of the sources.

12.5.10   Permissive Relaying

It is assumed here that the decision to operate and the starting of the prime mover are part of the control strategy in place. This discussion will center on the relay scheme. For successful parallel operation to result, the DG unit must be brought into synchronism with the grid-derived source. If the incoming source is an inverter or induction generator, synchronizing is quite readily achieved. Inverters' output circuits are connected to the grid and their solid-state switching circuitry is controlled to begin taking on load. For induction generators, a speed-sensing control initiates connection of the generator when it approaches synchronous speed. Upon connection, a loading control increases fuel flow to the prime mover. As the fuel flow increases, speed exceeds synchronous speed and the generator begins to produce output power. When this power flow reaches the desired level, fuel flow is held constant. For these two scenarios, the permissive relaying is built into the controls and usually consists of Dev. 27. The objective is to preempt the operation unless the grid-supplied voltage is acceptable. Since neither of these sources is capable of overpowering the grid, Dev. 81 is not required.

Where the power source is a synchronous generator, the permissive relay scheme will include a synchronizer or sync check relay, Dev. 25. The synchronizer will adjust the speed of the DG unit to closely match that of the grid, typically within 0.2 Hz, and will also adjust the voltage of the DG unit to match grid voltage, typically within 5%. When the frequency and voltage differences are acceptable, the synchronizer then reduces the phase angle difference between the two sources to an acceptable value, typically less than 5° (electrical). At $\Delta V = 5\%$, the voltage of the grid would be at 1.0 p.u. and that of the DG unit would be at 0.95 p.u. At $\Delta F = 0.2$ Hz, the speed of the angular difference would be 72° per second. Switching equipment used in this application would likely have a closing time of 5 cycles, 84 milliseconds, or less. Given these conditions and a closing initiated when crossing 5° going toward zero, the contacts of the closing device would meet when the angle was 1.04° past zero. Using the law of cosines to find the value of the resultant vector and given a subtransient reactance of 0.1 pu, the maximum transient current exchange could not exceed 0.53 pu. However, because the generator is excited at the no-load value, the synchronizing current would be considerably less than 0.53 pu. Thus, system disturbance would be minimal, if distinguishable at all. Typically, a sync check relay would monitor relative phase angle only and not provide any matching. Consequently, when a sync check relay is used, it is recommended that Devs. 27 and 81 be provided on both the grid- and DG-derived buses.

12.5.11 Protective Relaying

The primary function of a protective relay scheme for a grid-parallel DG unit should be to separate the power sources on the occurrence of an anomaly. It is taken on faith at this point in the discussion that both power sources will be adequately protected with a proper scheme (though it remains to be discussed herein). Therefore, if the sources are properly protected as isolated sources, then the only action to take when the sources are paralleled is to separate them so that each may initiate its appropriate protective device. When the anomaly is resolved, the affected circuit will be isolated and the remaining circuitry can restore operation automatically. Such a scenario ensures proper protection, power continuity, and minimum risk.

The protective relay scheme for islanded operation will be presented before exploring the parallel source scheme. Slightly different schemes will be utilized on low and medium voltage power systems, mainly due to the differences in equipment design. For example, instantaneous and time-delayed overcurrent trip functions are an integral part of the circuit breaker trip element in low-voltage equipment. These functions are provided by separate relay packages in medium-voltage equipment. The development of the protection scheme begins with the requirements for operating the power sources as islanded sources. In the islanded mode, the grid provides the source of power to the host facility. The protection scheme at the point of common connection is generally designed to limit exposure of the host facility loads to the damaging effects of excessive current flow from the grid, and provide protection against voltage configurations that could damage loads.

Figure 12.8 presents a single line diagram of the relay scheme at the point of common connection of the grid to the host facility. This is representative of both low-voltage and medium-voltage systems. For the low-voltage system, the wye point of the secondary of the transformer would likely be solidly grounded, as shown on the left of the diagram. In a medium-voltage system, the wye point would likely be resistance- or otherwise grounded, as shown on the right side of the diagram. Devs. 50 and 51 would provide for disconnection of the facility should a fault occur in its immediate zone. If the fault occurs further downstream in the facility, the timing of Dev. 51 would be coordinated with downstream overcurrent protection. Virtually all facilities of any appreciable size have three-phase motors downstream. It is, therefore, desirable to protect them against the heating effects of single phasing. For this reason, the service entrance protective scheme will include Dev. 47, negative sequence voltage function. This function is commonly available in combination with Dev. 27 in the same relay package. Dev. 27 would cause the main breaker to open on loss of voltage from the grid. In the permissive role, these functions would allow reclosure of the main breaker when grid voltage is again within acceptable limits.

Figure 12.8 also includes a single line diagram of a relay scheme for an in-house generator. This type of generator is typically installed for emergency, standby, or optional standby operation of critical loads. The diagram illustrates the same overcurrent protection as provided for the main service. Obviously, it serves the same purpose, although it is often somewhat more difficult to achieve coordination with downstream overcurrent protective devices. This is due to the very limited ability of the generator to supply the level of forcing current into a fault that is available from the grid. Usually, overcurrent protection on the generator will include a short time function to achieve the desired coordination. For generator circuits, Devs. 27 and 81 serve both permissive and protective functions. In the permissive role, these relays prevent the connection of the generator to a load until the output voltage and frequency are at least 95% of nominal. In the protective role, these relay functions will cause disconnection of the generator from a bus if these parameters fall below 90% of nominal. This is for protection in the islanded mode of operation. Sustained low voltage or frequency is indicative of an overload on the generator.

When these power sources are to operate in parallel for extended periods, additional relaying is required. Figure 12.9 illustrates what might be required. The circuit shown is for a soft-load system. A soft-load transfer system will synchronize and parallel the two sources. When in parallel, the loading controls will cause the load to shift from one source to the other at a preprogrammed rate. The objective of such a system is to minimize loading transients. In the case where extended parallel operation is desired, once the sources are paralleled the loading controls will control the amount of real and reactive load each source will carry. As can be seen in this figure, the protective relaying includes additional functions.

Beginning with the DG unit, two additional functions are required. Dev. 32R is applied in the reverse power mode. The reverse power relay serves the protective role to disconnect the DG unit from the power system. Power flow into this source is indicative of an unacceptable condition. Power flow into the source, reverse power, will occur when a malfunction in the prime mover or inverter occurs. For example, should a control or switching component in an inverter circuit fail, it may permit power to flow into the source. Where the prime mover is an engine, it can lose power and become motorized by the source. Under this condition, the DG unit must be separated from the power system. Dev. 40 provides a level of protection against failure of the excitation controls in the DG unit. Where the source of electricity is a synchronous generator, for the size range discussed here it will likely be brushless design. This design makes it difficult to monitor excitation so as to achieve true loss of excitation protection. However, when a brushless synchronous generator loses excitation while operating in parallel with a larger

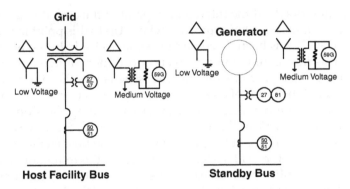

FIGURE 12.8
Single line diagram for grid and generation circuits.

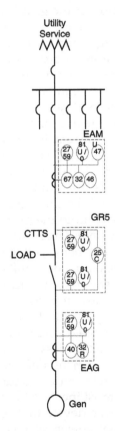

FIGURE 12.9
Power and relaying circuit for soft loading
and extended parallel operation.

power source, it will generally draw reverse VARs. Therefore, directional
power relays connected in quadrature are typically applied as loss of excitation protection on brushless generators.

At the CTTS, Devs. 27, 59, and 81 U/O act in the permissive role for permitting transfer and in the protective role for initiating transfer. Dev. 25 acts as

previously described. At the interconnect point of the circuit to grid-derived power, there is another set of protective relays. Up to this point in this discussion, only Dev. 27 has been discussed for voltage check. In the relay scheme shown at this interconnect point, over- and undervoltage and over- and underfrequency relaying are shown. The issue is whether the DG unit can drive the grid at the point of connection to overvoltage or over- or underfrequency. Under the most likely scenario for DG unit application, where the DG unit is less than 10% of the MVA rating of the grid, the only way the unit can cause a rise in voltage at the grid is by taking on host facility load so that the voltage drop in the grid system is reduced by the reduction in demand. However, many utility companies require these protective functions at this interface. Given that these functions are in the CTTS relaying, they are not required at the point of interconnect. Devs. 47 and 46 provide for protection of the DG unit generation process against single phasing of the grid connection. If this connection or the grid itself becomes single phased, the feeder breaker will be tripped, allowing the in-house source to carry the load. Negative sequence voltages and currents have detrimental thermal effects on induction and synchronous generators. While the decrement curve of these generators can tolerate this condition for some short period, it must still be cleared. Therefore, it is common to find the time setting of these relays to be in the two- to five-second range. In actuality, these devices will operate as backup to overcurrent relays for single-phase fault clearing most of the time.

Some discussion on the remaining protective devices is necessary. Only one or the other of Devs. 32 and 67 should be required at the point of interconnection. Dev. 32 is a directional overpower relay, and Dev. 67 is a directional overcurrent relay. It is this author's opinion that directional overcurrent, Dev. 67, should be used. This has to do with the fault current-driving ability of the unit in comparison to the grid. Recall that it has been demonstrated that the grid is many times larger than the DG unit. AC power is a vector dot product. It is the product of the voltage, current, and power factor at the point of measurement. The nature of a fault on an AC circuit is inductive. Since circuit conductors are selected to reduce voltage drop along their path, their resistance is quite low in comparison to their inductive reactance. As the ratio of XL:R increases, the power factor decreases. The voltage at the location of a fault is zero. The voltage drop in the power system to the point of fault is equal to the source voltage less the voltage drop across its internal impedance. Since power is a function of voltage and power factor and since both voltage and power factor reduce under fault conditions, it is therefore concluded that current must increase to accommodate these reductions in order to operate a protective device that operates at a fixed power setting. For example, a directional power relay is set at 5%, 0.5 pu. At unit voltage and current and 0.5 power factor, the relay will trip (.05 = 1.0 × .1 × .5; note that power relays may be less sensitive at power factors below 0.5). Under fault conditions in the grid, the power factor and voltage at the point of connection of the DG unit will be quite low. If the voltage were as high as 0.05 and the power factor as high as 0.2, 5 pu current flow through the measuring point would be required to cause the relay to trip. It is quite difficult for DG units to produce this current flow at the point of connection to the grid. Dev. 67 uses a small voltage sample for polarization and acts directly on current flow. A Dev. 67 set to the correct value of reverse current flow would be more reliable than a directional power relay when acting as the primary device to separate the power sources when an anomaly occurs in the combined power system.

In the control strategy where the DG unit acts to power in-house loads only while operating in parallel with the grid, Dev. 67 would be set for a very low current level, 0.05 pu. In a scenario where the DG unit will feed power to the grid, the Dev. 67 trip point is set at a current level equivalent to a slightly higher than expected value and at a slightly lower power factor. This relay is an instantaneous trip relay. Accordingly, on the occurrence of a fault, disturbance, or anomaly on the grid distribution radial to which the DG unit is connected, within six to nine cycles of that occurrence, the DG unit will be separated and disconnected. The recloser will not close before the DG unit is disconnected. Reconnection of the distribution radial feeder at the substation will clear the fault or open again and lock out. Whichever of those two states occurs, the DG unit cannot be reconnected to the grid until the voltage permissive relays enable the auto synchronizing control, Dev. 25, to produce synchronism and reconnection.

12.6 Summary

Despite the perception of DG as a suite of breakthrough technologies and applications that cannot be installed and operated safely without further analysis, the technical reality appears more sanguine. Available today are the requisite protective devices and relays to enable a variety of electrical schemes. Parallel codes and standards contain an understanding of fuel, electrical, fire, and life safety protection. The challenge is to bring all concerned parties to a common understanding so that the full economic benefits of DG can be realized.

References

Blume, F. L. et al., *Transformer Engineering*, John Wiley & Sons, New York, NY, 1992.
Borbely, A.M., U.S. Installation, Operation and Performance Standards for Microturbine Products, Pacific Northwest Laboratory, Richland, WA, 1999.
Conover, D.R., Energy Standards and Model Codes Development, Adoption, Implementation, and Enforcement, Pacific Northwest Laboratory, Richland, WA, 1994.

Conover, D.R., A foundation for success: building construction regulations and DG, E-Source, Boulder, CO, 1998.

Fink, D.G. et al., *Standard Handbook for Electrical Engineers*, McGraw-Hill, New York, NY, 1993.

Fitzgerald, A.E. et. al., *Electric Machinery*, McGraw-Hill, New York, NY, 1990.

13

Fuels

Branch Russell, Don Stevens, and Michael Godec

CONTENTS

This chapter begins with a discussion of power generation fuel types, sources, compositions, distribution infrastructure, and handling characteristics. North America's natural gas resources are summarized; future recoverable reserves are estimated. An overview of Fischer–Tropsch (FT) gas-to-liquids (GTL) and biomass fuel technologies completes the chapter.

Appropriate available fuel supply is a major factor in determining the suitability of a DG technology for a given site and will play a dominant role in the shape of the DG energy production industry. This chapter is intended to familiarize end users with the fuels available for on-site generation. Each alternative fuel has its own setup costs in terms of capital equipment, space requirements, safety, and environmental considerations.

Natural gas and diesel (fuel oil) are the most prevalent fuel types in use for small-scale power generation technologies today. Propane, LPG, naphtha, and kerosene are also important. At present, only small volumes of biomass fuels and synthetic fuels from FT GTL conversion, methanol, and hydrogen are produced and distributed as fuels. Active debate centers on what fuel will take the place of conventional petroleum fuels during the 21st century. Production of synthetic fuels compatible with the present distribution infrastructure is anticipated to grow significantly during the near to medium term. In the long term (>20 years), hydrogen is anticipated to become an important fuel.

Internal combustion engines can run on conventional liquid fuels or be converted to operate on gaseous hydrocarbon fuels. Turbines can operate on a variety of fuels including low-Btu vent gas from landfills or industrial operations such as steel furnaces. Natural gas, LPG, methanol, ethanol, gasoline, diesel, and synthetic fuels are under active evaluation as base fuels for fuel cells. The wide-scale direct production of hydrogen is also being studied.

The choice of fuel, like the choice of any commodity, is dictated by environmental benefits, equipment requirements, price, energy density, purity (contaminant content including sulfur, aromatics, and metals), reliability of supply, distribution infrastructure, and storage. If natural gas is available via pipeline, it will likely be the fuel of choice for DG applications. When this is not the case, the fuel choice becomes much more complicated.

13.1 Fuels Overview

Fuels most likely to be considered for use in DG applications fall roughly into four groups: (1) natural gas and natural gas derivatives, including pipeline natural gas, natural gas liquids (NGL), compressed natural gas (CNG), liquefied natural gas (LNG), propane, and liquefied petroleum gas (LPG), (2) conventional liquid petroleum fuels, including diesel, kerosene, and naphtha, (3) synthetic fuels, including FT gasoline and FT middle distillate (kerosene and diesel), and (4) non-petroleum alternative fuels, including methanol, ethanol, dimethylether (DME), biodiesel, and hydrogen.

13.1.1 Fuel Consumption

Fuel consumption is typically reported as the quantity of fuel consumed per hour based on a specified load. It is impossible for any DG developer to provide the customer with an exact calculation of fuel requirements unless the time-weighted load usage is specified. As such, the manufacturer usually reports the maximum fuel consumption at full (100%) load. Some generators at no load typically use about half of the fuel used at full load. This linear relationship between fuel usage and load allows a reasonable estimate of fuel consumption over an assumed range of operating conditions. Thus,

Fuel consumption = (estimated load amps/max. generator output) × 0.5 × max. fuel consumption + 50% of max. fuel consumption

Other technologies use only a small fraction of the full-load fuel. Each technology differs, and specific information should be used for correct economic and thermal analyses.

13.2 Natural Gas and Natural Gas Derivatives

Natural gas and associated gas, the gas produced in conjunction with oil production, is piped to field gas plants near the wellhead. Hydrogen sulfide, carbon dioxide, nitrogen, water, and other contaminates are removed, and a significant amount of NGL, essentially the C_3+ and C_2+ hydrocarbons, is stripped from the natural gas to meet pipeline specifications. There are over 700 field gas plants in Canada. The majority are located in Alberta. However, not all gas plants produce significant quantities of NGL. The average propane recovery is 0.04 units of NGL per unit of raw gas produced.

The processed natural gas is sent via pipeline from field gas plants to gas straddle plants to recover residual C_3+ and C_2+ components. The straddle plants produce one-third of the total propane produced from natural gas production. Separation facilities fractionate the NGL mixture into components for marketing. The separated fractions are moved by truck to local markets and through pipelines or by rail to distant markets. In countries with well developed gas pipeline infrastructures, such as the lower 48 states of the U.S. and western Canada, natural gas is processed and sold to customers who contract with pipelines to transport the gas.

In their pure form, NGLs are known as spec products. Ethane plus, known as C_2+, is a mixture of ethane, propane, butane, and a small amount of pentanes and higher molecular weight hydrocarbons called pentane plus. Propane plus, known as C_3+, is a mixture of propane, butane, and pentane plus. LPG is generally a byproduct of oil refinery processes and consists principally of propane and butane. As with NGL extraction processes, refinery propane extraction can be varied to meet demand. LPG is comprised of a mixture of hydrocarbons containing varying degrees of propane and butane.

The decision to process natural gas is dependent on both economics and pipeline specifications. Gas processing plants generally operate off the margin between the price of natural gas and price of NGLs. Through much of the 1980s and 1990s, this margin spread was either nonexistent or was too small to support the operating costs of processing plants. Consequently, gas processors tended to process only the minimum quantity of natural gas necessary to meet pipeline specifications. Energy companies have been generally unsuccessful in increasing market demand for natural gas liquids as rapidly as the demand for natural gas.

TABLE 13.1

Description of Natural Gas and Natural Gas Liquid Fuels
(Methane, Ethane, and Propane)

Ignition temperature, °F (°C)	920–1020° (493–549°)
Maximum flame temperature, °F (°C)	3600° (1982°)
Percentage of gas in air for maximum flame temperature	4.4%
Lower and upper limits of flammability (percentage of gas in air)	2.4 to 9.5%
Octane number (iso-octane = 100)	97 to 125

13.2.1   Compressed Natural Gas (CNG)

CNG is stored in welding bottle-like tanks at pressures over 3000 psig. CNG and LNG are both delivered to the internal combustion engines at low vapor pressure (< 300 psig). A complication with all natural gas and NGL fuels is that they are naturally odorless. A sulfur-bearing amyl mercaptan that is readily identified by smell is added for safety. Even trace amounts of the mercaptans are detrimental to the performance of advanced internal combustion (IC) engine emission reduction systems and fuel cells. Mercaptan removal systems or alternative additives will be required before these treated fuels are acceptable for use in sensitive generation equipment.

In the U.S., natural gas is typically transported by large FERC-regulated interstate pipeline companies, then delivered to local distribution companies that deliver low-pressure natural gas to consumers. Natural gas is stored by:

- Line pack — varying the pressure within the pipeline system to handle daily imbalances between supply and demand of natural gas

- Salt dome or underground cavern storage — used to provide large volumes of natural gas for heavy demand periods, such as extreme cold weather or supply outages caused by hurricanes.

- Depleted reserve storage — used to handle seasonal peaking requirements (this storage must typically be cycled each season)

13.2.2 Liquefied Natural Gas (LNG)

When natural gas is cooled to a temperature of approximately –260°F at atmospheric pressure, it condenses to LNG. One volume of this liquid takes up about 1/600th the volume of natural gas at a stove burner tip. LNG weighs less than one-half the weight of water, actually about 45% as much. LNG is odorless, colorless, noncorrosive, and nontoxic. When vaporized, it burns only in concentrations of 5 to 15% when mixed with air. Neither LNG nor its vapor explode in an unconfined environment.

The liquefaction process removes the oxygen, carbon dioxide, sulfur compounds, and water. The process can also be designed to purify the LNG to almost 100% methane.

Once liquefied, natural gas is much more compact and occupies only 1/600th of its gaseous volume. This makes it economical to ship over long distances.

LNG tanks are always double-wall insulated construction. Large tanks are low aspect ratio (height to width) and cylindrical in design, with domed roofs. Storage pressures are less than 5 psig. Smaller quantities, 70,000 gallons and less, are stored in horizontal or vertical vacuum-jacketed, medium pressure (up to 250 psig) vessels. LNG must be maintained below –117°F to remain liquid.

13.2.3 Liquefied Petroleum Gas (LPG)

Nationwide, approximately 45% of LPG comes from petroleum refining, with the balance from natural gas processing. Most residential gas appliances operate on LPG, i.e., barbecue grills, portable space heaters, and home heater units in rural areas that lack natural gas pipelines. Therefore, unlike CNG or LNG, LPG enjoys an extensive, mature distribution network nationwide. LPG is normally stored on-site under relatively low pressures. The bulk of LPG produced in North America is commonly used in central heating systems and as a feedstock in chemical plants. LPG is now aggressively marketed as a clean-burning fuel for internal combustion engines. LPG's main components, propane (C_3H_8) and butane (C_4H_{10}), have different boiling points, –45°C and 0°C, respectively. Moderate temperature reduction or pressure increase liquefies LPG. These properties make it transportable as a less explosive liquid that can be easily regasified on-site for combustion.

13.2.4 Propane

Propane in the U.S. is a mixture of approximately 85% propane and 15% butane; small amounts of pentane and isobutane are also incorporated. Outside the U.S. and Canada, the mixture can vary greatly from 80/20 to 30/70 propane to butane. Propane has a boiling point of –42°C (–44°F) at atmospheric pressure; therefore, when kept in a sealed vessel, propane has a vapor pressure that moves depending on outside temperature. For example, at 70°F, the vapor pressure of propane is 127 psi and of butane is 17 psi, so the vapor pressure is a weighted average of the two constituents. As the outside temperature drops, the vapor pressure also drops. Depending on where the fuel is being used, it may require some sort of active vaporization device to ensure that the fuel is in gaseous form when used. In liquid form, propane has a heat content of approximately 91,500 Btu/gal. In gaseous form, it has a heat content of approximately 2500 Btu/cu^3 (natural gas has a heat content of approximately 1000 Btu/cu^3).

Propane is gathered and processed from both natural gas and oil wells throughout the world. It is heavier than natural gas (96 to 98% methane) and than air, so in gaseous form leaks remain explosive for a longer period of time than natural gas because they take longer to dissipate. Propane can also displace oxygen, so there is danger of asphyxiation if care is not taken. For this reason, the U.S. government requires that propane be artificially odorized.

The primary market for propane is the residential and small business heating market, so propane prices are directly related to winter temperatures. During winter of the year 2000, the wholesale market price of propane was approximately $0.55/gal, with retail ranging from $0.65 to $1.35 per gallon. The price outlook for propane is flat for the next three to five years. Due to the four warmest winters on record in recent history, propane supplies are at an all-time high, and a very cold winter over much of the U.S. would be required to move prices a noticeable amount. Compared to natural gas, which usually costs U.S. residential consumers $3.00 to 7.00/mmBtu, propane costs about $7.50/mmBtu at current prices.

Propane is readily available in bulk and retail in all 50 United States and most areas of Canada and Mexico. In other markets, countries that are net exporters of petroleum products usually have an ample supply of propane, while those countries that are net importers tend to use very little and have little infrastructure for its use and delivery.

The propane industry is one of the best suited to deliver remote Btus to users who need a clean-burning fuel for a reasonable price. The average setup for basic home delivery costs about $500 and is usually leased to the home or business owner. Most power generation applications will require a larger tank, roughly 1000 to 2000 gallons at a price of $3000 to $6000. The tank must be certified and tested to the specifications required by local, state, and federal laws. These rules are fairly standard, and propane tanks are usually tested to two or three times their operating pressure, but, generally, the operating pressure will not exceed 250 psi. Each tank is equipped with a hydrostatic relief valve that will not allow the interior gas pressure to exceed the pre-set point. Propane installations are very safe, and the propane industry has a very good safety record. When these setups are used to fuel a DG installation, it is recommended that a propane-fired vaporizer be added to insure that the propane flows to the machinery in gaseous form. In colder climates, the temperature may be such that the propane will not gasify fast enough to supply the generator with sufficient gaseous propane.

Depending on electricity requirements, including location and availability of fuel, propane can and will be the fuel of choice in many situations. Since the propane industry is so well suited for fast and efficient delivery of remote fuel, propane will always be considered a good alternative for remote power applications where natural gas pipelines do not reach. Another attractive application of propane is for use as a backup power fuel. When the main purpose of a DG project is to provide power reliability, a higher confidence factor can be used for a fuel that is actually on-site, as opposed to natural gas, which is dependent on a system that could suffer disruption during a natural disaster.

13.3 Historical Gas Market Projections — Lessons from the Past

For the last two decades, virtually all of the participants in the North American natural gas market made major corporate decisions based on erroneous forecasts about future gas prices. Figure 13.1 compares forecasted average U.S. wellhead gas prices taken from the Annual Energy Outlook of the Energy Information Administration (EIA) in 1985, 1990, 1995, and 1999. As shown, in just the last 14 years, the EIA's forecast of gas prices has declined dramatically. During the last nine years, the decline was considerable, from a 1990 forecast of average wellhead gas prices in the year 2010 (in 1996 dollars) of $6.64/Mcf to its most recent forecast of wellhead prices in 2010 of approximately $2.45/Mcf (U.S. Department of Energy, 1998).

The EIA was by no means alone in its early 1990s' forecasts being so different from today's forecasts. A look at the predictions of any organization forecasting gas prices for the year 2010 in 1990 and today shows a similar trend.

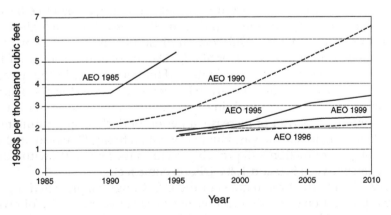

FIGURE 13.1
Natural gas future price estimates from 15 years of EIA reports.

This represents a fairly dramatic change in our collective view of the future for natural gas prices, and this change has occurred in just the last decade. In fact, today most forecasters predict gas prices in the year 2015 to be in the $2.20 to $2.70/Mcf range, i.e., not much different from today's prices (U.S. Department of Energy, 1998). (Transmission and distribution add $1.50 to $4.50 to the price of natural gas, creating delivered prices of $3.50 to $6.50 to consumers, given a wholesale price of $2.00.)

Why were historical forecasts about the future of U.S. gas markets wrong? What has caused this dramatic change in the perception of the future for natural gas? Many factors have been advanced as explanations. Several stand out:

- Our perceptions about the size of the North American gas resource base, which have evolved considerably from one of impending supply shortage to today's view of an enormous resource base

- Our increasing experience and optimism concerning the important role of technology and the ability of technology to advance rapidly and efficiently to ensure stable, reliable gas supplies in the future, despite relatively low gas prices

- Our understanding of the importance of the fundamental changes in the nature of the government's role in overseeing and regulating the U.S. energy industry; the dramatic impact that increased competition has had and will continue to have on driving greater efficiencies and continuous innovation in the marketplace

13.3.1   North America's Gas Resources

Our collective view of the size of the U.S. gas resource base has evolved considerably since the 1970s and early 1980s. Based on the work of Hubbert and others, the conventional wisdom was that the U.S. gas resource base was mature and was being depleted rapidly, and that the U.S. was a high-cost natural gas province. In 1967, Hubbert published his milestone work on the future of oil and gas resources, entitled, "Degree of Advancement of Petroleum Exploration in the United States" (Hubbert, 1967). In this paper, he stated that, "because oil and gas are exhaustible resources, the discovery and production history of these fuels in any particular area must be characterized by a beginning, a period of increase, a period of decline, and ultimately, an end." Pertaining to natural gas, Hubbert went on to say, "the peak in the rate of natural gas production will probably occur in the late 1970s, with ultimate production between 900 and 1,200 trillion cubic feet."

In 1977, Hubbert, at the request of the U.S. Congress, predicted a peak in U.S. gas production of 23 trillion cubic feet (Tcf) per year in 1977, followed by rapid decline. Hubbert predicted that by 1990 production would drop to 13 Tcf/year and would further decrease to 10 Tcf/year by 1995. Hubbert was not alone. Major oil companies were predicting gas resources sufficient for meeting only 15 to 30 years worth of supplies at historical rates of consumption. In 1976, the General Accounting Office declared that few additional gas supplies would be discovered at prices around $1.75 per Mcf.

In the case of natural gas, the fundamental issue, it was believed at the time, was how to slice an increasingly shrinking pie. Concerns about dwindling gas supplies prompted legislation restricting the use of gas in certain markets. In general, most industry experts believed that significantly higher gas prices would be required for any substantial increase in U.S. gas supplies to be realized, and that reserve additions and production would still decline despite increasing prices and increased levels of drilling and development. Given these concerns, the U.S. government began a process for decontrolling prices, with the primary objective of encouraging the development of U.S. gas resources and minimizing the impact of gas shortages that would result from rising wellhead prices.

The decontrol of gas prices did occur, and the U.S. market was opened to competition. But, instead of rapidly increasing, prices dropped and continued to stay relatively low; drilling levels declined dramatically. However, reserve additions and production remained relatively steady. This caused a considerable reassessment of our understanding of the dynamics of the North American gas market and the nature of gas supplies serving this market.

Figure 13.2 is modeled after work by Bill Fisher at the Bureau of Economic Geology at the University of Texas at Austin (Fisher, 1994) and shows how assessments of the U.S. gas resource base have changed over time and how rapid that change has been since 1985. Fifteen years ago, various organizations estimated the amount of potentially recoverable U.S. gas resources as ranging between 200 and 700 Tcf (those estimates were adjusted to account for the amount of gas produced since the estimates were made). Today, most estimates of recoverable gas resources in the U.S. range from 1200 to over 2000 Tcf. A similar perception change in the size of the gas resource base has occurred in Canada as well.

A supply view evolved that focused more on gas resources rather than gas reserves. Substantial quantities of gas are believed to exist in the traditional supply areas such as the mid-continent, Appalachia, and the Gulf Coast. Moreover, large quantities of (mostly undelineated) gas resources exist in Alaska. This evolving view of our gas resource endowment is characterized in terms of a resource pyramid, shown in Figure 13.3, and represents a more dynamic view of the resource base. This concept represents the gas resource base as successive layers of lower quality, less accessible, and/or more costly resources. However, the total volume of resource in each category tends to increase as you work your way down the pyramid to lower quality resource categories. As technology and our understanding of these resources advance, these lower quality, higher cost resources become more accessible and economical, allowing them to make a larger contribution to gas supplies.

The bulk of the resource endowment in North America, approximately one-third by today's estimates, exists in these lower quality, tight gas, coal bed, and gas shale settings. The emerging supply areas that are believed to

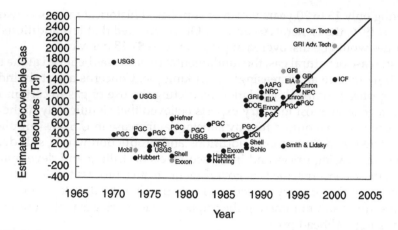

FIGURE 13.2
Estimates of recoverable gas reserves since 1970. Several estimates are for the lower 48 states only. All estimates are adjusted or production since that date of estimate, 1994. From Fisher, W. L., The U.S. Experience in Natural Gas, Proceedings of the Global Gas Resources Workshop, 1994 (with permission).

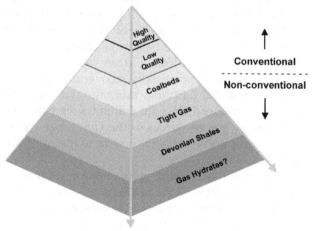

FIGURE 13.3
The resource pyramid concept.

contain the majority of these resources include portions of the Rocky Mountains and the Western Sedimentary Basin in Canada. However, it is important to realize that much of these so-called unconventional resources are quite accessible and conventional using available technology. A substantial amount of unconventional gas development activity continues in the U.S., even though some decline resulted when the Section 29 tax credits were discontinued. Approximately two-thirds of the successful gas wells in the U.S. are currently drilled into unconventional settings. Technology advances, numerous (although limited) high-quality prospects and sweet spots, and improving market conditions allow unconventional gas resources in the U.S. to remain economically attractive. In aggregate, unconventional gas resources contribute approximately 20% of U.S. production, and this percentage is growing. The nearly 50 Tcf of currently producing proven unconventional gas reserves in the U.S. represents about 30% of the total reserves.

In addition, substantial in-place volumes of other more speculative and/or challenging potential resources exist throughout North America. These areas include :

- Environmentally sensitive regions (offshore and federal leasing moratoria areas)
- Unconventional resources in the U.S. and Canada (beyond conventional areas in the San Juan and Warrior Basins for coal, and Michigan shale gas resources)
- Deepwater Gulf of Mexico (beyond where current activity is taking place)
- Resources in deep sediments (below 15,000 feet) that remain relatively untested
- Resources in the Northern Territories of Canada and Alaska

The majority of these emerging categories of resources are uneconomical to pursue given today's technology and market conditions. However, over the longer term, with improvements in technology and reduction of the costs associated with exploration, development, and production, the contribution of these currently emerging sources of supply is likely to grow considerably in the coming decades.

The North American continent has a natural gas resource endowment on the order of nearly 2700 Tcf when the significant gas resources in Canada and Mexico are included. Figure 13.4 illustrates this point.

The North American natural gas markets have access to diverse domestic and foreign sources of natural gas supplies:

- The U.S. (including Alaska) is the second largest producer of natural gas in the world and has over 1400 Tcf of remaining recoverable reserves.
- Canada has over 1100 Tcf of recoverable reserves that are mainly concentrated in the Western Canadian Sedimentary Basin (WCSB). Rapid growth in supply potential exists from the Canadian Eastern offshore and from the Northwest Territories. Although wide-scale exploitation of the frontier areas is not expected in the immediate future, they are believed to contain over 40% of the Canadian reserves.

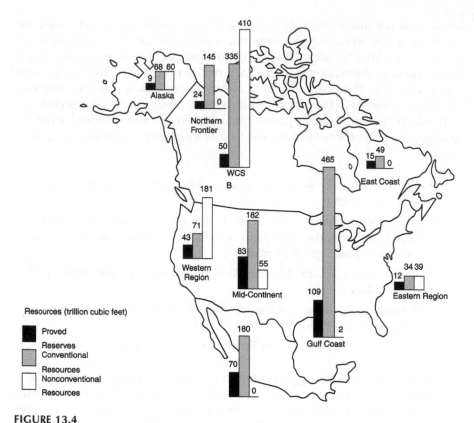

FIGURE 13.4
Technically recoverable gas resources in North America comprise almost 2700 trillion cubic feet.

- With 250 Tcf in recoverable resources and an extremely low extraction rate, Mexico has significant potential for production growth in North America.
- The four LNG facilities in the U.S. may provide access to the 6000 to 7800 Tcf of global gas reserves (see discussion below).

The size and geographical diversity of the North American gas resource base are significant.

13.3.2   Technology Advances in Exploration and Production

The second important reason why incorrect predictions were made about future gas markets and why today's views are quite different relates to how gas extraction technologies have advanced dramatically in the last two decades and how we now believe that these advances will continue to make an ongoing contribution to low-cost gas supplies in the future. A significant variety of technological advances contributed to increasing low-cost gas supplies. Flat oil and gas prices forced operators to implement programs to dramatically reduce expenditures and streamline operations. These technology advances include:

- The introduction of three-dimensional and four-dimensional seismic technology that reduces dry hole rates and improves overall reservoir performance. For example, recent work sponsored by the Gas Research Institute (GRI) showed that three-dimensional seismic tests in older onshore fields in the Gulf typically resulted in a three- to fourfold increase in production over that for fields where it was not applied (Caldwell, 1996).

- The application of horizontal drilling is making considerable improvements in cost effective recovery in a wide variety of settings. Since the mid-1980s, horizontal drilling has grown in application from almost nothing in 1990 to over 2700 horizontal wells drilled worldwide (Chambers, 1998).

- Developments in completion and stimulation technology in the past decade dramatically changed the economics of producing gas resources, especially in the unconventional gas resources in coal seams, shales, and low permeability sands.

- Improved drilling technologies including longer-life drill bits and improved control systems are dramatically reducing drilling costs and improving both initial productivity and ultimate recoveries per well.

- A significant contribution to the improved economics of offshore discoveries is the reduction in cycle time from exploration success to first production. Using Shell as an example, this time was reduced from seven to four years during the 1980s. Today, the average time between offshore discovery and first production is typically between one to two years depending on the size of the discovery (Funk, 1995).

A number of studies, including the 1992 natural gas study conducted by the National Petroleum Council, demonstrated that drilling costs in the U.S. have exhibited a 2 to 4% real decline per year after accounting for fluctuations in oil and gas prices, rig utilization rates, and other factors. This decline is believed to be largely attributable to technological advances (National Petroleum Council, 1992).

All of this resulted in an increase in the volume of gas reserves discovered per well. Average reserve additions from new gas wells increased from about 1 billion cubic feet (Bcf) per well in the 1980s to over 2 Bcf per well on average. In terms of gas discoveries per exploratory gas well drilled, reserve additions increased from between 5 and 10 Bcf in the 1980s to over 20 Bcf per well today. This obviously means that gas reserve additions, to a large extent, kept pace with production despite the fact that we are drilling half as many gas wells today as we were in the late 1970s and early 1980s. Since 1993, gas reserve additions in the U.S. have exceeded production. This is the first time in decades that production has been replaced on an annual and sustained basis.

Moreover, much of these reserve additions come from development drilling rather than new exploration. In fact, today over 40% of reserve additions in the U.S. are attributable to lower cost development drilling compared to an approximately 10% contribution in the early 1980s. This has led to significant increases in our estimates of remaining reserves in discovered fields. Some estimates suggest that initial discovery will grow by a factor of ten through additional delineation, infill development, and application of improved technology. These modern developments resulted in substantial reductions in finding and production costs. According to the EIA, real gas finding costs decreased by nearly 60% since the mid-1980s from about $2.00 to $2.25/Mcf in 1985 to $0.60 to $0.90/Mcf today. Production costs also dropped almost by half, from $1.50 to $1.75/Mcf in 1985 to $0.40 to $0.80/Mcf today.

It is likely that major technological developments will continue. These advances appear to be offsetting the impacts of resource depletion. Moreover, improved understanding of the resource base is expanding developments into more challenging settings. Finally, trends toward a more competitive and efficient energy industry are influencing a tremendous amount of innovation, not only in technologies but also fundamentally in how the North American gas market works now and, possibly, in the future.

13.4   Conventional Liquid Petroleum Fuels (Diesel and Kerosene, Naptha)

13.4.1   Diesel

Historically, diesel was the most common fuel source for stationary engine generator sets, but international and domestic concerns over the emissions generated by combusted diesel and a dramatic rise in the availability of piped natural gas is causing diesel to be phased out of the stationary power market. Some air-quality constrained portions of the U.S. and Canada have unilaterally halted all new permits for diesel-fired generators.

Diesel fuel stores almost as easily as kerosene and is becoming more and more popular among self-sufficient households in North America. It is difficult to ignite intentionally and almost impossible to ignite by accident. Two grades are available: No. 1 diesel which is actually kerosene, and No. 2 diesel, which is identical to No. 2 home heating oil. The most common diesel recommended by vendors for backup power diesel engines is grade No. 2-D or DF-2.

Diesel, however, brings with it a suite of issues related to constant, reliable power production, including:

- Poor or irregular fuel quality contributes to a high incidence of performance failure among diesel engines.
- Diesel degrades through oxidation and water condensation within the tank. Galvanized steel, zinc, and copper can serve as oxidation catalysts.
- Anaerobic bacteria in the water thrive on sulfur within the diesel, forming a sludge that renders the fuel unusable.
- Diesel-fired generators may experience problems if the temperature of the fuel falls below the cloud point (temperature at which paraffin in the fuel crystallizes).
- Biocides and stabilizers are used commonly throughout the industry to prevent degradation of the fuel when stored for longer than six months.

Due to the susceptibility of diesel to degradation, the best maintenance program is consumption. The contents of the fuel storage tanks should be turned over at least twice annually.

13.4.2 Kerosene

Kerosene is a refined middle-distillate petroleum product that finds considerable use as a jet fuel and as a fuel for cooking and space heating around the world. When used as a jet fuel, some of the critical qualities are freeze point, flash point, and smoke point. The boiling point ranges for commercial and military jet fuels are about 375 to 525 and 130 to 550°F. Kerosene, with less critical specifications, is used for lighting, heating, solvents, and blending into diesel fuel.

Kerosene is one of the easiest fuels to store. It does not evaporate as readily as gasoline and will remain stable in storage with no special treatment, unlike diesel. Many pre-1950 farm tractor engines were designed to run on kerosene, and diesel generators can run on kerosene if necessary.

13.5 Fischer–Tropsch Synthetic Fuels

The Fischer–Tropsch (FT) gas-to-liquid (GTL) process essentially involves three catalytic processes. In the first reaction, natural gas is combined at specific temperature, pressure, and ratio with air, oxygen-enriched air, or oxygen and a small amount of water to manufacture synthesis gas. The resulting synthesis gas consists primarily of carbon monoxide and hydrogen that is diluted with nitrogen to the extent that air was used in the feed. Table 13.2 illustrates the conversion process.

TABLE 13.2

Conversion of Natural Gas to Synthesis Gas

Natural Gas	Air	Steam	Catalyst	Synthesis Gas (diluted with nitrogen if air instead of O_2 is used)	Water
$CH_4 +$	$O_2 +$	$(N_2 +) H_2O$	=====>	$CO + H_2 + (N_2 + ...)$	H_2O

In the second process, commonly referred to as the Fischer–Tropsch reaction, the synthesis gas flows into a reactor containing either an iron or cobalt-based catalyst. As the synthesis gas passes over the catalyst, it is converted into hydrocarbons of various molecular weights, with byproduct water and carbon dioxide also produced. The produced hydrocarbons and water are subsequently separated.

Cobalt catalysts are well suited for the conversion of natural gas because they work most efficiently with hydrogen to carbon monoxide ratios of approximately 2:1, which happens to be the ratio of synthesis gas from methane. Iron FT catalysts are better suited for synthesis gas feedstocks with hydrogen to carbon ratios of less than 2:1, such as refinery coke and coal. The many differences between iron-based and cobalt-based FT processes are beyond the scope of this discussion. However, one drawback of iron-based FT processes is that when natural gas is used as a feedstock, iron-based FT processes produce significantly more CO_2 per unit volume of synthetic fuel produced than cobalt-based FT processes. Additionally, synthetic oils from iron-based FT processes are undersaturated with respect to hydrogen and, thus, must undergo additional hydroprocessing if saturated paraffins are the desired products.

The following chemical notation illustrates the FT reaction:

Synthesis Gas (if diluted with nitrogen)	Catalyst	Hydrocarbons	Nitrogen	Water
$H_2 + CO (+ N_2)$	=====>	$CnH(2n + _2) +$	$(N_2 +)$	H_2O

FT catalysts typically produce a very waxy synthetic crude oil. More than 50% of a barrel of synthetic crude oil is solid at room temperature due to the high wax content. Thus, the third process to make FT synthetic fuels is the conversion of these waxy hydrocarbons into fuels (gasoline, kerosene, and diesel) using conventional refining technologies (hydroisomerization). Just as in a conventional refinery, yields of diesel, kerosene, and naphtha are controlled by hydrocracker severity — the more severe the hydrocracker reaction, the lighter (smaller) the average molecule in the product from the hydrocracker. The hydrocracker product is fractionated into constituent fuels, diesel, kerosene/jet fuel and naphtha, which are defined by product boiling point ranges. Fuel yields are dictated by fuel and oil market conditions.

Gasolines from FT processes have unique characteristics when compared to typical gasoline that consists of the C_5 to C_9 fraction. Without further upgrading, the octane of FT gasoline is too low for internal combustion engines. However, it has been demonstrated to be an excellent fuel cell fuel. Additionally, FT diesel and perhaps gasoline may be a viable fuel when emulsified with water and minor additives to yield a diesel substitute that is anticipated to significantly reduce emissions and improve engine efficiency. This technology, analogous to water injection, is in the engine testing phase.

The FT distillate (C_{10} to C_{20}) stream includes both the jet/kerosene range (C_{10} to C_{14}) as well as traditional diesel (C_{13} to C_{20}). Due to its strong paraffinic character, FT distillate has a very high cetane number, which lies in the mid-70s. Also due to its chemical nature, FT distillate has a relatively high pour point that may restrict its appeal in more northern regions. Further isomerization of the paraffin and the use of additives can reduce pour point and cloud point temperatures.

Blending of low sulfur level FT distillate with cheaper, high sulfur crude-derived stocks can allow a refiner to both meet finished product specifications and reduce refining costs. The level of allowable sulfur in diesel is dropping in many parts of the world. Future changes in sulfur specifications are imminent including 0.005% in Western Europe and parts of the Asia/Pacific region. A low sulfur blending-stock is attractive in these countries as an alternative to more severe and costly hydrotreating to remove the sulfur. .

With the primary exception of the U.S. (except California), almost all countries have relatively high minimum cetane number/ index requirements for diesel. In the Asia Pacific and Western Europe regions, cetane specifications range from 45 to 53. Latin America cetane requirements run from 42 to 47. The U.S. requires a 40 to 42 cetane depending on the pipeline, except for California where CARB specifications call for a cetane of 48. Numerous recent engine testing programs have demonstrated that FT diesel significantly reduces several criteria emissions (particulate matter, NO_x, CO, HC).

FT synthetic oil may also become a viable alternative to LNG for new and existing power generation. It can be produced and delivered for less cost per Btu, less safety risk, fewer handling and capital investment issues, and still provide all of the advantages of LNG for power generation. Virtually all power generation facilities that use LNG can also use alternate fuel oil to cover shortages of LNG. FT synthetic oil contains no sulfur, no aromatics, and no particulates or metals. Where local markets offer high prices for premium quality diesel fuel or kerosene/jet fuel, either component may be separated from the commingled product stream for sale, with the balance of production shipped on to power plants.

13.6 Biomass

In the U.S., biomass-fueled facilities account for about 7000 MWe of installed capacity. A diverse range of interests including utilities, independent power producers, the pulp and paper industry, and other forest product businesses operate facilities at more than 350 locations.

Essentially all electricity from biomass is generated using conventional steam-cycle processes. Biomass is burned, the heat is used to generate steam, and the steam is used as a working fluid in an expansion turbine to provide mechanical power for electrical generators. While these processes are effective and commercially available, they operate at relatively low efficiencies. Typical biomass steam-cycle processes are approximately 20 to 25% efficient at converting the energy in biomass into electricity. The newer systems have efficiencies of at least 25 to 30%.

In the U.S., commercial biopower facilities are typically in the size range of 20 to 30 MWe, with a few as large as 50 MWe. This size range is dictated by the economies of scale of the steam-cycle process and the availability of biomass. Smaller facilities are not cost-effective for commercial power generation. Larger-scale facilities are also difficult to build because of feedstock availability. Present facilities use waste as a source of biomass. Obtaining reliable supplies of these feedstocks at scales above 50 MWe may be difficult. The upper size limit may be substantially increased by using dedicated energy feedstocks. However, those fuels cost more than waste.

In situations where there are needs for both heat and power, biomass facilities may be profitable at smaller scale. Many CHP (combined heat and power) facilities exist in the range of 1 to 10 MWe in size, both in industrial applications and in district heating facilities in Europe.

13.6.1   Advanced Biomass Conversion Technologies

Throughout the world, there is extensive interest in increasing the use of biomass for power generation. This interest arises primarily from environmental considerations related to global warming. To encourage the expanded use of biomass for power, governments and industries are sponsoring ongoing research and development programs. Current efforts include the following:

- Cofiring biomass with coal — cocombustion of biomass with coal reduces SO_x emissions and reduces net CO_2 emissions. By adding a biomass feed-handling system, existing coal-fired power generation facilities can use biomass. This approach allows the amount of biomass to be rapidly increased and creates the infrastructure for future biopower from dedicated energy crops. This is appropriate for existing, large-scale facilities.

- Advanced conversion technologies — research is being conducted on systems to improve the efficiency of power generation from biomass, which would be suitable for small-scale applications. In most cases, biomass would first be *gasified* to produce a clean fuel gas. The gasification process can either be thermochemical or biological in nature The gas would then be used in applications such as gas turbines (250 kWe to, perhaps, 50 MWe), fuel cells (less than 1 MWe), Stirling engines (less than 1 MWe), and IC engines (a few hundred kWe).

It may also be possible to produce a low-cost liquid "biocrude" through emerging pyrolysis technologies. These fuels may be suitable for fueling gas turbines, IC engines, or Stirling engines.

13.6.2 Biomass Fuel Characteristics

Biomass is produced by complex photochemical and biological reactions that involve the conversion of carbon dioxide and water to stored chemical energy. The simplified reaction scheme is:

$$6CO_2 + 6H_2O = C_6H_{12}O_6 \text{ (hexose)} + 6O_2$$

Because biomass resources are very diverse, their characteristics as energy feedstocks also vary significantly. As shown in Table 13.3, the energy content of woody biomass is typically about 19 to 20 MJ/kg (HHV, dry basis, or about 8300 BTU/lb), while that of other biomass or wastes is usually less. In practice, the energy content of the biomass is also highly dependent on both the moisture content of the biomass and the bulk density. Woody biomass typically contains about 50% moisture (total weight basis), while other feedstocks and wastes range anywhere from 5 to 95% moisture as harvested. The lower-moisture feedstocks can be dried for thermal conversion systems such as gasifiers by a combination of storage practices and the use of waste heat without significant loss of overall thermal efficiency. Higher-moisture feedstocks can be used without drying in biological conversion systems such as anaerobic digesters. The bulk density of the feedstock will also vary depending on whether it is stored as chips, powder, hogged fuel, etc.

Biomass contains inorganic material (ash) that varies by species, as shown in Table 13.4. Woody biomass typically has about 1 to 2% ash, while herbaceous species or wastes such as rice hulls may contain 15 to 20%. This ash is composed primarily of salts and oxides of calcium, potassium, and silicon. Sulfur in harvested biomass is typically less than 0.5%, although the sulfur content of selected wastes such as animal manures can be high.

It is important to note that, despite the variability of the characteristics, biomass is used extensively worldwide as an energy resource for power generation. Effective feedstock preparation, handling, and conversion systems have been designed and deployed by considering the types of biomass a specific facility is likely to use. Hundreds of facilities worldwide routinely use a wide range of biomass feedstocks and wastes. Each system is typically capable of handling a range of feedstocks (such as forest residues, orchard prunings, nut shells, cocoa bean pressings, and similar materials at the same site).

TABLE 13.3

Energy Content and Inorganic Content
of Biomass Feedstocks

Material	Inorganic Material, dry wt%	HHV, MJ/dry kg
Woody		
Cottonwood	1.1	19.5
Hybrid poplar	1.0	19.5
Loblolly pine	0.5	20.3
Eucalyptus	2.4	18.7
Herbaceous		
Switchgrass	10.1	18.0
Sweet sorghum	9.0	17.6
Miscellaneous		
Rice straw	19.2	15.2
Paper	6.0	17.6
Cattle manure	23.5	13.4
Brown kelp	45.8	10.3
Pine bark	2.9	20.4

Source: Klass, D.L., *Biomass for Renewable Energy, Fuels, and Chemicals*, Academic Press, San Diego, 1998 (with permission).

TABLE 13.4

Composition of Ash from Selected Biomass

	Ash Composition by Constituent, Dry wt%							
	CaO	K_2O	P_2O_5	MgO	Na_2O	SiO_2	SO_3	Other
Hybrid Poplar	47.2	20.0	5.0	4.4	0.2	2.6	2.7	17.9
Pine	49.2	2.6	0.3	0.4	0.4	32.5	2.5	12.1

Source: Klass, D.L., *Biomass for Renewable Energy, Fuels, and Chemicals*, Academic Press, San Diego, 1998 (with permisison).

13.6.3 Biomass Costs and Availability

The costs of biomass feedstocks vary significantly. Low-value wastes such as animal manures can command tipping fees, while very high-value feedstocks such as clean wood chips (suitable for paper production) may cost $75 to $90/ton. Power producers attempt to obtain the lowest priced mix of feedstocks compatible with their generation system.

The paper and forest products industries generate power using processing wastes that otherwise must be disposed. In that sense, they have feedstock costs of $0 or less. Some areas have wood wastes or food processing wastes that are abundant. These feedstocks can have relatively low costs of $5 to $10/dry ton ($0.30 to 0.60/mmBtu). In the broader market, biomass feedstock costs for power generation in the U.S. cover a wide range, perhaps $15 to $30/dry ton ($0.90 to 1.80/mmBtu) delivered to the site for typical combustion facilities.

Biomass is widely available in North America, but careful studies must be made by potential power generators to ensure that sufficient quantities are available at specific sites. This is particularly crucial with larger-scale facilities of 20 to 50 MWe capacity. Biomass can typically be gathered and transported within a radius of 50 miles before transportation costs become prohibitive. Sufficient biomass is currently available in a few localities for facilities up to about 100 MWe. In the future, the establishment of short-rotation forestry and the development of energy crops may provide abundant reliable supplies of biomass for even larger facilities. The costs of these energy crops are estimated to be in the range of $1.20 to 2.25/mmBtu on a dry basis when delivered to a large facility.

Smaller facilities, below about 5 MWe in scale, typically have little problem with feedstock availability. These small facilities frequently utilize local point sources of low-cost wastes, such as sawdust, to reduce their feedstock costs. The power producer, however, must exercise due diligence to ensure that feedstocks will be available on a continuing basis at reasonably predictable costs.

Biomass is generally available throughout the world for small-scale facilities, but larger-scale applications have more limitations. For instance, larger facilities can be sited in many places in North America and northern Europe, but higher population densities in southern Europe limit the amounts of biomass there. For smaller-scale distributed generation systems, biomass availability is significantly constrained only in those situations of climate extremes or population densities that limit biomass production.

13.6.4 Transportation and Storage Costs

Biomass is delivered by trucks or rail transportation to power generation facilities as a solid fuel. Individual power producers establish specifications for the material they will accept. The specifications define the average size of the material (i.e., chips 1 to 3"), the acceptable limits for fines, the maximum inorganic content, average moisture, and similar parameters. The necessity of meeting these requirements falls upon the feedstock producer, who is usually independent of the power producer. Thus, the power producer typically agrees to buy defined types of feedstock at prices determined by contracts or the market. The cost of initial preparation and transportation of the material is included in the delivered feedstock costs described above.

Individual facilities store sufficient feedstock to last for a few days or months depending on location. Areas where weather may prevent year-round harvesting require longer storage periods than those with regular feedstock availability. Biomass storage is usually in outdoor piles with minimal cover. Due to biological activity, loss of the feedstock in the piles is typically 0.5 to 1.0% per month, a level of little significance.

When needed at the conversion facility, the bulk feedstock is transferred by front-end loaders to mechanized feed systems. These typically include steps to sort off-sized material, separate remaining metal magnetically, and, perhaps, dry the material. The costs of the feedstock preparation and handling steps are included in the capital and operating costs of the conversion facility.

13.6.5   Gasification

The fact that biomass is a solid fuel limits its direct use in advanced power generation systems. Biomass can readily be burned to provide heat for conventional steam-cycle generation, but that is inefficient at a small scale. To use biomass as a fuel for advanced power generation systems such as gas turbines, microturbines, or fuel cells, it is necessary to convert the biomass into an intermediate gaseous or liquid fuel. Biomass thermal gasification produces a low- or medium-energy fuel gas (5 to 15 MJ/Nm^3) that can be cleaned and used as a substitute for natural gas. The product gas contains mixtures of combustible gases including CO, H_2, CH_4, and others, as well as CO_2 and sometimes N_2. The gas cleaning process must remove unwanted ash and tars from the product. Biomass gasifiers are commercially available on a small scale (~10 MWt output or less) and have recently become available on a larger scale (~60 MWth). Gas cleanup technologies are available to clean the product gas for high-demand applications such as gas turbines. The gasification process allows biomass to be used in advanced-technology power generation technologies with little or no modification. The cost of gasifying the biomass will vary depending on many factors, but the overall cost of high-efficiency gasification/power generation systems is projected to be less than lower efficiency, conventional combustion systems at equivalent scales. For the very small scale required for microturbines (~100 kWe), the biomass preparation/gasification/cleanup steps are roughly estimated to require capital costs of about $800/kWh of installed generating capacity.

Biological gasification processes are more appropriate for converting wet biomass and wastes to fuel gases. Biological conversion changes the carbohydrate portion of the biomass to a medium gas consisting primarily of CH_4

and CO_2. The lignin portions of the feedstock (10 to 40%, depending on species) cannot be converted biologically, and must be disposed of as a sludge. Biological conversion of low-sulfur feedstocks produces a readily usable product that can directly replace natural gas in most applications. Wet wastes such as animal manures and MSW usually contain sulfur that is partially converted to H_2S in the digestion reaction. In some cases, equipment must be included to remove the H_2S prior to use in electrical generation systems.

Anaerobic digestion systems using manures are commercially mature but are seldom used in the U.S. because they are not economically competitive. The collection and cleaning of the gas costs more than natural gas in the U.S., although some systems are used in situations that mandate environmental cleanup. Some digestors are also used in developing countries. Landfill gas collection sites are a specialized instance of biological gas generation. The landfills provide a localized source of biomass, and environmental/safety regulations may require methane recovery. Commercial power generation from landfill gases is widely practiced in Europe, and in North America to a lesser extent.

Biomass can also be converted to a liquid product using pyrolysis technology. The liquid is a highly oxygenated "biocrude" with a heating value of about 60% of diesel fuel. The technology is compatible with power generation systems in the 0.5 to 10 MWe size range. The compatibility of this product with advanced power generation systems has not yet been shown, although the technology is used on a small scale for the production of liquid smoke, a food additive. The oil contains some particulates and alkali salts that may impact advanced generation technologies. Research on these products is ongoing.

13.7 Hydrogen

The potential for hydrogen to fuel electric power generators has been gaining popularity in recent years. The principal thrusts behind this move include the advancement of hydrogen production, transport, and storage technologies, with a concomitant rise in environmental concerns over the accelerating use of conventional hydrocarbon fuels. Hydrogen is not expected to play a significant role in electrical power generation in the near future and is therefore not given extensive treatment in this book.

13.7.1 The Current State of H_2 Technology

Various hydrogen technologies have, to some extent, been tested and have been in use for decades. This is particularly the case in the chemical and petrochemical industries, where large amounts of hydrogen are used for the synthesis and upgrading of crude oil. Hydrogen is currently produced from reformed natural gas, partial oxidation of heavy fuel oil (or diesel), and coal. Small reformers and partial oxidizers are being developed to provide hydrogen for fuel cells at the customers' sites or in vehicles.

13.7.2   Storage and Transport

Depending on the end use, hydrogen must be either compressed or liquefied prior to handling. Hydrogen compression is similar to natural gas but is energy intensive. It is frequently possible to use the same compressors with slight modifications. Small volume stationary storage is handled with aboveground tanks at 5 MPa. Bottle storage can also be used if the volume is sufficient. Alternative storage media are under development, including metal hydrides, iron oxides, and adsorption on carbon microfibers.

Compressed hydrogen is delivered in compressed tanks, currently via truck or train. Germany has developed several small-scale pipeline distribution networks for hydrogen that have operated without major incident for decades. A large-scale hydrogen pipeline network is in operation along the Texas Gulf Coast to supply refineries and chemical plants. Liquid hydrogen is distributed entirely by tank to date.

References

Caldwell, R.H., Merkel, J.L., and Hansen, J., Study to gauge impact of technology advances, *Oil Gas J.*, Feb. 12, 1996.

Chambers, M.R., Multilateral technology gains broader acceptance, *Oil Gas J.*, November 23, 1998.

Fisher, W.L., How technology has confounded U.S. gas resource estimators, *Oil Gas J.*, October 24, 1994.

Funk, J.M., The Gulf of Mexico: Economics of the U.S. Gas "Breadbasket," presentation given at the 1995 Natural Gas Summit, Calgary, Alberta, Canada, November 1995.

Hubbert, M.K., Degree of advancement of petroleum exploration in the United States, *AAPG Bull.*, v. 51(11), 2207, 1967.

Klass, D.L., *Biomass for Renewable Energy, Fuels, and Chemicals*, Academic Press, San Diego, 1998.

National Petroleum Council, *The Potential for Natural Gas in the United States*, Volume II, 1992.

U.S. Department of Energy, Energy Information Administration, Annual Energy Outlook — 1999, DOE/EIA – 0383 (99), December 1998.

Additional Reading

Government Working Group on Sulfur in Gasoline and Diesel Fuels, Preliminary Report of the Government Working Group on Sulfur in Gasoline and Diesel: Setting A Level for Sulfur in Gasoline and Diesel, March 27, 1998.

Government Working Group on Sulfur in Gasoline and Diesel Fuels, Final Report of the Government Working Group on Sulfur in Gasoline and Diesel Fuel: Setting A Level for Sulfur in Gasoline and Diesel Fuel, July 14, 1998.

Masters, C.D., World Resources of Natural Gas — A Discussion in the Future of Energy Gases, U.S. Geological Survey Professional Paper 1570.

Tushingham, M., International Activities Directed at Reducing Sulfur in Gasoline and Diesel: A Discussion Paper, Environment Canada, 1997.

U.S. Geological Survey, Ranking of the World's Oil and Gas Provinces by Known Petroleum, Open-File Report 97-463.

Index

A

Activation polarization, 162–163
Advanced Reciprocating Gas Engine
        Technology, 63–64
Air pollution
    federal regulations regarding,
        39–41
    fossil fuels as cause of, 37
Air pollution control districts, 42
Air quality
    federal regulations regarding, 41
    state regulations regarding, 41
Alkaline fuel cell, 156
Amyl mercaptan, 358
Automatic generation control, 306

B

Band gap, 99
Banking, 244–245
Basis and location swaps, 241
Bearings, for single-shaft gas-powered
        microturbines, 125
Best available control technology, 40
Biomass fuel
    characteristics of, 373–374
    commercial facilities for, 372
    description of, 372
    electricity generated from, 372
    feedstocks
        availability of, 375
        costs of, 375
        energy content, 373–374
        gasification of, 376–377
        storage of, 376
        transportation of, 375–376
    production of, 373
    technologies, 372–373
BMEP, *see* Brake mean effective pressure

Brake mean effective pressure, 62, 65
Brayton cycle
    inverted, 123–124
    recuperated, 84–85
Breaker-reclosing schemes, 113
Buildings
    commercial, *see* Commercial
        buildings
    residential, 31–33
Buyback priority, 187–189

C

CAA, *see* Clean Air Act
Calendar swaps, 241
Call options, 235, 241
Capacity factor, 283
Capital recovery factor, 211–212
Caps, 239–240
Carbon dioxide emissions
    combined heat and power system
        benefits, 292–293
    electricity industry production, 43
    global warming caused by, 43
Cash flows
    continuous, 215
    discounting of, 207–210
    discrete, 215
    equivalent, 210–215
    levelizing of, 210–215
Casten, Thomas R., 266–267
Cells, *see* Fuel cells; Solar cells
Chlorofluorocarbons, 44
CHP systems, *see* Combined heat and
        power systems
Clean Air Act, 38–40
Closed-transition transfer, 69, 337,
        345–347
Coal-fired facilities
    description of, 11

Nomenclature

A	Annual payment
A	Area, m^2
A_{life}	Levelized annual cost
A_m	Annual cost for maintenance, first-year $
$(A/P, r, N)$	Capital recovery factor
AFUE	Annual fuel utilization efficiency, %
C_{base}	Base level energy (hot water, cooking, lights, etc.), kWh/yr
C_{cap}	Capital cost, first-year $
C_d	Draft coefficient for resistance to airflow between floors
C_{eff}	Effective heat capacity of building, J/°C
$[C_{SO_2}]$	Concentration of pollutant (for example, SO_2), $\mu g/m^3$
C_{life}	Life cycle cost
C_{salv}	Salvage value, first-year $
C_{yr}	Normalized annual consumption
CLF_t	Cooling load factor at time t
$CLTD_t$	Cooling load temperature difference at time t, °C
COP	Coefficient of performance
CPI	Consumer price index
c	Cost
c	Heat capacity, kJ/°C
c_p	Specific heat, kJ/(kg – K)
d	Distance, km
d	Diameter, meters
$CDD(T_{bal})$	Cooling degree-days for base T_{bal}, °C-days
$HDD(T_{bal})$	Heating degree-days for base T_{bal}, °C-days
E	Radiation emissive power, W/m^2
E_b	Blackbody emissive power, W/m^2
Epol	Emission rate of pollutants, g/kWh
f_{dep}	Present value of total depreciation as fraction of C_{cap}
$f_{dep, n}$	Depreciation during year n as fraction of C_{cap}

f_l	Fraction of investment paid by loan
g	Acceleration due to gravity = 9.81 m/s^2
H_o	Extraterrestrial daily irradiation, MJ/m^2
$H_{glo, hor}$	Daily global irradiation at earth's surface, MJ/m^2
$H_{glo, vert}$	Daily global irradiation on vertical surface, MJ/m^2
HV	Heating values, J/kg
h	Enthalpy, kJ/kg
h	Height, meters (use appropriate subscripts)
h	Hydraulic head referring to pressure, meters
h_{con}	Convection heat transfer coefficient, W/(m^2 – °C)
h_i	Indoor surface heat transfer coefficient, W/(m^2 – °C)
h_o	Outdoor surface heat transfer coefficient, W/(m^2 – °C)
I_o	Extraterrestrial irradiance, W/m^2
I_{dif}	Diffuse irradiance on horizontal surface, W/m^2
I_{dir}	Beam (direct) irradiance at normal incidence, W/m^2
$I_{glo, hor}$	Global horizontal irradiance, W/m^2
$I_{glo, p}$	Global irradiance on tilted plane, W/m^2
J	Joules
K_{cond}	Conductive heat transmission coefficient, W/°C
K_T	Daily solar clearness index
$\overline{K_T}$	Monthly average solar clearness index
K_{tot}	Total heat transmission coefficient of building, W/°C
k	Thermal conductivity, W/(m–°C)
k_T	Instantaneous or hourly clearness index
L	Load, kW
Lat	Latitude, deg
Long	Longitude, deg
M	Mass, kg
$\dot{M}$	Mass flow rate, kg/sec
N	System life, yr
N_{bin}	Number of hours per bin of bin method
N_{dep}	Depreciation period, yr
p	Pressure, Pa
(P/F, r, N)	Present worth factor
P_{max}	Peak demand, kW
P_{dem}	Demand charge, $/kW/month
PLR	Part load ratio

p_e	Price of energy, \$/GJ
p_e	Levelized energy price, \$/GJ
p_{ins}	Price of insulation, \$/m²
Q	Energy consumption, Joules
$\dot{Q}$	Heat flow, watts
Q_{annual}	Annual energy, kWh
SC	Shading coefficient
SEER	Seasonal energy efficiency ratio
SHGF	Solar heat gain factor, W/m²
SPF	Seasonal performance factor
s	Seconds
s	Entropy, kJ/(kg –°C)
T	Temperature, R or °C
T_{db}	Dry-bulb air temperature, °C
T_{bal}	Balance-point temperature of building, °C
T_i	Indoor air temperature, °C
T_{tstat}	Thermostat setpoint temperature, °C
T_o	Outdoor air temperature °C
$T_{o,\,ave}$	Average outdoor temperature on design day, °C
$T_{o,\,max}$	Design outdoor temperature, °C
T_{OS}	Sol-air temperature, °C
$T_{OS,t}$	Sol-air temperature of outside surface at time t, °C
$T_{o,t}$	Average outdoor temperature for any hour t of month, °C
$T_{o,\,yr}$	Annual average temperature, °C
t_{sol}	Solar time, h
t_{ss}	Sunset time, h
U	Overall heat transfer coefficient, W/(m² – °C)
u	Wind speed, m/s
V	Flow rate, m³/s or liters/s
v	Volume, liters
v	Velocity, m/s
W	Work, kJ
w	Thickness of wall, ft (in)
x	Distance, meters
Y	Annual yield

Greek

α	Absorptivity for solar radiation
β	Grid penetration, fraction
β_s	Altitude angle of sun (= $90° - \theta_s$)
Δp	Pressure differential
ΔT	Indoor/Outdoor temperature difference, $T_i - T_o$, °C
Δt	Time step h
Δx	Thickness of layer, meters
δ	Solar declination, degrees
η	Efficiency
θ_i	Incidence angle of sun on plane, degrees
θ_p	Zenith angle of plane (tilt from horizontal, up > 0), degrees
θ_s	Zenith angle of sun, degrees
λ	Latitude
ρ	Density, kg/m^3
ρ_g	Reflectivity of ground
ϕ_p	Azimuth angle of plane, degrees
ϕ_s	Azimuth angle of sun, degrees

Conversion Factors

1 meter = 3.2808 ft = 39.37 inches

1 km = 0.621 miles

1 m^2 = 10.76 ft^2 1 bar $\equiv$ 105 N/m^2 = 14.504 lbf/in^2

1 cm = 0.155 in^2

1 gal $\equiv$ 0.13368 ft^3 = 3.785 liters

1 kg = 2.2046 lbm

1 lbf = 4.448 N

1 Btu = 252 cal = 1055 joules

1 Atm $\equiv$ 14.696 lbf/in^2 = 101325 Pa

1 mm Hg = 0.01934 lbf/in^2

1 bar = 10^5 N/m^2 = 14.504 lb/in^2

1 Pa $\equiv$ 1 N/m^2 = 0.00014504 lbf/in^2

1 in Hg $\equiv$ 3376.8 Pa

1 in water $\equiv$ 248.8 Pa

1 W/m^2 = 0.3170 Btu/ft^2·hr

1 W/m·°C = 0.5778 Btu/hr·ft·°F

1 kJ/kg = 0.4299 Btu/lbm

1 kJ/kg·°C = 0.23884 Btu/lbm·°F
1 kW = 3412 Btu/hr
1 watt = 1 joule/second
1 HP ≡ 550 ft·lbf/s = 746 watts
1 Quad = 10^{15} Btu
°F = °C·1.8 + 32
0°C = 32°F, 273.16 K and 491.69 R